CMOS Image Sensors

Online at: https://doi.org/10.1088/978-0-7503-3235-4

CMOS Image Sensors

Konstantin D Stefanov
Centre for Electronic Imaging, The Open University, Milton Keynes, UK

IOP Publishing, Bristol, UK

ISBN 978-0-7503-3235-4 (ebook)
ISBN 978-0-7503-3233-0 (print)
ISBN 978-0-7503-3236-1 (myPrint)
ISBN 978-0-7503-3234-7 (mobi)

DOI 10.1088/978-0-7503-3235-4

Version: 20221101

IOP ebooks

British Library Cataloguing-in-Publication Data: A catalogue record for this book is available from the British Library.

Published by IOP Publishing, wholly owned by The Institute of Physics, London

IOP Publishing, No.2 The Distillery, Glassfields, Avon Street, Bristol, BS2 0GR, UK

US Office: IOP Publishing, Inc., 190 North Independence Mall West, Suite 601, Philadelphia, PA 19106, USA

To David Burt, from whom I have learnt many of the things in this book, and to my family, for their support and encouragement.

Contents

Preface

Image sensors are fascinating devices that straddle the boundary between semiconductor physics and electronic engineering. Nowadays, they are used in almost everything, come in bewildering varieties, and are mostly made with complementary metal-oxide semiconductor (CMOS) technology, like the billions of other integrated circuits (IC) manufactured every year. And they look good! They are among the handful of IC types which can be easily seen through their transparent glass cover.

To understand how they work, we need to know a fair bit of semiconductor physics, especially when dealing with high performance imagers designed for science applications. However, this is not always enough; there are a lot of electronic circuits inside a CMOS image sensor (CIS), and even the simple ones can show subtle behaviour and throw up surprises. Without claiming to cover everything, this book strives to cover both the semiconductor physics and the essential electronics found inside a CMOS image sensor.

A relatively small number of concepts from semiconductor physics form the backbone of image sensors' operation—depletion, drift, diffusion, recombination, charge conversion, to name a few. Knowing them well provides the foundations to understand practically all image sensors and helps with the more complex structures that exist or are yet to be invented.

It is impossible to imagine doing any serious work into image sensors without semiconductor technology CAD (TCAD). Very often it is the only way to 'see' what is happening inside, and this book offers many examples of device simulations. A successful TCAD simulation is a good result, but does not guarantee that something will work. However, if something doesn't work in TCAD, it's probably not going to work in silicon either.

To make full use of this book, some basic knowledge of electronics is essential. Knowing what an amplifier is, being familiar with gain, bandwidth, and noise, can be very advantageous. Freely available electronic simulation tools, such as SPICE, are great for designing and verifying the performance of various circuits. Throughout my career, my hobby in electronics has helped me enormously when working with image sensors. Building my first electronic circuit at around 14 years of age, I was fascinated but I only had a faint understanding of how it worked. In a few years, electronics gradually started to make more sense, and after learning semiconductor physics at university it was clear to me that this is what I wanted to do.

I have often found that many important 'bread and butter' topics in image sensors and their operation are difficult to find in books and papers, and sometimes are not there at all. Some of those I have only been able to find out in discussions with more experienced colleagues who have been longer in the field. This book is an attempt to put some of this 'unofficial knowledge', some of which could be simply due to my ignorance, in one place.

This book is intended to be used as a tutorial and has many examples, taken mostly from practice. Solved examples are essential for proper understanding of the theory and bring 'life' to the formulas. They also help with appreciating the parameters in real-world applications. Very often, a good enough grasp of the phenomena can be obtained from relatively simple formulas. They may not be super-accurate, but can give a decent approximate answer and can serve as a 'sanity check' for more detailed results derived from TCAD and SPICE.

Some of the contents of this book are derived from a practical course on CMOS image sensor operation and characterisation techniques that we deliver at the Open University to students and staff.

Chapter 1 covers the fundamentals of image sensors, starting with the photoeffect and charge generation, and including charge collection and transfer, drift and diffusion, recombination, and carrier lifetime. The chapter also describes the fundamental building blocks of CIS—diodes, MOS capacitors and transistors, and the basic MOSFET circuits—source followers and analogue switches.

Chapter 2 deals with 3T, 4T and other CMOS pixel architectures. Most of the material in this chapter is dedicated to the pinned photodiode (PPD) because of its importance for image sensor technology. This includes the operating principles of the PPD, doping profiles, charge transfer, full well capacity and image lag. Other PPD-based designs, such as the 5T, pixels with charge domain signal storage, and several high dynamic range pixels are also covered, as well as hybrid and 3D-intgerated sensors.

In chapter 3, some of the more specialised subjects in CIS performance are discussed, such as the collection of photogenerated signal in *pn* junctions, the sources of dark current, reflective barriers, the backside interface and its effect on the quantum efficiency in backside-illuminated sensors, potential gradients and punch-through.

Chapter 4 is dedicated to the different sources of noise in electronics components and MOSFETs, the readout techniques used for CIS and their noise performance. The two main correlated double sampling (CDS) methods, based on the double sampling and the dual slope integrator are discussed in detail and their noise performance is compared. In addition, the chapter deals with digital CDS, noise in the column readout in CIS, and MOSET noise optimisation.

Chapter 5 begins with the principles of electro-optical characterisation and readout modes in CIS. It describes the measurement methods for obtaining the most important sensor parameters: photoresponse, linearity, system gain, readout noise, dynamic range, full well capacity, dark current, image lag, quantum efficiency and modulation transfer function. Special attention is paid to the photon transfer curve (PTC) because of the wealth of information it provides, and many experimental tips are given.

Chapter 6 covers the on-chip and off-chip electronics used to control and readout the pixels and the sensor. On-chip amplifiers, correlated double sampling circuits, buffers, switches, drivers and logic are some of the circuits described here. Off-chip electronics providing power, bias and amplification is shown with practical examples and performance calculations.

I hope that this book will be useful to all who are using, characterising, or designing CMOS image sensors, beginners and experts alike.

Acknowledgement

The stimulating research environment at the Centre for Electronic Imaging (CEI) at the Open University is one of the main reasons for the existence of this book. Many of the topics and ideas presented here came up from discussions with my colleagues, students, industrial collaborators, and external partners, and by the challenges and the difficult questions they often had.

I am grateful for the inspiration and the knowledge I have gained thanks to David Burt, Andrew Holland, Chris Damerell, Ray Bell, Jérôme Pratlong, Paul Jerram, Doug Jordan, Neil Murray, Pete Turner, Dave Barry, David Hall, Matthew Soman, Julian Heymes, Martin Prest, James Ivory, Chiaki Crews, Nathan Bush, Steve Bowring and Giulio Villani.

Finally, I would like to thank IOP Publishing, and in particular John Navas for overseeing the production and for making this book possible.

Konstantin Stefanov
August 2022

Author biography

Konstantin D Stefanov

Konstantin D Stefanov was born in Rousse, Bulgaria, and has received his MSc in applied physics from Sofia University 'St. Kliment Ohridski'. He received his PhD degree in physics from Saga University, Japan, in 2001. As a research scientist at the Rutherford Appleton Laboratory in Oxfordshire, UK, Dr Stefanov has worked on the development of CCD and CMOS sensors for particle physics. Since 2012 he has been working at the Centre for Electronic Imaging at the Open University, Milton Keynes, UK, where he is developing CMOS image sensors for scientific and space applications. His research interests are in the areas of physics, technology, and design of CMOS image sensors for science applications, semiconductor device simulations, device characterisation, radiation damage effects, detector electronics and data acquisition systems. He has published over 80 research papers, has co-written two book chapters and holds several patents on CMOS image sensors.

List of frequently used abbreviations

AC	Alternating current
ADC	Analogue-to-digital converter
ADU	Analogue-to-digital unit
APS	Active pixel sensor
BSI	Back-side illumination
CCD	Charge coupled device
CDS	Correlated double sampling
CG	Conversion gain
CIS	CMOS image sensor
CMOS	Complementary metal-oxide-semiconductor
CVF	Charge-to-voltage conversion factor
DC	Direct current
DN	Digital number
DR	Dynamic range
DSNU	Dark signal non-uniformity
DUT	Device under test
EO	Electro-optical
ENC	Equivalent noise charge
ETF	Electrical transfer function
FPN	Fixed pattern noise
FSI	Front-side illumination
FWC	Full well capacity
HDR	High dynamic range
IR	Infrared
LED	Light emitting diode
MOS	Metal-oxide-semiconductor
MOSFET	Metal-oxide-semiconductor field effect transistor
MTF	Modulation transfer function
MVC	Mean-variance curve
NIR	Near infrared
NMOS	N-channel MOSFET
PD	Photodiode
PMOS	P-channel MOSFET
PPD	Pinned photodiode
PRNU	Photo response non-uniformity
PTC	Photon transfer curve
RMS	Root mean square
RTN	Random telegraph noise
RTS	Random telegraph signal
SF	Source follower
SNR	Signal-to-noise ratio
QE	Quantum efficiency
UV	Ultraviolet

Table of common symbols and units

Symbol	Description	Value/units
α	Photon absorption coefficient	cm^{-1}
B	Signal bandwidth	Hz (Hertz)
B_n	Noise power bandwidth	Hz
C	Capacitance	F (Farad)
C_{ox}	Area oxide capacitance	F cm^{-2}
C_{GS}	MOSFET gate-source capacitance	F
D_n, D_p	Diffusion coefficient for electrons or holes	$cm^2\ s^{-1}$
e_n	Voltage noise density	$V/\sqrt{Hz}$
e_{nw}	Voltage white noise density	$V/\sqrt{Hz}$
e_{nf}	Voltage $1/f$ noise density	V (Volts)
E	Electric field	V cm^{-1}
E_a	Activation energy	eV
E_c	Conduction band energy	eV
E_g	Bandgap energy	eV
E_v	Valence band energy	eV
E_i	Intrinsic Fermi level	eV
E_F	Fermi level	eV
E_t	Trap energy	eV
E_e	Irradiance	W cm^{-2}
E_{ph}	Photon energy	eV
E_w	Ionisation energy	eV
ε_0	Dielectric permittivity of vacuum	8.85×10^{-14} F cm^{-1}
ε_{Si}	Relative dielectric permittivity of silicon	11.9
f	Frequency	Hz
f_{nc}	$1/f$ noise corner frequency	Hz
f_c	Cut-off frequency	Hz
Φ	Photon flux	$cm^{-2}\ s^{-1}$
φ_T	Thermal potential, kT/q	V (Volts)
G	Carrier generation rate	$cm^{-3}\ s^{-1}$
G_c	Conversion gain	$\mu V/e^-$
G_{SF}	Source follower gain	–
g_m	MOSFET gate transconductance	A V^{-1}
h	Planck's constant	6.62×10^{-34} J s
I	Current	A (Amperes)
i_n	Current noise density	$A/\sqrt{Hz}$
J	Current density	A cm^{-2}

k	Boltzmann constant	1.38×10^{-23} J K^{-1}
K	System gain	e^{-}/ADU
L_n, L_p	Diffusion length for electrons or holes	cm
λ	Photon wavelength	nm
μ_n, μ_p	Mobility for electrons or holes	$cm^2V^{-1}s^{-1}$
n	Electron concentration	cm^{-3}
n_p	Electron concentration in p-type semiconductor	cm^{-3}
n_{p0}	Electron concentration in p-type semiconductor in equilibrium	cm^{-3}
N_A	Acceptor concentration	cm^{-3}
N_D	Donor concentration	cm^{-3}
N_{ss}	Surface trap energy density	$cm^{-2}eV^{-1}$
N_{st}	Surface trap density	cm^{-2}
N_t	Trap concentration	cm^{-3}
p	Hole concentration	cm^{-3}
p_n	Hole concentration in n-type semiconductor	cm^{-3}
p_{n0}	Hole concentration in n-type semiconductor in equilibrium	cm^{-3}
P_{ph}	Optical power	W (Watts)
Q	Charge	C (Coulomb)
q	Elementary charge	1.6×10^{-19} C
R	Resistance	Ω (Ohm)
σ, σ_n, σ_p	Trap capture cross section for electrons or holes	cm^{-2}
S_{ph}	Photosensitivity	A W^{-1}
S_n	Surface recombination velocity for electrons	cm s^{-1}
t	Time	s (seconds)
τ_n	Electron lifetime	s
T	Temperature	K, °C
U	Recombination rate	cm^{-3} s^{-1}
U_s	Surface recombination rate	cm^{-2} s^{-1}
v_{th}	Electron or hole thermal velocity	cm s^{-1}
v_d	Drift velocity	cm s^{-1}
V_{GS}	MOSFET gate-source voltage	V
V_T	MOSFET threshold voltage	V

IOP Publishing

CMOS Image Sensors

Konstantin D Stefanov

Chapter 1

The fundamentals

1.1 Introduction—what is an image sensor and what does it do?

An image sensor has the job to convert an image, consisting of photons emitted or reflected by an object, into an electronic signal. To register a photon, it must be *absorbed* by the sensor and converted to an electric signal. For most sensors, this happens using the internal photoelectric effect. Once the electric signal is registered and processed, we obtain an electronic image, a representation of the incoming photons. The photon energy can span from the far-infrared up to x-rays.

Image sensors come in a bewildering variety of types, shapes and sizes. Most commonly, image sensors are segmented into individual sensitive elements, called pixels. Each pixel is intended to register photons independently of its neighbours, however, in most real-world sensors there is some electrical and optical crosstalk. An array of pixels forms a 2-D image sensor, which is by far the dominant type, and a line of pixels is a linear array, or 1-D image sensor. A simple photodiode has no pixel structure and can therefore be called a 0-D image sensor.

Typically, we expose the image sensor to photons for a certain time, called integration time. During that time, each pixel absorbs photons and registers an electric signal. The job of the image sensor, broadly speaking, is to determine *how many photons have been received in each pixel during the integration time*. Ideally, every photon should be registered separately, and some image sensors can do exactly that. More often, however, the pure photon-induced signal is mixed with the intrinsic noise of the sensor, and we can find out only the *average* number of registered photons. A linear photoresponse is often desired in an image sensor, so that the electrical output signal is proportional to the number of registered photons. Most sensors have small nonlinearity in their response not exceeding a few percent, but some types have intentionally much stronger nonlinearity, for example with logarithmic response, helping to achieve wide dynamic range.

The output signal from a sensor can be a continuous current proportional to the illumination, and this is the way the photodiodes are normally operated. However,

doi:10.1088/978-0-7503-3235-4ch1

in low light conditions the photogenerated currents can be extremely small and be counted in few electrons *per second*. Such currents are impossible to measure directly, therefore the method we use is to *integrate* the charge over a certain time on a collection element because this creates a signal that is much easier to measure. Most image sensors, including the ones described in this book, are of the integrating type.

1.2 Charge generation

1.2.1 Photoeffect

For a semiconductor image sensor to detect light, the photons impinging on it must interact with the sensitive regions of the sensor and be converted to an electric charge, which is then collected and recorded.

The dominant process of photon conversion in image sensors is the internal photoeffect. An incident photon with sufficient energy can liberate a valence electron from an atom, which becomes a free electron in the conduction band and can move about in the crystal lattice. At the same time, the missing valence electron becomes a hole, which is also mobile, as shown in figure 1.1. In this way, electrons and holes are created in pairs, and the photon is absorbed and disappears. The minimum photon energy for the photoeffect to occur is the bandgap of the semiconductor E_g. The excess photon energy above this threshold is dissipated as crystal vibrations or by generating secondary electron–hole pairs as the primary pair dissipates its kinetic energy.

Photons with energy lower than the bandgap E_g are not able to create electron–hole pairs, and since there is no other mechanism for photons to lose energy, silicon appears transparent at their wavelength. The photon energy E_{ph} is given by:

$$E_{ph} = \frac{hc}{\lambda} \tag{1.1}$$

where h is Planck's constant, c is the velocity of light and λ is the photon's wavelength. Equation (1.1) can be more conveniently written as (1.2) where the photon energy E_{ph} is in electron-volts ($1\ \text{eV} = 1.6 \times 10^{-19}$ J) and the wavelength λ is in nanometres:

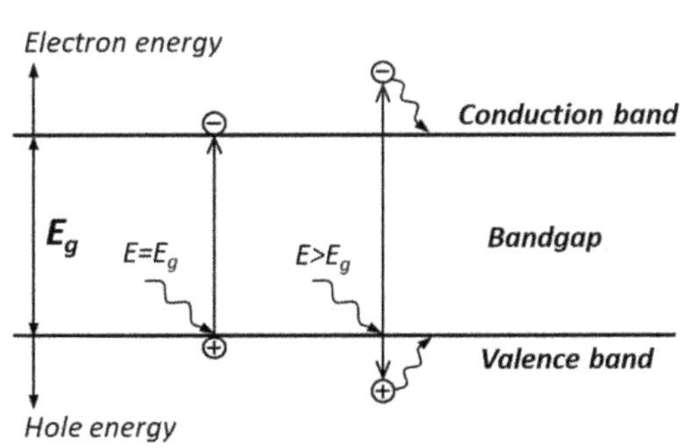

Figure 1.1. Photoeffect in semiconductors.

$$E_{ph} = \frac{1240}{\lambda} \tag{1.2}$$

Using (1.2) we can calculate that a photon with energy equal to the bandgap of silicon ($E_{ph} = E_g = 1.12$ eV at 300 K) has a wavelength of 1107 nm, and this is commonly referred to as the cut-off wavelength. The bandgap increases slightly at lower temperatures [2] leading to shorter cut-off wavelength and weaker absorption in the near-IR band.

In silicon, as an indirect bandgap semiconductor, the excitation of a valence electron into the conduction band requires that lattice vibrations (phonons) are involved to obey both energy and momentum conservation laws [2]. As the photon energy increases, the amount of momentum transfer must increase too, and the energy required to generate one electron–hole pair gradually increases. Because of this, a photon with energy equal to double the silicon bandgap (2.24 eV, or 554 nm) still generates one electron–hole pair, and not two.

The band structure of silicon has a direct bandgap of 3.1 eV [3] (400 nm) as well, allowing an electron–hole pair to be created directly, without the assistance of a phonon. However, it is thanks to its indirect bandgap that silicon is sensitive to visible light; if it only had the 3.1 eV direct bandgap it would have been sensitive only to wavelengths shorter than 400 nm and unusable for mainstream imaging.

A single electron–hole pair, corresponding to internal quantum yield of unity, is created up to a photon energy equal to three times the bandgap (3.36 eV, or 369 nm) [4]. Above this energy two e–h pairs begin to be created, but at a very low rate. Multiple e–h pair creation becomes significant only when the photon energy exceeds 4 eV (310 nm) [5, 6].

As the photon energy increases further, the ionisation energy E_w needed for the creation of one e–h pair reaches a peak of approximately 4.5 eV [4, 5]. For $E_{ph} > 10$ eV the pair creation energy E_w levels off to around 3.65 eV, as shown in figure 1.2.

An incoming beam of photons with flux Φ_0 (number of photons per unit area per second) and energy higher than the bandgap is gradually absorbed in the

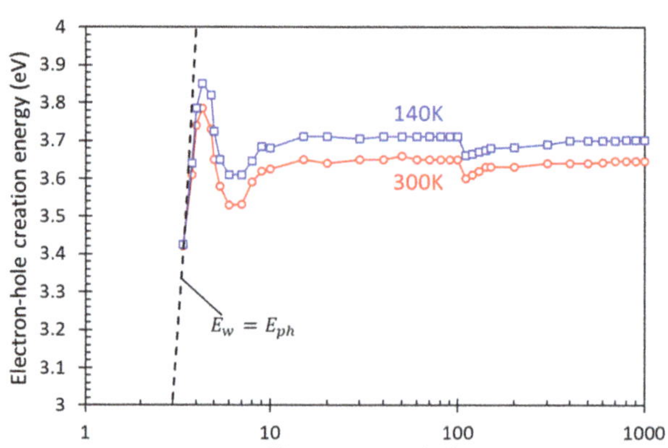

Figure 1.2. Electron–hole pair creation energy in silicon at 140 and 300 K (data from [1]). The dashed line marks $E_w = E_{ph}$.

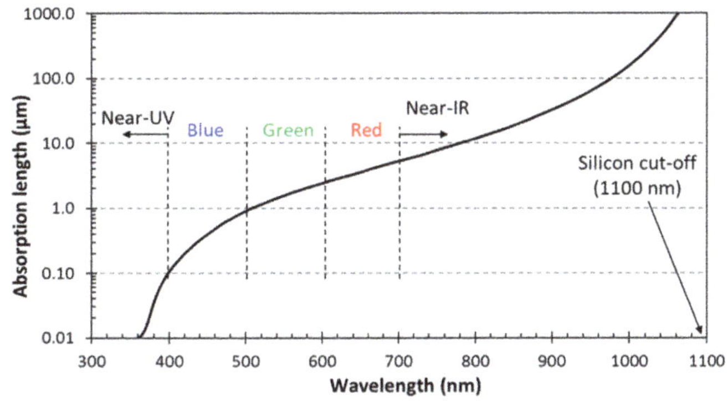

Figure 1.3. Photon absorption in silicon for low energy photons from near-UV to near-IR at 300 K. The data is from [7].

semiconductor. Ignoring any reflections, the flux $\Phi(x)$ at a distance x away from the illuminated surface is given by the Beer–Lambert law:

$$\Phi(x) = \Phi_0 e^{-\alpha x} \tag{1.3}$$

where α is the absorption coefficient, typically measured in units of cm^{-1}. At distance $x_0 = 1/\alpha$ the incoming photon flux is attenuated by $1/e$, which means that 63% of the light has been absorbed. The distance x_0 is called absorption length and is often more practical to use than the absorption coefficient because it allows straightforward comparison with the dimensions used in image sensors.

The absorption length depends strongly on the wavelength of light and changes by a factor of 50 between the lower end (400 nm) and the top end (700 nm) of the visible light range[1], as shown in figure 1.3 and table 1.1. As the bandgap increases at low temperatures the absorption length increases too, especially at near-IR wavelengths [7]. It is worth noting that photon absorption does not depend on the doping concentration or the free carrier concentration (either electrons or holes) in silicon for most practical cases.

Very often we would like to know what the silicon thickness should be to achieve certain level of photon absorption.

Example 1.1. Calculate the silicon thickness for 95% photon absorption for light with 400, 700 and 900 nm wavelength.
Solution: 95% absorption means that only 5% of the light is left. From formula (1.3) we have $e^{-\alpha x} = 0.05$ and therefore $x = -\ln(0.05)/\alpha = 3.0/\alpha$. From table 1.1 we get $1/\alpha = x_0 = 0.105$ µm for 400 nm wavelength, therefore the thickness is $x = 0.31$ µm. For 700 nm we have $1/\alpha = 5.263$ and 15.8 µm silicon thickness; and for 900 nm $x = 97.9$ µm.

---

[1] In the CIE (The International Commission on Illumination) luminous efficiency functions [8] the wavelength range of visibility is 380–700 nm.

Table 1.1. Absorption length in silicon for light wavelengths from 300 to 1100 nm at 300 K. Data from [7].

Wavelength (nm)	Absorption length (μm)
300	0.006
350	0.010
400	0.105
450	0.392
500	0.901
550	1.565
600	2.415
650	3.559
700	5.263
750	7.692
800	11.77
850	18.69
900	32.68
950	63.69
1000	156.3
1050	613.5
1100	2857

The light attenuation with distance for the wavelengths used in example 1.1 is plotted in figure 1.4 and illustrates the huge differences in silicon thickness required for the same absorption. This example shows the extremes of the visible range; in practice silicon thickness of 5 μm is often sufficient for visible light imagers because it allows acceptable absorption in the red end of the spectrum around 600–650 nm.

The fact that in the visible wavelength range each photon creates one electron–hole pair can be used to calculate the total light-generated charge in a volume of silicon. This charge, if collected, is the electrical output of the image sensor. Knowing the incident optical power P_{ph} (measured in watts) and the photon energy we can calculate the number of e–h pairs ΔN_{e-h} generated per unit time Δt based on the energy conservation law, simply as this:

$$\frac{\Delta N_{e-h}}{\Delta t} = \frac{P_{ph}}{E_{ph}} \tag{1.4}$$

As we can see the number of generated e–h pairs is inversely proportional to the photon energy, therefore lower energy photons, corresponding to near-IR and red light generate more carriers at the same optical power. While this is true, e–h pair generation requires that the photons are absorbed; for those long wavelengths the silicon must be very thick to ensure full absorption as figure 1.4 tells us.

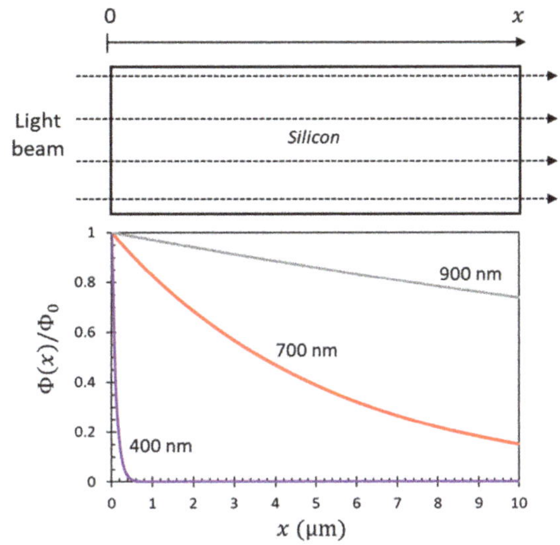

Figure 1.4. Photon absorption in silicon for violet (λ = 400 nm), red (λ = 700 nm) and for 900 nm light in the near-infrared.

Example 1.2. Calculate the number of e–h pairs generated per second by red light (λ = 650 nm) with an irradiance (power per unit area) of $E_e = 1$ W m^{-2} in a square pixel with a size a = 10 μm. Assume that the light is fully absorbed in the pixel's volume.
Solution: From equation (1.2) we find that for λ = 650 nm the photon energy is $E_{\text{ph}} = 1.91$ eV. The energy deposited in the pixel per second is the irradiance multiplied by the pixel area a^2. The number of generated e–h pairs per second is the energy deposited in the pixel per second, divided by the energy to create one e–h pair:

$$\frac{\Delta N_{\text{e-h}}}{\Delta t} = \frac{E_e a^2}{qE_{\text{ph}}} = \frac{1 \times 10 \times 10^{-6} \times 10 \times 10^{-6}}{1.6 \times 10^{-19} \times 1.91} = 3.27 \times 10^{8}\ \text{s}^{-1}$$

Here we have multiplied E_{ph} by the elementary charge q to convert the photon energy from eV to Joules. Irradiance of 1 W m^{-2} at 650 nm corresponds to approximately 68.3 lux, or a dimly lit room. This example shows that even meagre illumination manages to create a third of a billion e–h pairs every second in a tiny 10 μm pixel.

If all the e–h pairs were collected, a steady state *photocurrent* I_{ph} will flow; it is given by multiplying formula (1.4) by the elementary charge to convert the number or e–h pairs per second to coulombs per second, which is current:

$$I_{\text{ph}} = \frac{q\Delta N_{\text{e-h}}}{\Delta t} \tag{1.5}$$

Example 1.3. Calculate the photocurrent flowing in the pixel in example 1.2.
Solution: Multiplying the answer by the elementary charge gives:

$$I_{ph} = \frac{q\Delta N_{e-h}}{\Delta t} = 1.6 \times 10^{-19} \times 3.27 \times 10^{8} = 5.235 \times 10^{-11}\text{A} = 52.4\ \text{pA}$$

Very often the sensitivity of a semiconductor material is given as the photocurrent per watt of incident light energy. To compare devices irrespective of their structure or pixel size the current can be expressed as *current density* J_{ph}, i.e. amperes per unit area. If we use the optical power per unit area E_e (irradiance), equation (1.5) can be rewritten by dividing both sides by the device (or pixel) area, and using (1.4) we get

$$J_{ph} = \frac{qE_e}{E_{ph}}. \tag{1.6}$$

From (1.6) we arrive at the astonishingly simple expression (1.7) for the sensitivity S_{ph} in terms of detector current density per unit of optical irradiance. S_{ph} is measured in units of ampere per watt (A W^{-1}) because the area cancels from both J_{ph} and E_e.

$$S_{ph} = \frac{J_{ph}}{E_e} = \frac{q}{E_{ph}} \tag{1.7}$$

Equation (1.7) gives the theoretical maximum photosensitivity of an ideal image sensor having perfect light absorption, without any light losses due to reflection, and of course with complete charge collection. The ratio E_{ph}/q gives the photon energy in units of eV. From here we can calculate that the theoretical maximum photosensitivity for $\lambda = 650$ nm (as in example 1.2) is $S_{ph} = 1/1.91 = 0.52$ A W^{-1}. Using (1.2), expression (1.7) can also be written as $S_{ph} = \lambda/1240$, where λ is in nanometres.

1.2.2 Ionisation

Silicon makes an excellent sensor material not just for the near-IR, visible and UV light, but also for much more energetic photons, such as x-rays. The absorption length covering photon energies from 1.2 eV to 10 keV in figure 1.5 shows what happens at the higher end of silicon's usable range: similarly to the near-IR end, silicon becomes transparent beyond photon energy of about 10 keV. The absorption above 100 eV shows discontinuous absorption edges at the energy levels of the L-shell ($\approx$100 eV) and the K-shell (1839 eV) of the silicon atom. As the photon energy exceeds the binding energy of a shell, the electrons occupying it can be excited and the photon absorption sharply goes up. This corresponds to a stepwise *decrease* in the absorption length, most clearly seen at the K-edge.

For photons with $E_{ph} > \approx 50$ eV the ionisation energy is $E_w = 3.65$ eV at 300 K [11] and is nearly constant. This allows us to calculate the number of electron–hole pairs N_{e-h} generated by x-rays and gamma-rays using this simple expression:

$$N_{e-h} = \frac{E_{ph}}{E_w} \tag{1.8}$$

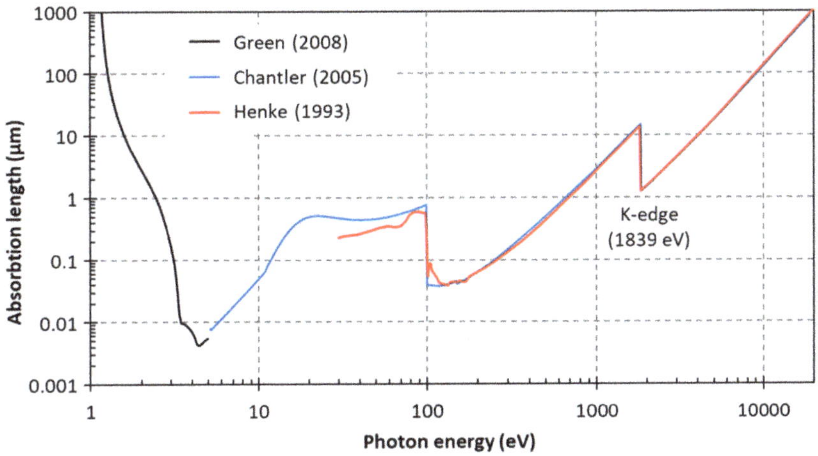

Figure 1.5. Photon absorption due to photoeffect in silicon for a wider energy range. Data for <5 eV from [7], 5 – 20000 eV from [9], and >30 eV from [10].

The ionisation energy is temperature dependant; this is due to the reduction of the bandgap as the temperature increases [1, 12]. X-rays in the range 1–10 keV, emitted by radioactive sources and x-ray fluorescence from various materials are particularly useful for sensor characterisation. They are widely used for calibration due to the well-known x-ray energies and the amount of charge created in silicon by them. Also, the initial charge cloud created by low energy x-rays is very compact [13, 14] and this allows the charge to be considered a point source.

One very popular calibration source is the ^{55}Fe isotope which decays via electron capture to manganese (^{55}Mn) with a half-life of 2.737 years. ^{55}Mn emits characteristic K-shell x-rays with energies 5.89 keV (Mn-K$_\alpha$) and 6.49 keV (Mn-K$_\beta$), with probabilities of 24.4% and 2.9%, correspondingly [15]. Figure 1.6 shows an example of a ^{55}Fe spectrum obtained by a CMOS image sensor.

The absorption length for 5.9 keV photons in silicon is approximately 28 µm [10]. This length is much larger than the depth of the active silicon in the typical optical sensor, therefore only a small fraction of the incoming x-rays is absorbed and converted to charge.

Example 1.4. Calculate the number of electron–hole pairs generated by the dominant Mn-K$_\alpha$ x-ray, and the current in a pixel receiving one hundred Mn-K$_\alpha$ x-rays per second, assuming that all the charge is collected.

Solution: Using formula (1.8) the number of generated electron–hole pairs per x-ray is

$$N_{\text{e–h}} = \frac{E_{\text{K}\alpha}}{E_{\text{w}}} = \frac{5890}{3.65} = 1614.$$

The current for 100 x-rays per second is calculated using (1.5):

$$I_{\text{ph}} = \frac{qN_{\text{e–h}}}{t} \times 100 = \frac{1.6 \times 10^{-19} \times 1614}{1} \times 100 = 25.8 \text{ fA}$$

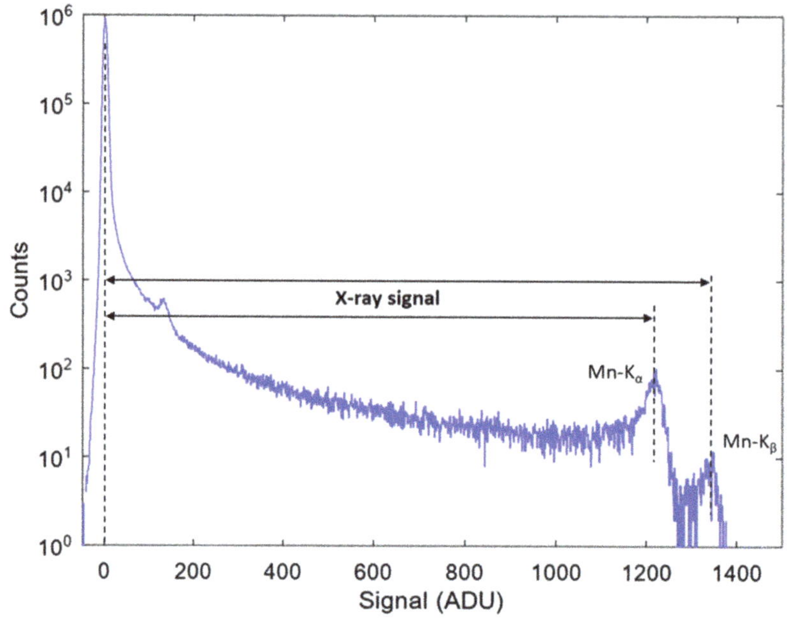

Figure 1.6. ^{55}Fe spectrum obtained by a CMOS image sensor. The peak at 0 ADU is due to pixels without x-ray signal, and the continuum leading to the x-ray peaks is caused by charge collected by more than one pixel.

At much higher photon energies two other mechanisms overtake the photoeffect and begin to play an increasing role—Compton effect and electron–positron pair creation [16]. Besides the detection of optical and x-ray photons, silicon is widely used for the detection of high energy, ionising particles. Traversing the material, charged particles lose energy due to several mechanisms, and most of that energy loss is due to ionisation. The ionisation loss is very high at low particle energies but decreases and flattens off at higher energies in a logarithmic dependence [16]. Particles with energies at and above the plateau, which usually lies at hundreds of MeV, are called minimum ionising particles (MIP).

The average number of the created electron–hole pairs is calculated by the most probable energy loss divided by the ionisation energy as in (1.8). The energy loss due to ionisation has large statistical fluctuations. For particles which lose only a small part of their energy in the material (i.e. the material is 'thin') the energy loss is described by the Landau distribution [17]. In a couple of microns of silicon there is a significant probability that a traversing high energy particle will cause no ionisation at all [18]. At the same time, the probability that the energy loss can far exceed the most probable value is also significant, due to the long tail in the Landau distribution.

The most probable number of e–h pairs increases with the thickness, as shown in figure 1.7, and above 10 μm the energy loss begins to approach the Landau distribution.

1.3 Charge collection

In the previous section we looked at how photons generate free charge, consisting of electrons and holes, and how the amount of charge depends on the photon energy. The question now is how to collect this charge and measure it.

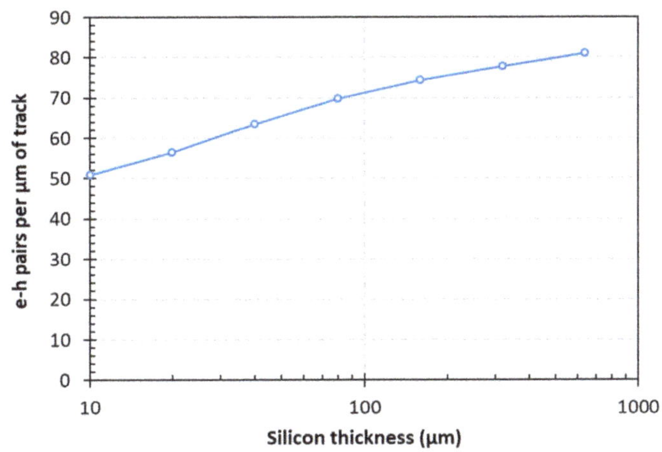

Figure 1.7. Most probable number of electron–hole pairs per micrometre of track for high energy charged particles, data from [17].

Left on its own, the charge will simply diffuse out, never to be seen again. Diffusion is fundamental in nature and always occurs when there is a difference in carrier concentration. The charge generated in pixels receiving more illumination will diffuse towards pixels receiving less illumination, until we get nearly uniform charge 'blob' everywhere. Obviously, this is not what we want to happen in an image sensor.

We need a charge collecting element—something that is electrically attractive to either electrons or holes (but obviously cannot be attractive to both). To make the charge move in a particular direction for collection we need to create an *electric field*; within it the charge experiences an electrostatic force and begins to accelerate in a direction opposite to the field (for electrons) and along the field (for holes). This charge movement in the presence of electric field is called *drift* and is the primary mechanism for charge collection. During drift the charge continues to diffuse due to its concentration gradient, regardless of the presence of any electric field; this is unavoidable but not always undesirable.

1.3.1 Carrier lifetime

An important point in image sensor operation, which often goes without much mention, is that the generated e–h pairs must survive, i.e. not recombine or get trapped, for sufficiently long time so that they can be collected. The characteristic describing the 'life duration' of electrons and holes is called *carrier lifetime* [19] and is widely used in semiconductor physics. What 'sufficiently long' means in practice will be explored in the next two sections. As a rough indicator the charge collection time rarely exceeds a few hundreds of nanoseconds, and this is how long the carriers must survive. Carrier lifetime can be many orders of magnitude longer, especially in high quality epitaxial silicon.

Whenever electrons and holes are created, for example by illumination with light, the excess carrier concentration will decay back to equilibrium after the source of e–h pair generation is turned off. In silicon the dominant physical mechanism for the decay is trap-assisted recombination. Direct e–h recombination occurs too, but at a much smaller rate. The rate of decay towards the equilibrium concentration, expressed as the change of carrier concentration per unit time $U = \Delta n/\Delta t$, is called *recombination rate*. The simplest possible mathematical description of this process is to assume that the recombination rate U is proportional to the excess carrier concentration. For example, if the electron concentration in p-type silicon is n_p and the equilibrium concentration is n_{p0}, the recombination rate can be written as $U \propto (n_p - n_{p0})$. Since proportionality is assumed, we need a proportionality constant in units of seconds, so that the recombination rate is measured in units of carrier concentration over time (cm^{-3} s^{-1} in semiconductor physics). This constant is called the carrier lifetime (τ_n for electrons, τ_p for holes) and can be thought of as the characteristic time over which the carrier concentration decreases. Now, the recombination rate can be written in its familiar form:

$$U = -\frac{dn_p}{dt} = \frac{n_p - n_{p0}}{\tau_n} \tag{1.9}$$

We have added a negative sign in (1.9) because due to recombination the carrier concentration decreases ($dn_p/dt < 0$) when $n_p - n_{p0} > 0$, and $U > 0$. Because n_{p0} is constant, we can write that

$$\frac{d(n_p - n_{p0})}{dt} = -\frac{n_p - n_{p0}}{\tau_n} \tag{1.10}$$

and after separating the variables we can integrate both sides:

$$\int \frac{d(n_p - n_{p0})}{n_p - n_{p0}} = -\frac{1}{\tau_n} \int dt \tag{1.11}$$

$$\ln(n_p - n_{p0}) + \text{const} = -\frac{t}{\tau_n} \tag{1.12}$$

The final equation can be written by using the initial conditions: at $t = 0$ the excess electron concentration is $n_p(0) - n_{p0}$ and at $t \to \infty$ naturally $n_p - n_{p0} = 0$, with only the equilibrium concentration n_{p0} left. Therefore, the constant in equation (1.12) must equal $-\ln\left(n_p(0) - n_{p0}\right)$ and we arrive at the time dependence of n_p:

$$n_p(t) = n_{p0} + (n_p(0) - n_{p0})e^{-\frac{t}{\tau_n}} \tag{1.13}$$

Equation (1.13) tells us that the electron lifetime τ_n is the characteristic time over which the excess carrier concentration decreases by $1/e$, i.e. only 37% of the excess carriers remain. After three times the lifetime only 5% of the initial charge will be left.

1.3.2 Recombination

This is a good place to answer an important question: why don't the electrons and the holes recombine immediately after they are generated? After all, they are created together and close to each other, and it would be reasonable to expect that they should recombine at high rate. Fortunately, such direct (band-to-band) recombination in silicon is very rare because it is an indirect bandgap semiconductor. Carrier lifetime controlled by band-to-band recombination is very long; in high purity silicon the electron and hole lifetimes can be many milliseconds. This long lifetime allows the charge to diffuse out a long distance from the place it was generated without recombining, unless it is quickly collected with the help of an electric field.

Figure 1.8 shows two direct, and very rare band-to-band recombination mechanisms: radiative with the emission of a photon, and Auger recombination where the excess energy is transferred to another carrier, such as a hole in highly doped p-type silicon.

The third one, trap-assisted recombination via a mid-band trap, is by far the dominant mechanism in silicon, described by the Shockley–Read–Hall (SRH) theory [20]. Traps are produced by imperfections or impurities in the crystal lattice, which introduce energy levels deep in the bandgap. Traps take part in both capture and emission of carriers and are also called generation-recombination centres.

The recombination rate from a trap with concentration N_t and energy level E_t above the valence band is given by

$$U = \frac{\sigma_n \sigma_p v_{\text{th}}(pn - n_i^2)N_t}{\sigma_n\left[n + n_i \exp\left(\frac{E_t - E_i}{kT}\right)\right] + \sigma_p\left[p + n_i \exp\left(-\frac{E_t - E_i}{kT}\right)\right]} \tag{1.14}$$

Here E_i is the intrinsic Fermi level (approximately the mid-band energy level);
k is the Boltzmann's constant;
T is the absolute temperature;
v_{th} is the carrier thermal velocity;
n_i is the intrinsic carrier concentration;
n and p are the electron and hole concentrations, respectively;
σ_n and σ_p are the electron and hole capture cross-sections, respectively.

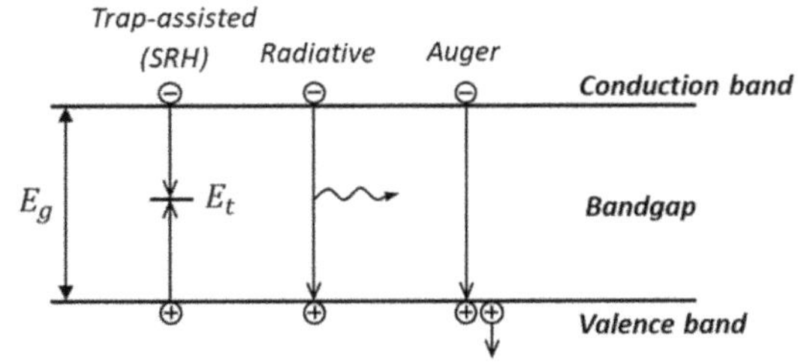

Figure 1.8. Recombination mechanisms.

The maximum recombination rate occurs when the denominator is at its minimum, achieved when $E_t = E_i$. This indicates that mid-band traps are the most effective in limiting the carrier lifetime.

In a p-type semiconductor we have $p \gg n_i$ and $p \gg n$, and if the two capture cross-sections are similar ($\sigma_n \cong \sigma_p$), the second term in the denominator of (1.14) becomes much larger than the first. Therefore, we can write that

$$U \approx \frac{\sigma_n v_{\text{th}}(pn - n_i^2)N_t}{p} \tag{1.15}$$

Introducing a small excess concentration of electron–hole pairs does not change significantly the equilibrium hole concentration p_{p0} in p-type silicon. Using that $p \approx p_{p0}$ and $n_i^2 = p_{p0}n_{n0}$ [21], equation (1.15) can be written as

$$U \approx \frac{\sigma_n v_{th}(p_{p0}n_p - p_{p0}n_{n0})N_t}{p_{p0}} = \sigma_n v_{\text{th}}N_t(n_p - n_{n0}) \tag{1.16}$$

Comparing with (1.9) we see that the term multiplying the concentration difference is the inverse of the electron lifetime (also called recombination lifetime)

$$\tau_n = \frac{1}{\sigma_n v_{\text{th}}N_t} \tag{1.17}$$

Typical capture cross-sections are in the range 10^{-16} to 10^{-14} cm^2.

Example 1.5. Calculate the electron lifetime due to a mid-band trap ($E_t = E_i$) with $\sigma_n = 10^{-15}$ cm^2 and concentration 10^{12} cm^{-3}, using that the electron thermal velocity is $v_{\text{th}} = 1.4 \times 10^7$ cm s^{-1}.
Solution: Using (1.17) we get

$$\tau_n = \frac{1}{10^{-15} \times 1.4 \times 10^7 \times 10^{12}} = 71.4\ \mu\text{s}$$

To put this into perspective, trap concentration of 10^{12} cm^{-3} corresponds to an average of one trap per cubic micrometre, or one trap per 50 billion Si atoms.

Measurements show that in high quality, lightly doped (<10^{15} cm^{-3}) silicon the minority carrier lifetime can be tens of milliseconds [22]. Considering the calculation in example 1.5, this implies that the trap density responsible in the SRH model must be less than 10^{10} cm^{-3}, or one trap per 5 trillion atoms. As the dopant concentration increases above 10^{16} cm^{-3} the lifetime begins to decrease and this is taken into account as concentration-dependent SRH lifetime [23].

At high carrier density, such as along a dense ionisation track or in solar cells, Auger and band-to-band radiative recombination can begin to limit the lifetime even in low-doped silicon. Auger recombination involves a direct recombination between an electron and a hole, with the excess energy transferred to another electron or hole. The Auger lifetime for high excess carrier concentration in a lightly doped semiconductor is [22]:

$$\tau_{\text{Auger}} = \frac{1}{C_a \Delta n^2} \tag{1.18}$$

Here C_a is the ambipolar Auger coefficient (1.66×10^{-30} cm^6 s^{-1} in silicon [24]) and $\Delta n = n_p - n_{n0}$ or $\Delta n = p_n - p_{p0}$ is the excess carrier concentration.

The lifetime due to band-to-band radiative recombination is given by

$$\tau_{\text{rad}} = \frac{1}{B \Delta n} \tag{1.19}$$

where $B = 4.7 \times 10^{-15}$ cm^3 s^{-1} is the radiative coefficient in silicon at 300 K [25]. It is easy to see that the Auger and radiative lifetimes are much larger than τ_{SRH} and can become comparable to the SRH lifetime only at excess concentration above 10^{16}–10^{17} cm^{-3}. In a typical image sensor, the excess carrier concentration due to illumination rarely exceeds 10^8–10^{10} cm^{-3}, therefore the direct recombination mechanisms have negligible influence.

The total lifetime τ_{tot} can be calculated from Matthiessen's rule for the three recombination processes as in [19]

$$\frac{1}{\tau_{\text{tot}}} = \frac{1}{\tau_{\text{SRH}}} + \frac{1}{\tau_{\text{Auger}}} + \frac{1}{\tau_{\text{rad}}} \tag{1.20}$$

Equation (1.20) is analogous to the one used to calculate the resistance of parallel resistors; physically it means that the different recombination mechanisms work independently and in parallel.

1.3.3 Drift

We are going to consider a hypothetical collection element without specifying what it is and how it is made; then in the following section we will talk about two real charge collection elements—the *pn* junction and the MOS capacitor.

Figure 1.9 shows our hypothetical collection element. From the surface down to depth d there is a constant electric field E, and below d the field is zero. This may look artificial but is not far off from reality.

In this structure only the electrons are collected, and the holes are discarded never to be seen again, as is typical for most image sensors. Electrons are preferred because they move much faster than the holes, resulting in shorter collection times. Holes are allowed to diffuse until they reach the backside substrate electrode, or they recombine after travelling a long distance away from the charge collection element.

Within a region having an electric field E electrons experience a force $F = -qE$ and begin to accelerate. Holes experience the same force but with the opposite sign and move in the other direction. As mentioned previosly, this movement under the influence of an electric field is called *drift*. In semiconductors it is experimentally

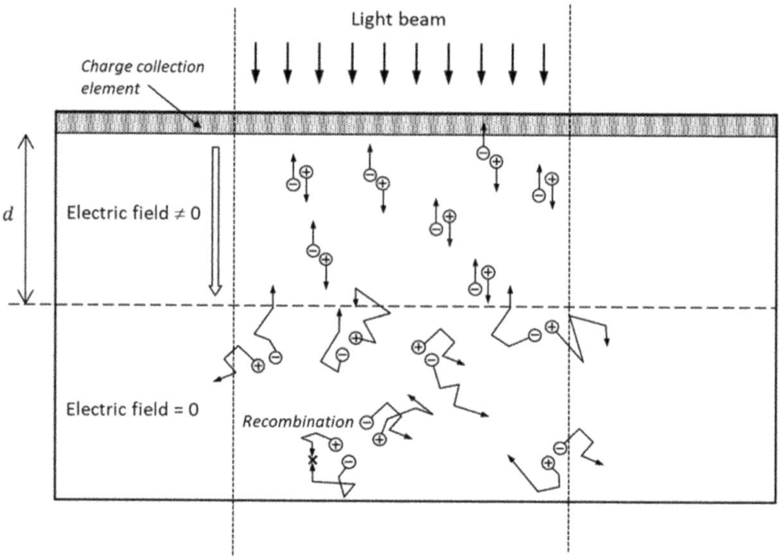

Figure 1.9. Charge collection of electrons created by a light beam and experiencing drift and diffusion.

observed that at low electric fields ($<\sim 10^4$ Vcm^{-1}) the charge carriers acquire drift velocity v_d proportional to the electric field E:

$$v_d = \mu E \tag{1.21}$$

The proportionality factor μ is called *mobility* and is not a constant—it decreases as the temperature and the doping concentration increase [26]. The electron mobility μ_n at low fields is about three times higher than the hole mobility μ_p; the values in low-doped silicon at 300 K are $\mu_n = 1400$ cm^2 V^{-1} s^{-1} and $\mu_p = 470$ cm^2 V^{-1} s^{-1} [26]. The higher electron mobility and velocity is one of the main reasons why we prefer collecting and transferring electrons, rather than holes.

The drift velocity decreases below the value calculated by (1.21) at higher electric fields (above 10^4 V cm^{-1}), and stops increasing altogether (i.e. saturates) at $E > 10^5$ V cm^{-1}. The saturation drift velocity v_d^{sat} for both electrons and holes is approximately 10^7 cm s^{-1}.

Drift velocity is directional and determined by the applied electric field; it is also superimposed on the thermal carrier velocity caused by their random movement in the crystal lattice. The thermal velocity for electrons is given by

$$v_{th} = \sqrt{\frac{3kT}{m_0^*}} \tag{1.22}$$

where k is the Boltzmann constant, T is the absolute temperature and $m_0^* = 0.26m_0$ is the effective electron mass [2]. At 300 K formula (1.22) gives $v_{th} = 2.3 \times 10^7$ cm s^{-1}. This is similar to the experimentally observed saturation velocity v_d^{sat}, and is an indication that the simple proportionality in formula (1.21) is valid only when $v_d \ll v_{th}$.

Example 1.6. Calculate the electron drift velocity for $E = 1000\ \text{V cm}^{-1}$ and electron mobility $\mu_n = 1400\ \text{cm}^2\ \text{V}^{-1}\ \text{s}^{-1}$ (value for Si at 300 K and for low doping concentration), and compare it to the thermal velocity v_{th}.
Solution: From formula (1.21) we get

$$v_\text{d} = \mu_n E = 1400 \times 1000 = 1.4 \times 10^6\ \text{s}^{-1}$$

This velocity is about 20 times lower than the random thermal electron velocity.

Knowing the drift velocity allows us to calculate the time it takes to collect the charge. For simplicity we can consider that v_d is much smaller than the saturation velocity. The travel distance x is simply the velocity multiplied by the time:

$$x = v_\text{d} t = \mu_n E t \tag{1.23}$$

The charge collection time t is the device thickness divided by the drift velocity, and substituting the drift velocity from (1.21) we arrive at:

$$t = \frac{d}{v_\text{d}} \cong \frac{d}{\mu_n E} = \frac{d^2}{\mu_n V} \tag{1.24}$$

Here we have used that the electric field is the applied voltage V divided by the thickness d. This is an approximate and simple, but very useful formula; we will refine it further in the following sections.

Example 1.7. Calculate the charge collection time in silicon with thickness $d = 5\ \mu\text{m}$ and applied voltage across it $V = 1$ V. The electron mobility is $\mu_n = 1400\ \text{cm}^2\ \text{V}^{-1}\ \text{s}^{-1}$. Compare with the charge collection time under velocity saturation ($v_\text{d}^\text{sat} \cong 10^7\ \text{s}^{-1}$).
Solution: First, we need to see how the electron drift velocity compares to the saturation velocity. Using (1.21)

$$v_\text{d} = \mu_n E = \mu_n \frac{V}{d} = 1400 \times \frac{1}{5 \times 10^{-4}} = 2.8 \times 10^6\ \text{s}^{-1}$$

we see that it is about a factor of 3 lower than v_d^sat, therefore formula (1.24) can be used and gives:

$$t = \frac{(5 \times 10^{-4})^2}{1400 \times 1} = 0.18\ \text{ns}$$

The collection time under velocity saturation is calculated as

$$t = \frac{d}{v_\text{d}^\text{sat}} = \frac{5 \times 10^{-4}}{10^7} = 0.05\ \text{ns}$$

This example shows that for the typical sensor thicknesses the charge collection time by drift is very short and may become an issue only if very fast operation is required. However, in a much thicker sensor, made so that it can have higher absorption at near-IR wavelengths, the charge collection time could be substantially longer due to the quadratic dependence on the device thickness.

1.3.4 Diffusion

Charge generated in a field-free semiconductor diffuses out from the point at which it is created and can travel large distances. It can reach a region with an electric field, where it is swept away, or it can recombine with the assistance of bulk or surface traps.

If we generate a point-like sphere of electron–hole pairs, they will expand stochastically in a cloud described by the Gaussian distribution. After time t, the RMS cloud radius r_n for electrons and r_p for holes in one dimension is given by:

$$r_n = \sqrt{2D_n t} \tag{1.25}$$

$$r_p = \sqrt{2D_p t} \tag{1.26}$$

where D_n is the diffusion coefficient for electrons and D_p for holes, measured in $\text{cm}^2\ \text{s}^{-1}$. The diffusion coefficients are connected to the mobility via the Einstein relationship:

$$D_n = \mu_n \frac{kT}{q} \tag{1.27}$$

$$D_p = \mu_p \frac{kT}{q} \tag{1.28}$$

The diffusion radii are the standard deviations of the charge density spread; from the properties of the Gaussian distribution, we know that 95% of the charge is contained within radius $2r_n$ for electrons and $2r_p$ for holes.

Figure 1.10 shows the diffusion spread with time of a point-like charge generated at $t = 0$ without any recombination. The electron density is described by a Gaussian with standard deviation given by (1.25), with the peak moving by a distance x determined by (1.23) when the electric field is not zero.

Electrons spread $\sqrt{3}$ times faster than holes due to their diffusion coefficient being three times higher; the values can be calculated from (1.29) and (1.30) and are $D_\text{n} = 36\ \text{cm}^2\ \text{s}^{-1}$, $D_\text{p} = 12\ \text{cm}^2\ \text{s}^{-1}$ at 300 K.

During diffusion the charge carrier concentration decreases because they spread out; at the same time their concentration decreases also because they are subjected to various recombination processes, with their combined influence reflected in the carrier lifetime. The longest distance the carriers can travel is naturally limited by their lifetime and is called *diffusion length*. It is an important parameter in semiconductors and enters numerous formulas describing image sensor operation. The diffusion length L_n for electrons is defined by:

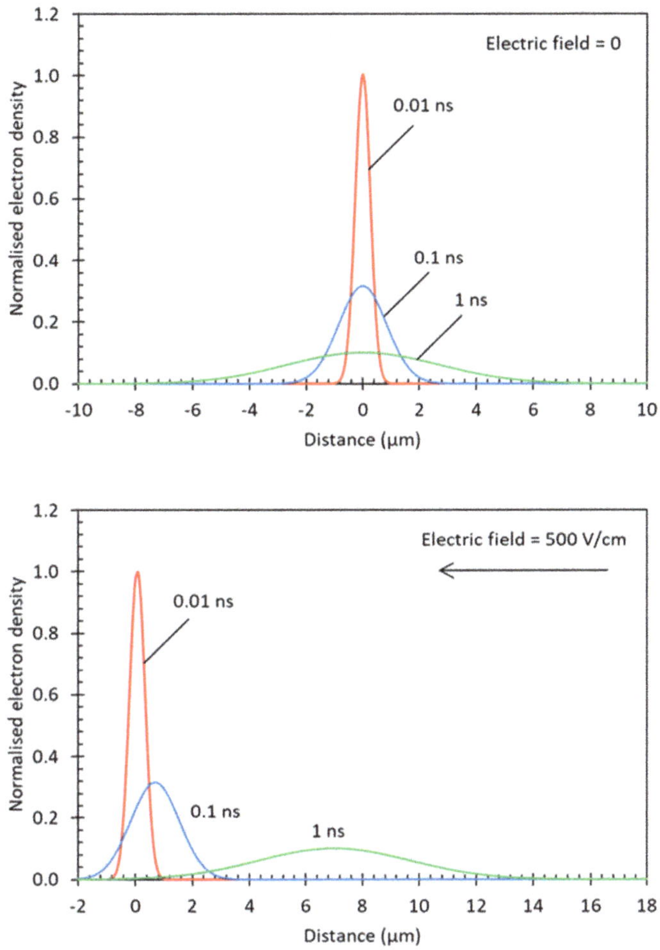

Figure 1.10. Electron spread only due to diffusion (at zero electric field) and due to drift and diffusion with electric field = 500 V cm^{-1} (after [2], p 55).

$$L_n = \sqrt{D_n \tau_n} \tag{1.29}$$

and a similar expression can be written for hole diffusion length L_p. The diffusion length is usually much larger than the typical pixel sizes due to the long carrier lifetime.

Example 1.8. Calculate the diffusion length for electrons in silicon with electron lifetime $\tau_n = 1$ ms (fairly typical for low-doped, high quality epitaxial silicon). The diffusion coefficient is $D_n = 36\ \text{cm}^2\ \text{s}^{-1}$.

Solution:

$$L_n = \sqrt{36 \times 10^{-3}} = 1897\ \mu\text{m}$$

This is a *really long* distance; some image sensors are physically smaller than this!

Since charge travels so well on its own, can we collect it using only diffusion? The short answer is 'No'—collection entirely by diffusion can lead to significant losses because the charge is not going to stay in the confines of the pixel where it was generated. Diffusion is isotropic, and charge moves in all directions. To capture it, we need the collection element to surround the charge on all sides, or a special structure that forces the charge to go in one predominant direction.

Drift is a far better choice for charge collection because it is directional towards the source of electric field. It is also much faster and minimises the chance of charge loss. The distance travelled under drift (1.23) is proportional to time, while the diffusion radius (1.25) increases much slower as a square root. Also, electrons move three times faster than holes in electric field and not only $\sqrt{3}$ times faster as in diffusion.

Example 1.9. Calculate the electron collection time if the charge moves only by diffusion for the values in example 1.7, assuming that the charge is forced to travel in one direction, and there is no charge loss.

Solution: Using equation (1.25) we can calculate the time it takes the electrons to travel the longest distance, equal to the depth of the sensor $r_n = d = 5\ \mu\text{m}$:

$$t = \frac{r_n^2}{2D_n} = \frac{(5 \times 10^{-4})^2}{2 \times 36} = 3.5 \text{ ns}$$

Because of the stochastic nature of the diffusion this time can be considered as a time constant; three time constants would suffice for 95% charge collection, and equals to 10.5 ns. This time is two orders of magnitude longer than in charge collection by drift.

1.4 Charge transfer

After the charge is collected it needs to be converted to a voltage or a current that can be measured using electronic circuits. This could happen at the collection element itself, but very often the tasks of collection and conversion are physically separated for good reasons. The charge needs to travel to a dedicated place where it is converted to an electrical signal; therefore, the task is to perform an efficient *charge transfer*.

Figure 1.11 shows schematically three cases of charge transfer. Figure 1.11(a) corresponds to CMOS image sensors using pinned photodiode as a collection element and is probably the most widely used. Here the photogenerated electrons are directed to the collection element by an electric field. The conversion element is shielded from direct charge collection and receives only the charge transferred to it. This is usually accomplished by either an optical shield over the conversion element, or electrically—by steering the electrons away from it.

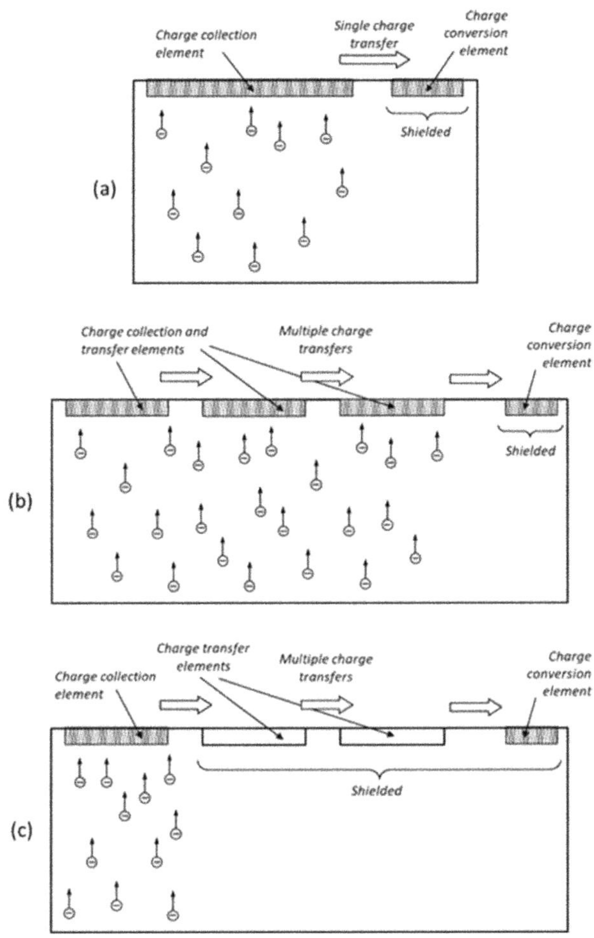

Figure 1.11. Charge transfer in image sensors: (a) single transfer; (b) multiple transfers with charge collection elements capable of charge transfer; (c) multiple transfer with dedicated transfer elements which do not collect charge.

Figures 1.11(b) and 1.11(c) show examples where the charge travels larger distances, and the transfer is done is many steps. The transfer elements can either be the same as the charge collection elements, as in full frame CCDs, or they can be dedicated to charge transport only, as in interline transfer CCDs.

1.5 Charge conversion

Charge conversion is a crucial step that occurs at a *sense node*, where the charge is converted to an electrical signal—most frequently a voltage. This can happen *destructively*—meaning that after the conversion the charge cannot be recovered or returned to its original state. This is the typical conversion in image sensors.

The conversion can also occur *non-destructively*, which is when the charge remains unaltered and intact, and the same charge can be converted multiple times.

A circuit allowing non-destructive conversion would normally couple capacitively to the charge without getting in physical contact with it.

Why is destructive conversion preferred? Because it is more sensitive and a larger electrical signal can be obtained for the same charge, it is usually sufficient to convert the signal into voltage only once. Multiple signal measurements requiring non-destructive conversions are used only in some specialised sensors where they offer some advantages, such as charge measurement during collection, or for noise reduction.

Figure 1.12 shows four methods of charge conversion: (a)–(c) are destructive because there is an electrode attached to the collection element; once the electrons reach it, they exit the device through that external connection and there no easy way of reversing this process.

In figure 1.12(a) the photogenerated current is continuously flowing out of the device through a resistor; the voltage drop across it (i.e. the difference between the bias voltage and the output) is the measure of the photocurrent.

Figure 1.12(b) shows the typical charge conversion circuit in the vast majority of image sensors. A transistor act as an electronic switch to momentarily connect the output to a stable bias voltage, and after that it is disconnected. The output terminal has an equivalent capacitance C to ground, and when the charge Q reaches the output a voltage step is produced:

$$\Delta V = \frac{Q}{C} \tag{1.30}$$

This voltage step is proportional to the photogenerated charge and is the output signal. Figure 1.12(c) shows a more sophisticated version which uses a Charge

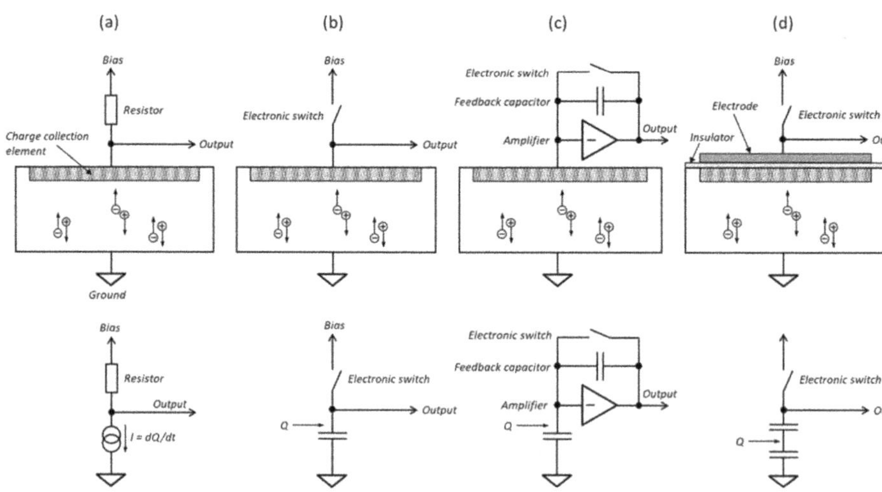

Figure 1.12. Charge conversion types: (a) continuous current on a resistor; (b) on the effective capacitance of the charge collection element; (c) using a charge-sensitive amplifier; (d) non-destructive by electrostatic induction.

Sensitive Amplifier (CSA). Due to the negative feedback the CSA 'moves' the incoming charge to the feedback capacitor and the voltage step (1.30) appears across it, while the input stays at nearly constant voltage. The output signal can be large because the feedback capacitor can be made very small.

Figure 1.12(d) illustrates the fourth method of charge conversion discussed here, where the charge is sensed non-destructively by capacitive coupling. The output is the top electrode of a floating capacitor, separated by an insulator from the structure underneath. The collection element has capacitance to both the output and to ground, so the equivalent circuit has two capacitors connected in series. After the switch is disconnected, the collected charge couples by electrostatic induction to the output electrode and produces a voltage change across the effective capacitance. This method does not interfere with the collected charge because there is no connection to it.

The common feature between the charge conversion methods in figures 1.12(b) and (c) is that the charge is converted to voltage on a capacitance. This may not be a physical capacitor as the feedback capacitor in figure 1.12(c), but an effective capacitance to ground formed by the structure of the charge collection element.

The conversion to voltage is characterised by the *conversion gain* G_c, expressed as the voltage change at the output per one collected electron. Using (1.30), the conversion gain is written as:

$$G_c = \frac{q}{C} \tag{1.31}$$

where q is the elementary charge and C is the *conversion capacitance*. The term charge to voltage factor (CVF) is also frequently used as a synonym for the conversion gain.

1.6 *pn* junction

The *pn* junction is arguably the most important element in image sensor technology. It can be used for both charge collection and charge-to-voltage conversion and is used in practically all image sensors. Many excellent books describing the *pn* junction have been written (for example, [2] and [21]) and we are going to spend a great deal on it too, due to its importance to image sensors.

1.6.1 *pn* junction in equilibrium

The *pn* junction is one of the types of charge collection elements with internal electric field, discussed in the previous section. To form a *pn* junction a donor dopant is implanted into *p*-type silicon (or the other way around), and then the device is annealed to activate the implant[2]. The free electrons from the *n*-side start diffusing into the *p*-type silicon and the free holes from the *p*-side into the *n*-type silicon; this occurs naturally because of the concentration differences. Donor and acceptor atoms on both sides of the junction are left without their electrons and holes to keep

[2] Nobody is forming junctions by bringing *p*- and *n*-type silicon bars together.

them neutral, and this creates a region of fixed *space charge*—positively charged donor atoms on the n-side and negatively charged acceptor atoms on the p-side. This fixed charge creates a built-in electric field which counteracts the diffusion of both majority carriers. The same field, however, is accelerating the minority holes in the n-side and the minority electrons in the p-side towards the opposite sides of the junction. As the space charge region grows, the electric field increases, slowing down the diffusion until an equilibrium is reached. The two opposing currents balance each other for both electrons and holes, the net flow becomes zero and the width of the space charge region settles to x_n on the n-side to x_p on the p-side as shown in figure 1.13.

The concentration of majority electrons and holes in the space charge region is nearly zero due to the electric field which forces them away. The space charge region is depleted from majority carriers and this is why it is more often called the *depletion region*. The depletion behaves essentially as an insulator, separating the electrically neutral (and conducting due to the large carrier concentration) n and p regions on both sides.

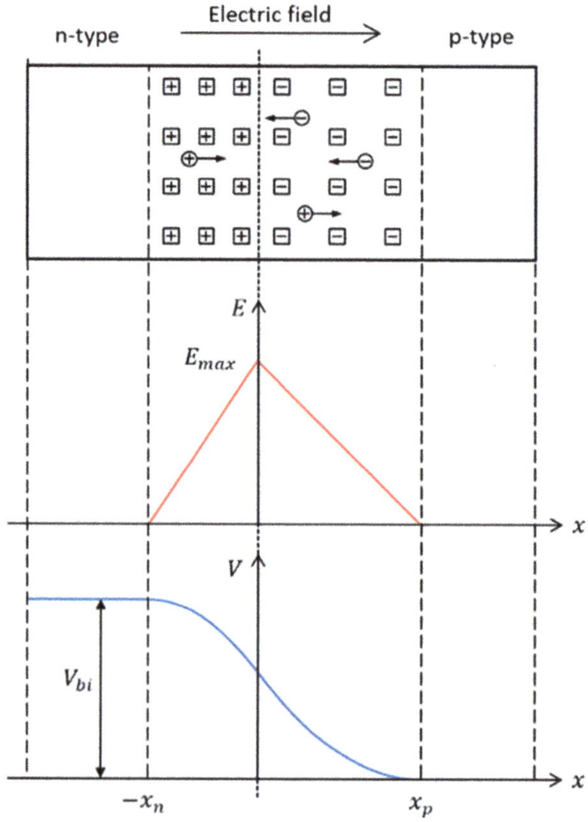

Figure 1.13. Structure, electric field and potential in a pn junction in equilibrium.

It is this electric field that makes the *pn* junction so useful in image sensors. Electrons and holes generated by the photoeffect get separated upon entering the depletion region and create current, which can be measured. Also, only a negligible current will flow in the absence of light due to the depletion behaving as an insulator. This is what was described in the previous sections when we discussed the idealised charge collection.

In equilibrium the whole of the *pn* junction must remain electrically neutral regardless of the fixed space charge. If we consider an *abrupt junction*, where the doping concentrations N_D (in the *n*-side) and N_A (in the *p*-side) are uniform and change in a stepwise fashion at the junction at $x = 0$, the condition of electrical neutrality can be written as

$$N_D x_n = N_A x_p \tag{1.32}$$

Equation (1.32) mathematically means that in the space charge region the number of donor and acceptor atoms per unit area is equal. Despite its idealistic appearance, the abrupt junction is a very good approximation to real *pn* junctions found in image sensors.

From the condition of thermal equilibrium, it follows that the Fermi level through the device is constant. Therefore, the valence and the conduction bands must bend so that the Fermi level stays flat, creating the built-in potential V_{bi} in figure 1.13. Because the electric field is the gradient of the potential distribution, this is just another way to describe the presence of electric field in the space charge region. The built-in voltage can be calculated from the uniformity of the Fermi level and is [2]

$$V_{bi} = \frac{kT}{q}\ln\left(\frac{N_A N_D}{n_i^2}\right) \tag{1.33}$$

where n_i is the intrinsic carrier concentration.

Integrating the Poisson equation for both sides [2], we can calculate the electric field on the *n*-side ($-x_n \leqslant x \leqslant 0$)

$$E(x) = \frac{qN_D}{\varepsilon_0 \varepsilon_{Si}}(x + x_n) \tag{1.34}$$

and on the *p*-side ($0 \leqslant x \leqslant x_p$)

$$E(x) = \frac{qN_A}{\varepsilon_0 \varepsilon_{Si}}(x_p - x) \tag{1.35}$$

Here ε_0 is the dielectric permittivity of vacuum and ε_{Si} is the dielectric constant of silicon. The electric field has a linear dependence on distance because of the constant doping concentrations in the abrupt junction approximation. We can see in figure 1.13 that the electric field has a characteristic triangular shape and decreases linearly from its maximum E_{max} at $x = 0$, where the formulas (1.34) and (1.35) join up.

$$E_{\max} = \frac{qN_A x_p}{\varepsilon_0 \varepsilon_{Si}} = \frac{qN_D x_n}{\varepsilon_0 \varepsilon_{Si}} \tag{1.36}$$

The field becomes zero in the neutral silicon at $x \leqslant -x_n$ and $x \geqslant x_p$.

Furthermore, we can calculate the potential distribution in the junction by integrating (1.34) and (1.35), using the boundary conditions $V_n(-x_n) = V_{bi}$ and $V_p(x_p) = 0$. This gives the expected quadratic dependence on distance, since we are integrating a linearly changing electric field:

$$V_n(x) = V_{bi} - \frac{qN_D}{2\varepsilon_0 \varepsilon_{Si}}(x + x_n)^2 \tag{1.37}$$

$$V_p(x) = \frac{qN_A}{2\varepsilon_0 \varepsilon_{Si}}(x - x_p)^2 \tag{1.38}$$

Here we must clarify an important point about the built-in potential—it does not appear across the terminals of the junction. Anybody who has measured the voltage across a *pn* diode with a voltmeter will testify that the voltage is zero, unless the diode is illuminated. Obviously the *pn* junction is not a battery! The reason we do not see the built-in voltage is the contact potential between the silicon and the electrodes used to connect it to the outside world. The contact potentials between the metal electrodes on the *n* and the *p*-side precisely cancel the built-in voltage, and there is no potential difference between the two external electrodes. If that were not the case, a shorted diode would generate a continuous current through itself, which obviously does not happen.

The total width of the depleted region $W = x_n + x_p$ can be found from (1.37) and (1.38) by using that $V_n(0) = V_p(0)$, which gives

$$V_{bi} - \frac{qN_D}{2\varepsilon_0 \varepsilon_{Si}} x_n^2 = \frac{qN_A}{2\varepsilon_0 \varepsilon_{Si}} x_p^2 \tag{1.39}$$

From here, using (1.32) we can write

$$V_{bi} = \frac{q}{2\varepsilon_0 \varepsilon_{Si}} \frac{(N_A + N_D)N_D}{N_A} x_n^2 = \frac{q}{2\varepsilon_0 \varepsilon_{Si}} \frac{(N_A + N_D)N_A}{N_D} x_p^2 \tag{1.40}$$

and finally, we get

$$x_n = \sqrt{\frac{2\varepsilon_0 \varepsilon_{Si}}{q} \frac{N_A V_{bi}}{(N_A + N_D)N_D}} \tag{1.41}$$

$$x_p = \sqrt{\frac{2\varepsilon_0 \varepsilon_{Si}}{q} \frac{N_D V_{bi}}{(N_A + N_D)N_A}} \tag{1.42}$$

The depletion width W is the sum of (1.41) and (1.42)

$$W = \sqrt{\frac{2\varepsilon_0\varepsilon_{Si}}{q}\left(\frac{N_A + N_D}{N_A N_D}\right)V_{bi}} \tag{1.43}$$

Most of the *pn* junctions used in image sensors are one-sided (also called asymmetric), i.e. one of the dopants has much higher concentration than the other. Typical examples are n^+p junctions used as photodiodes and for charge conversion.

In a one-sided n^+p junction $N_D \gg N_A$, which leads to almost all of the depletion being on the *p*-side because $x_p \gg x_n$. From (1.42) and (1.43) we see that $x_p \cong W$ and the expression for the depletion width simplifies to

$$W = \sqrt{\frac{2\varepsilon_0\varepsilon_{Si}}{qN_A}V_{bi}} \tag{1.44}$$

Why are we using one-sided junctions? If one dopant has significantly higher concentration than the other, for example by a factor of 100, the depletion width is determined entirely by the low-doped side, and only its concentration has to be precisely controlled. The built-in voltage depends on both dopant concentrations, but thanks to the logarithmic dependence in (1.33) it has much weaker effect on the depletion width.

1.6.2 *pn* junction under reverse bias

The *pn* junction is very useful even without any voltage applied to it. Whether its terminals are shorted or left floating, the internal electric field is still there and can separate photogenerated electron–hole pairs. However, the depletion width in equilibrium can be quite small, unless very low doping concentrations are used. Increasing the depletion is often desired as it reduces the size of the field-free regions and the extent of charge diffusion. This can happen when a reverse bias is applied across the junction with the same polarity as the built-in voltage, i.e. positive on the *n*-side (cathode) relative to the *p*-side (anode).

When a reverse bias V_r is applied, the total voltage across the junction is $V_{bi} + V_r$ and the depletion width in a n^+p junction becomes

$$W = \sqrt{\frac{2\varepsilon_0\varepsilon_{Si}}{qN_A}(V_{bi} + V_r)} \tag{1.45}$$

Due to the asymmetric doping, the depletion width is almost entirely contained in the *p*-side of the junction, and so is the electric field.

Example 1.10. Calculate the depletion depth of a silicon one-sided n^+p junction with $N_A = 6.7 \times 10^{14}$ cm^{-3} (resistivity $\rho = 20$ Ωcm) and $N_D = 10^{18}$ cm^{-3} for $V_r = 0$ and $V_r = 5$ V and 300 K. Use that $n_i = 1.45 \times 10^{10}$ cm^{-3}, $\varepsilon_0 = 8.85 \times 10^{-14}$ F cm^{-1} and $\varepsilon_{Si} = 11.9$. Also calculate E_{max} and x_n at $V_r = 5$ V.

Solution: First, from (1.33) we calculate V_{bi}:

$$V_{\rm bi} = \frac{kT}{q}\ln\left(\frac{N_{\rm A}N_{\rm D}}{n_i^2}\right) = \frac{1.38\times10^{-23}\times300}{1.6\times10^{-19}}\ln\left(\frac{6.7\times10^{14}\times10^{18}}{2.1\times10^{20}}\right) = 0.74\ \rm V$$

Next, from (1.45) we calculate the depletion widths

$$W(0{\rm V}) = \sqrt{\frac{2\times8.85\times10^{-14}\times11.9\times0.74}{1.6\times10^{-19}\times6.7\times10^{14}}} = 1.21\ \mu{\rm m}$$

$$W(5{\rm V}) = \sqrt{\frac{2\times8.85\times10^{-14}\times11.9\times(0.74+5)}{1.6\times10^{-19}\times6.7\times10^{14}}} = 3.36\ \mu{\rm m}$$

The maximum electric field and x_n can be calculated from (1.36) using that $x_p \cong W$:

$$E_{\rm max} = \frac{qN_{\rm A}x_p}{\varepsilon_0\varepsilon_{\rm Si}} = \frac{qN_{\rm A}W}{\varepsilon_0\varepsilon_{\rm Si}} = \frac{1.6\times10^{-19}\times6.7\times10^{14}\times3.36\times10^{-4}}{8.85\times10^{-14}\times11.9} = 34\,200\ {\rm Vcm^{-1}}$$

$$x_n = \frac{\varepsilon_0\varepsilon_{\rm Si}E_{\rm max}}{qN_{\rm D}} = \frac{8.85\times10^{-14}\times11.9\times34\,200}{1.6\times10^{-19}\times10^{18}} = 2.3\ \rm nm$$

This example shows how much smaller the depletion is on the *n*-side in a one-sided n^+p junction compared to the *p*-side; in this case x_n is entirely negligible. The characteristic right-triangular shape of the electric field and its drop to zero at the edge of the depletion are important features with implications to charge collection.

Figure 1.14 shows the calculated potential and electric field from example 1.10 using (1.34)–(1.38), and a simulation using commercial TCAD software [27].

It is useful to compare the analytic formulae with a finite element device simulation using the same parameters; this provides a necessary cross-check even if both cannot be made exactly the same. The matching in figure 1.14 is good, considering the approximations in the analytical calculation and the discrete structure of the simulation model.

The approximation of ideal abrupt *pn* junction assumes that the space charge region and the majority carrier concentration have infinitely sharp edges. This is of course an idealisation and is the reason why the electrical field falls to zero at both ends of the depletion. In practice infinitely sharp edges do not exist and the majority carrier concentration changes smoothly.

The finite element analysis (FEA) TCAD simulation solves the Poisson equation without this assumption and reveals the fine details at the edge of the depletion. Plotted on a logarithmic scale, the electric field in figure 1.15 continues to be above zero for about a micron beyond the calculated edge of the space charge region. This field is small but can make a sizeable effect on the charge collection time due to drift being much faster than diffusion, as discussed in section 1.3.3.

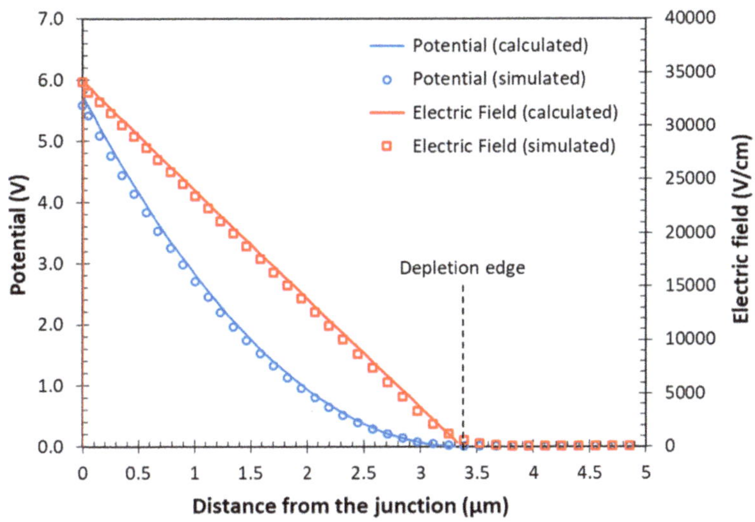

Figure 1.14. Potential and electric field calculated and simulated in example 1.10 for $V_r = 5$ V.

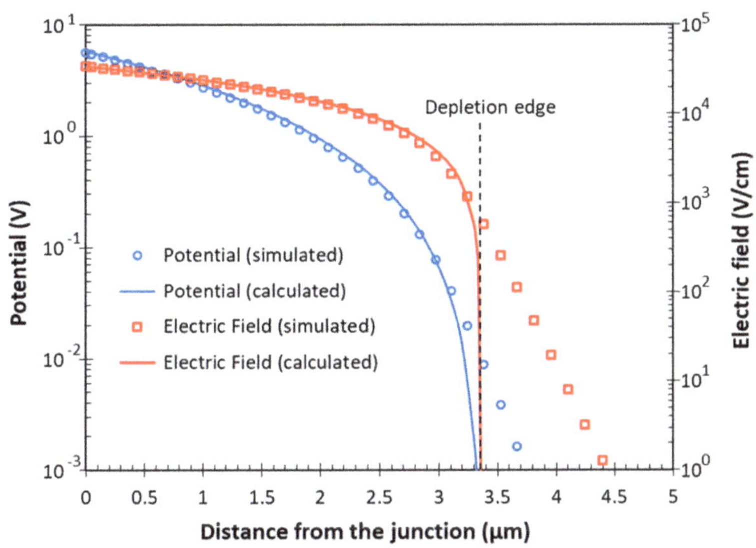

Figure 1.15. A comparison between the electric field calculated using the abrupt n^+p junction approximation in example 1.10 for $V_r = 5$ V, and an FEA device simulation which does not make this approximation.

The calculations in this section used the abrupt one-sided *pn* junction model, which is a good approximation to most practical devices. This gives the familiar square root dependence (1.45) of the depletion width on the reverse bias. A different doping profile would produce a different voltage dependence, with the linearly graded junction [2] the most notable example, giving $W \propto (V_{bi} + V_r)^{1/3}$.

1.6.3 Charge collection

Due to its electric field, the *pn* junction collects electrons on the cathode, biased or not. Looking back at figure 1.9, showing our then hypothetical charge collection element, this is exactly what the *pn* junction does.

Without a reverse bias, the depletion is usually much smaller than the depth of the device. In this case some of the charge will diffuse before collection, as shown in figure 1.16(a). To minimise charge spread the field-free region must be eliminated, so that *full depletion* is achieved. This is shown in figure 1.16(b), and the mathematical condition is that the depletion width becomes equal to the device thickness. Applying that $W = d$ to (1.45) we can calculate the reverse bias V_{fd} at which full depletion is achieved:

$$d = \sqrt{\frac{2\varepsilon_0\varepsilon_{Si}}{qN_A}(V_{bi} + V_{fd})} \tag{1.46}$$

which gives

$$V_{fd} = \frac{qN_A d^2}{2\varepsilon_0\varepsilon_{Si}} - V_{bi} \tag{1.47}$$

Equation (1.47) tells us that the full depletion voltage is proportional to the dopant concentration, and this is why it is easier to reach in high resistivity, low-doped semiconductors.

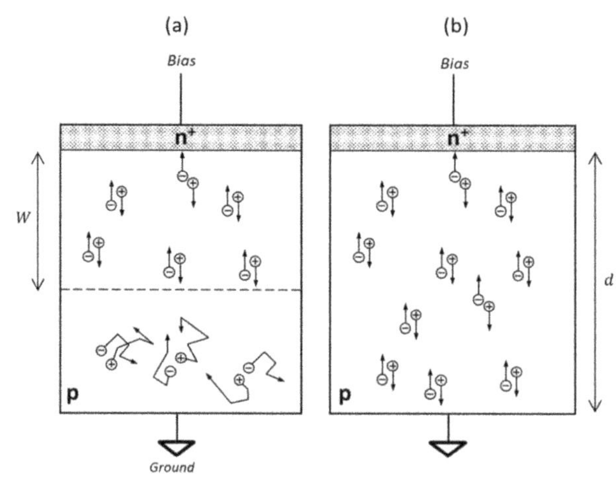

Figure 1.16. Charge collection in partially depleted (a); and in a fully depleted *pn* junction (b).

Example 1.11. Calculate the full depletion voltage of the n^+p junction in example 1.10 for a device thickness d = 5 μm.

Solution: Using (1.47) and the previously calculated V_{bi}

$$V_{\rm fd} = \frac{1.6 \times 10^{-19} \times 6.7 \times 10^{14} \times (5 \times 10^{-4})^2}{2 \times 8.85 \times 10^{-14} \times 11.9} - 0.74 = 11.98 \text{ V}$$

For very thick devices the full depletion voltage is large and we can ignore $V_{\rm bi}$ because $V_{\rm fd} \gg V_{\rm bi}$.

We can now refine the treatment of the charge collection time in (1.24) for an abrupt one-sided *pn* junction. Figure 1.17 shows the familiar triangular shape of the electric field, which in full depletion is described by (1.35) with $x_p = W = d$.

$$E(x) = \frac{qN_{\rm A}}{\varepsilon_0 \varepsilon_{\rm Si}}(d - x) \tag{1.48}$$

To avoid the electric field dropping to near zero at the back of the device, the reverse bias can be increased above that required for full depletion so that the junction becomes *over-depleted*. This creates an additional electric field $E_{\rm od} = V_{\rm od}/d$, where $V_{\rm od}$ is the voltage top-up above $V_{\rm fd}$. Adding $E_{\rm od}$ to $E(x)$, and using (1.47) with $V_{\rm fd} \gg V_{\rm bi}$ the electric field becomes

$$E(x) = \frac{2V_{\rm fd}}{d^2}(d - x) + \frac{V_{\rm od}}{d} \tag{1.49}$$

Due to the linear dependence of the electric field the electron velocity v_n will change linearly too, depending on the position in the device according to $v_n(x) = \mu_n E(x)$. The travel time can be obtained by solving equation (1.50) as in [28]

$$v_n(x) \equiv \frac{dx}{dt} = -\mu_n \left[\frac{2V_{\rm fd}}{d^2}(d - x) + \frac{V_{\rm od}}{d} \right] \tag{1.50}$$

Here the minus sign takes into account that the electrons move opposite to the direction of the x-axis, and we ignore velocity saturation. The solution for charge generated at coordinate $x_0 < d$ is:

$$t = \frac{d^2}{2\mu_n V_{\rm fd}} \ln \left[\frac{1}{1 - \left(\frac{2V_{\rm fd}}{2V_{\rm fd} + V_{\rm od}} \right) \frac{x_0}{d}} \right] \tag{1.51}$$

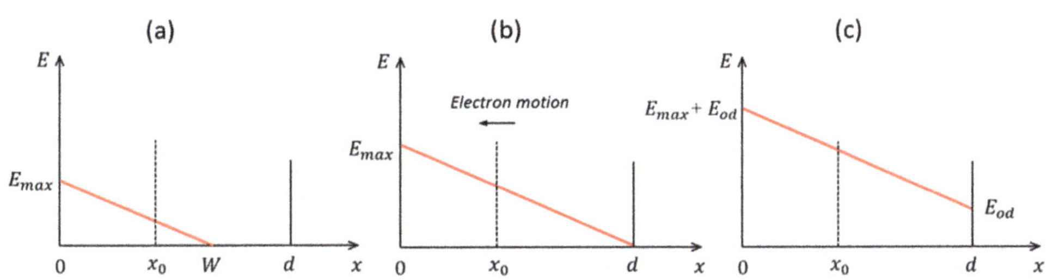

Figure 1.17. Electric field in a partially depleted (a), fully depleted (b) and over-depleted (c) abrupt *pn* junction.

If $V_{od} = 0$ (1.51) simplifies to

$$t = \frac{d^2}{2\mu_n V_{fd}} \ln\left(\frac{d}{d - x_0}\right) \tag{1.52}$$

and shows that when charge is generated near the back of the device ($x_0 \approx d$) the charge collection time tends to infinity. This is because we are considering only charge drift and the electric field at the back of the device for $V_{od} = 0$ is zero, therefore an electron will stay there forever and will not be collected. Fortunately, diffusion is ever-present and takes care of this 'unphysical' situation—the charge will diffuse to regions with non-zero field will be quickly swept away.

The maximum charge collection time t_{max} for over-depleted junction ($V_{od} > 0$) can be obtained from (1.51) for $x_0 = d$

$$t_{max} = \frac{d^2}{2\mu_n V_{fd}} \ln\left(\frac{2V_{fd} + V_{od}}{V_{od}}\right) \tag{1.53}$$

Comparing (1.53) to (1.24), the differences are a factor of two in the denominator and the logarithmic term, which can become significant for low over-depletion voltages. The two formulas give the same result when $V_{od} = 2V_{fd}/(e^2 - 1) = 0.31V_{fd}$.

Example 1.12. Calculate the maximum charge collection time for the *pn* junction in example 1.11 for $V_{od} = 1$ V, ignoring V_{bi}.
Solution: Using formula (1.53) we get

$$t_{max} = \frac{(5 \times 10^{-4})^2}{2 \times 1400 \times 11.98} \ln\left(\frac{2 \times 11.98 + 1}{1}\right) = 24 \text{ ps}$$

However, this is wrong because we have ignored velocity saturation in the derivation of the formula. A quick check using (1.49) for $x = 0$ gives $v_n = 6.7 \times 10^7$ cm s^{-1}, well above the saturation velocity. Therefore, this calculation gives too short charge collection time; a better estimation as in example 1.7 would give around 50 ps.

You may ask the question: do we need a specialised semiconductor structure (e.g. *pn* junction or a MOS capacitor) to generate an electric field? Can we just apply some voltage across silicon to collect the charge? Let's investigate this, picking a square pixel with size $a = 10$ μm in silicon substrate $d = 5$ μm thick. Applying voltage V across the pixel will force current to flow because silicon has finite resistance. This current will look like photogenerated signal and must be reduced as much as possible; therefore, we have to choose intrinsic (i.e. undoped and pure) silicon, which has the highest resistivity $\rho = 230$ kΩcm at room temperature. If we now apply one volt across the pixel, the current can be calculated from the Ohm's law:

$$I = \frac{V}{R} = \frac{V}{\rho d/a^2} = \frac{1}{230 \times 10^3 \times 5 \times 10^{-4}/(10 \times 10^{-4})^2} = 8.7 \text{ nA}$$

Comparing with examples 1.3 and 1.4 we see that this current is many orders of magnitude higher than the photogenerated current; obviously a pixel made like this will be a very poor image sensor. It may have a chance in very bright illumination conditions, or when the sensor is cooled down (to increase the resistivity and reduce the current), but not as a normal image sensor we are all used to.

1.6.4 Junction capacitance

Besides for charge collection, the *pn* junction can be used for charge-to-voltage conversion as in the diagrams shown in figures 1.12(a)–(c). Two of the circuits rely on the conversion of the photocurrent on an external resistor (figure 1.12(a)), or on an external capacitor (figure 1.12(c)). The diagram in figure 1.12(b) shows the most popular use, where the junction capacitance itself is used for the conversion.

As mentioned before, the depletion region is an insulator separating the conducting *p* and *n*-type field-free regions. This is exactly the situation in a parallel plate capacitor with a distance between the electrodes W and capacitance $C = \varepsilon_0 \varepsilon_{Si} A / W$, where A is the electrode area. Using (1.45) the capacitance of the *pn* junction is

$$C = \frac{\varepsilon_0 \varepsilon_{Si} A}{W} = A\sqrt{\frac{\varepsilon_0 \varepsilon_{Si} q N_A}{2(V_{bi} + V_r)}} \tag{1.54}$$

Example 1.13. Calculate the capacitance of the *pn* junction with an area $A = 25\ \mu m^2$ with the parameters given in example 1.10 for $V_r = 1$ V.
Solution: From (1.54)

$$C = 25 \times 10^{-8} \times \sqrt{\frac{8.85 \times 10^{-14} \times 11.9 \times 1.6 \times 10^{-19} \times 6.7 \times 10^{14}}{2 \times (0.74 + 1)}} = 1.42\ \text{fF}$$

After the junction has been biased to V_r and left floating by disconnecting the switch, photogenerated electrons will collect at the cathode and reduce its potential according to formula (1.30). However, because the junction capacitance depends on the applied voltage, it is not an ideal parallel plate capacitor. As electrons are collected, the voltage on the junction goes down, and the capacitance goes up. The change of the capacitance is nonlinear, and its dependence on the reverse bias can be found by differentiating (1.54) by V_r:

$$\frac{dC}{dV_r} = -\frac{A}{2}\sqrt{\frac{\varepsilon_0 \varepsilon_{Si} q N_A}{2(V_{bi} + V_r)^3}} = -\frac{C}{2(V_{bi} + V_r)} \tag{1.55}$$

The nonlinear capacitance means that the conversion from charge to voltage will be nonlinear too, which is not what we normally want from an image sensor. As figure 1.18 shows, the change of capacitance can exceed 20% when the voltage across the junction changes by 1 V. This change is large but can be counteracted by connecting the junction in parallel with a larger, linear capacitance, so that the nonlinearity is much reduced. In image sensors the role of this additional capacitance is performed by the readout circuitry, as well as by actual capacitors.

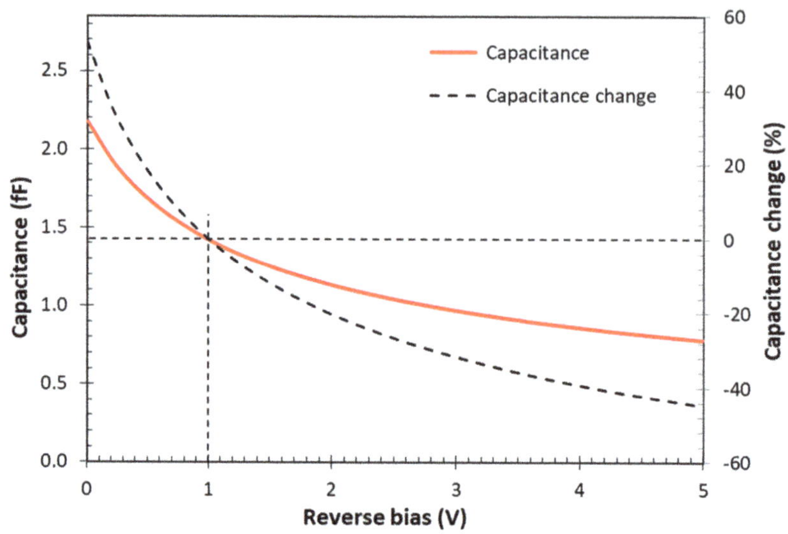

Figure 1.18. Junction capacitance and the change of the capacitance relative to $V_r = 1$ V for the *pn* junction in example 1.13.

1.7 MOS capacitor

1.7.1 Depletion

The MOS capacitor is a structure that combines a metal electrode (usually called a 'gate') deposited on top of an insulator (typically SiO_2), which is grown on silicon, as shown in figure 1.19. Normally the gate is not made of metal but of heavily doped, highly conductive polycrystalline silicon because this greatly improves the quality of the Si–SiO_2 interface and device yield. The gate was made of metal in the early days of semiconductor technology and hence the term 'MOS' was coined back then, but it remains in use to this day.

The MOS capacitor is the basic building block of the CCD and also of certain type of CIS using photogates instead of a photodiode. Photogate-based CIS find applications where fast or multiple charge transfer is needed, such as in some time-of-flight (ToF) sensors.

Similarly to the *pn* junction, the MOS capacitor can be used to create an electric field and depletion region suitable for charge collection. We will consider the example in figure 1.19 with a n^+-doped polysilicon gate and a *p*-type substrate. Intuitively, a positive gate potential with respect to the substrate should force the holes away from the Si–SiO_2 interface and create a depletion region with depth W. In the band diagram in figure 1.19 this is shown as downward bending of the conduction and valence bands. Downward bending indicates increased potential near the interface relative to the neutral bulk, which is connected to ground.

In an electrically neutral semiconductor, the difference between the Fermi level E_F and the mid-band (intrinsic) Fermi level E_i is the same everywhere, including at the

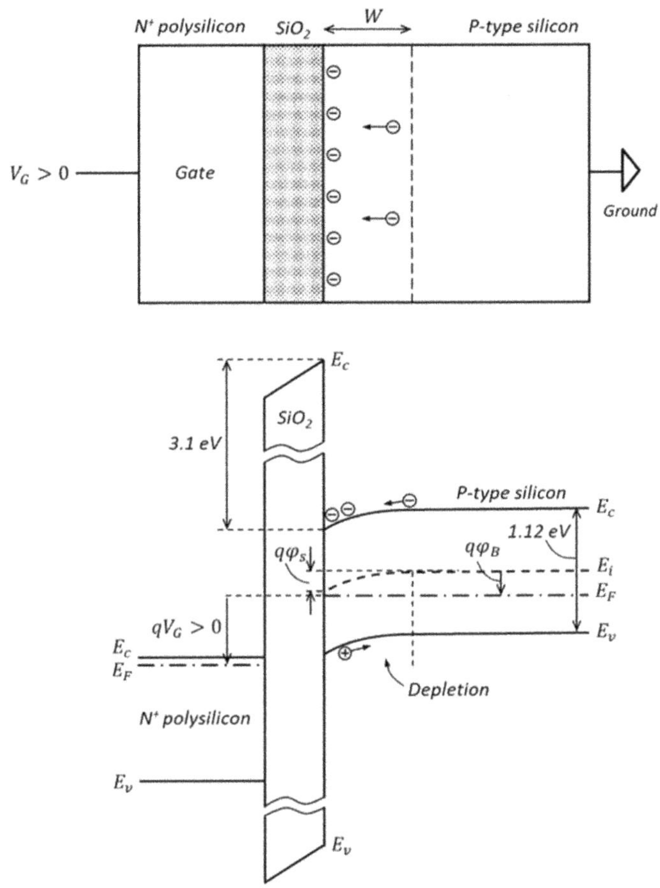

Figure 1.19. Energy-band diagram of a MOS capacitor in depletion.

Si–SiO_2 interface. The difference $E_i - E_F$ in the bulk of a neutral *p*-type semiconductor can be expressed as:

$$q\varphi_B = E_i - E_F = kT\ln\left(\frac{N_A}{n_i}\right) \tag{1.56}$$

When a gate voltage V_G is applied so that the bands bend downwards, the surface potential φ_s increases, and the surface hole concentration decreases according to

$$p = N_A\exp\left(-\frac{q\varphi_s}{kT}\right) \tag{1.57}$$

When $\varphi_B > \varphi_s > 0$ the semiconductor is depleted because the hole concentration at the surface is lower than in the bulk. When $\varphi_s = \varphi_B$ the surface hole concentration becomes the intrinsic concentration n_i, as can be verified by using equations (1.56) and (1.57). If the gate voltage is increased even further, so that $\varphi_s > \varphi_B$, the surface becomes inverted because there are more electrons than holes. Strong inversion happens when $\varphi_s > 2\varphi_B$ and is the condition used in enhancement mode MOS transistors.

The depletion region in figure 1.19 looks very similar to the *p*-side of an abrupt *pn* junction (figure 1.14) if we take the voltage at the junction as φ_s. Therefore, the depletion depth can be calculated with equation (1.45)

$$W = \sqrt{\frac{2\varepsilon_0\varepsilon_{Si}\varphi_s}{qN_A}} \tag{1.58}$$

and from it, the surface potential is

$$\varphi_s = \frac{qN_A W^2}{2\varepsilon_0\varepsilon_{Si}} \tag{1.59}$$

The potential in the depletion region away from the surface (which is at $x = 0$) is quadratic as in equation (1.38) describing the *pn* junction:

$$\varphi(x) = \varphi_s\left(1 - \frac{x}{W}\right)^2 \tag{1.60}$$

Since the substrate is at ground, we can write that the gate voltage is the sum of the voltage across the oxide V_{ox} and the surface potential, ignoring for the moment the flat-band voltage offset (described in the next section):

$$V_G = V_{ox} + \varphi_s \tag{1.61}$$

The oxide voltage is equal to the space charge in the depleted region divided by the oxide capacitance per unit area C_{ox}:

$$V_{ox} = \frac{qN_A W}{C_{ox}} \tag{1.62}$$

Substituting (1.59) and (1.62) into (1.61), we get the equation (1.63)

$$V_G = \frac{qN_A W}{C_{ox}} + \frac{qN_A W^2}{2\varepsilon_0\varepsilon_{Si}} \tag{1.63}$$

which can be solved to give the depletion depth as

$$W = -\frac{\varepsilon_0\varepsilon_{Si}}{C_{ox}} + \sqrt{\left(\frac{\varepsilon_0\varepsilon_{Si}}{C_{ox}}\right)^2 + \frac{2\varepsilon_0\varepsilon_{Si}}{qN_A}V_G} \tag{1.64}$$

We see that the depletion depth changes as the square root of the gate voltage, similarly to the reversed-biased *pn* junction.

When $\varphi_s > 2\varphi_B$ and in the absence of a ready source of electrons, as in figure 1.19, inversion takes some time to take hold because the electrons needed to populate the interface have to be generated thermally ([26], chapter 5). This allows the surface to be taken far beyond $\varphi_s > 2\varphi_B$ into *deep depletion* (figure 1.20) without surface inversion, so that photogenerated electrons can be collected. This condition is what

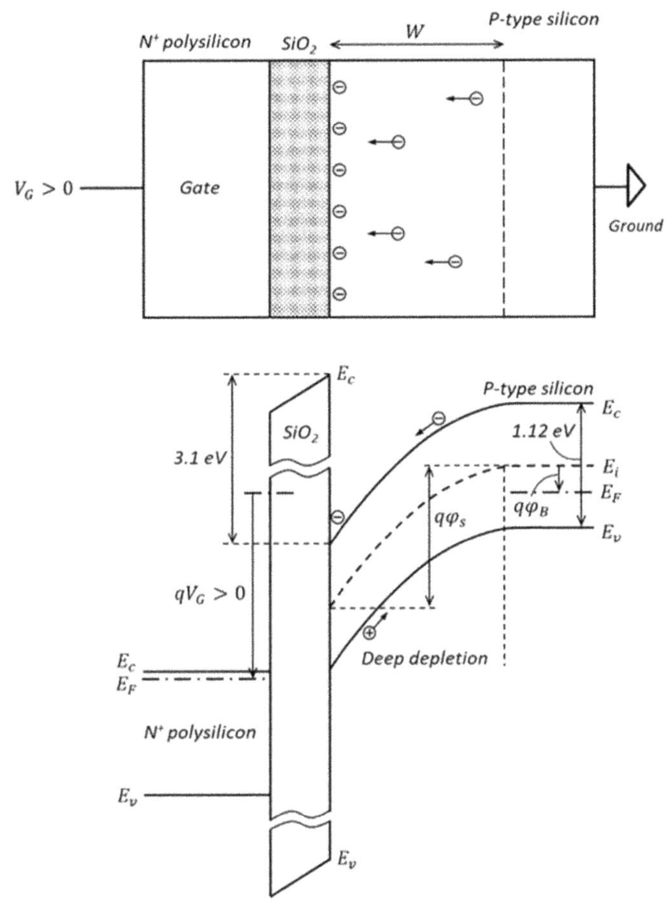

Figure 1.20. Energy-band diagram of a MOS capacitor in deep depletion.

makes image sensors using photogates work. If inversion were to be established immediately as in MOSFETs, the photogenerated charge would be swamped by the many more electrically induced carriers.

Once collected, the signal electrons have nowhere else to go because they cannot cross over the oxide, therefore signal detection via current is not an option. However, the charge can be measured electrostatically (non-destructively) as in figure 1.12(d), or it can be transferred to a separate charge conversion element. In CCDs, for example, the charge is transported over large distances to a reverse-biased *pn* junction for detection.

Regardless of how the signal is measured, it must be cleared from the MOS capacitor before the next signal is collected. Unlike a *pn* junction the photogenerated charge cannot be drained away simply by re-connecting the bias since the MOS capacitor has no conductive path. The possibilities for clearing the charge are to either: (a) transfer it in a controlled manner; (b) drain it by using an additional structure next to the MOS capacitor, or (c) to recombine the signal electrons by flooding the interface with holes.

1.7.2 Gate capacitance

Similarly to the *pn* junction, the depletion depth in the MOS capacitor is nonlinear due to the square root in (1.64). This makes the MOS capacitance voltage dependent too. The difference is that the total gate capacitance is determined by the oxide and the depletion capacitances connected in series, as illustrated in figure 1.21:

$$\frac{1}{C_G} = \frac{1}{C_{ox}} + \frac{1}{C_{dep}} \tag{1.65}$$

The oxide capacitance per unit area is given by

$$C_{ox} = \frac{\varepsilon_0 \varepsilon_{ox}}{t_{ox}} \tag{1.66}$$

where ε_{ox} is the dielectric permittivity of SiO_2 and t_{ox} is the gate oxide thickness. The depletion capacitance per unit area can be determined by using the formula for the parallel plate capacitor and (1.64):

$$C_{dep} = \frac{\varepsilon_0 \varepsilon_{Si}}{W} = \frac{1}{-\frac{1}{C_{ox}} + \sqrt{\frac{1}{C_{ox}^2} + \frac{2V_G}{\varepsilon_0 \varepsilon_{Si} q N_A}}} \tag{1.67}$$

Due to the thin gate oxide C_{ox} is usually very large, and in deep depletion V_G is large too, so that $C_{ox} \gg C_{dep}$ and $1/C_{ox}$ becomes negligible in (1.67). In this condition the gate capacitance can be approximated with

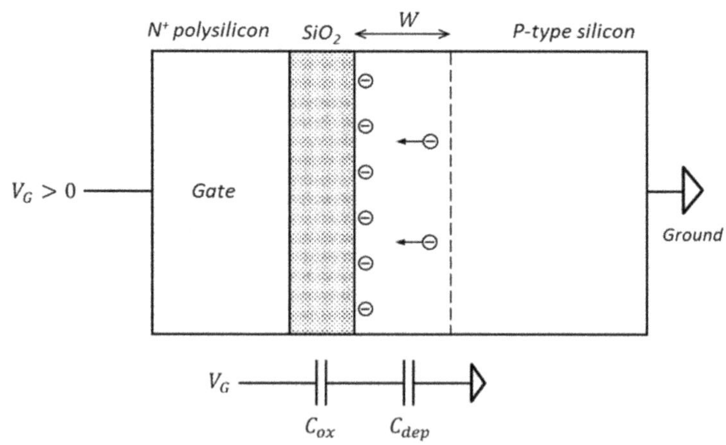

Figure 1.21. MOS gate capacitance.

$$C_G \approx \sqrt{\frac{\varepsilon_0 \varepsilon_{Si} q N_A}{2V_G}} \tag{1.68}$$

which is almost identical to the capacitance of the *pn* junction in (1.54).

Example 1.14. Calculate the oxide capacitance per square centimetre and square micron of 7 nm thick SiO_2, using that $\varepsilon_{ox} = 3.9$. Also calculate the gate capacitance in deep depletion for $N_A = 6.7 \times 10^{14}$ cm^{-3} (resistivity $\rho = 20$ Ωcm) and $V_G = 3.0$ V.
Solution: From (1.66)

$$C_{ox} = \frac{3.9 \times 8.85 \times 10^{-14}}{7 \times 10^{-7}} = 4.93 \times 10^{-7}\ \text{F cm}^{-2} = 4.93\ \text{fF}\ \mu\text{m}^{-2}$$

Next, using (1.68)

$$C_G = \sqrt{\frac{8.85 \times 10^{-14} \times 11.9 \times 1.6 \times 10^{-19} \times 6.7 \times 10^{14}}{2 \times 3.0}} = 4.34 \times 10^{-9}\ \text{F cm}^{-2}$$

$$= 0.04\ \text{fF}\ \mu\text{m}^{-2}$$

The gate capacitance is about two orders of magnitude smaller than the oxide capacitance, which should not come as a surprise because the depletion depth in silicon is 2.4 μm (as can be verified from (1.64)), compared to just 0.007 μm oxide thickness. In this case, ignoring C_{ox} in the approximate formula (1.68) is fully justified.

The gate capacitance in deep depletion is dominated by the depletion depth in silicon because the oxide capacitance is normally much larger. However, if the depletion is to shrink to zero, the silicon under the oxide becomes conductive and acts as an electrode to the gate oxide. This makes the gate capacitance equal to the very large C_{ox}, which is the upper limit. The way to reach this condition is to operate the MOS capacitor in *accumulation* by biasing the gate sufficiently negative relative to substrate, so that holes gather at the Si–SiO_2 interface. Another way to increase the capacitance to C_{ox} is to invert the surface and populate it with electrons, which act as the second electrode. This is used in capacitors based on MOSFETs and is described in the following section.

1.8 MOS transistor

1.8.1 Structure

MOS field effect transistors (MOSFETs) are the staple of microelectronics and are the fundamental building block of nearly every integrated circuit. In image sensors MOSFETs are used as buffers for signals with high output impedance, such as the photogenerated voltage in a photodiode, and as amplifiers, current sources and sinks, active loads, switches and capacitors.

MOSFETs work using the *field effect*, describing the strong change of conductivity in a semiconductor (or another material such as graphene) under the influence of an electric field. In *p*-type semiconductor the electric field induced by the gate–substrate voltage can create a thin inversion layer of electrons at the Si–SiO_2 interface. Devices operating like this are called surface channel transistors, shown in figure 1.22(a), and comprise the vast majority of MOSFETs. They are also called enhancement mode, or 'normally off' transistors, because at zero voltage on the gate the transistor does not conduct.

Another type of MOSFET, called buried channel, or depletion mode transistor, is shown in figure 1.22(b). In the buried channel *n*-MOSFET a *n*-type dopant is implanted in the channel, which becomes conducting so that the transistor is normally on without a voltage applied to the gate. To turn it off, a *negative gate voltage* with respect to the source must be applied. This depletes the channel of electrons until the conduction is cut off.

Described in simple terms, when the gate–source voltage V_{GS} is above the threshold voltage V_T the transistor is conducting current, and when below threshold the transistor is off, representing an infinite resistance. For MOSFETs in digital circuits this description is very accurate—the transistors are either 'fully on' or 'fully off', with no intermediate state. Analogue circuits use precise ways to smoothly control the drain current so that the MOSFETs work as amplifiers of small signals, or as current sources and sinks.

The source and the drain of the transistors shown in figure 1.22 are symmetrical and interchangeable. The source and the drain are *pn* junctions in their own right, and so is the whole of the buried channel in figure 1.22(b). They are always reverse biased with respect to the substrate, and in common with any *pn* junction can be photosensitive.

MOSFETs are essential for image sensor operation—the main reason is their extremely high input impedance. As we saw already, the currents involved in imaging can be very low; as an example, the dark current though a pixel could be below 1 electron per second at room temperature—this is 1.6×10^{-19} A (or 160 zepto amps), a phenomenally low current. Only a good insulator such as the SiO_2 used as a gate oxide in MOSFETs has a chance of low enough leakage that is not disturbing this tiny current. For comparison, a very good discrete *pn* junction (e.g. a low

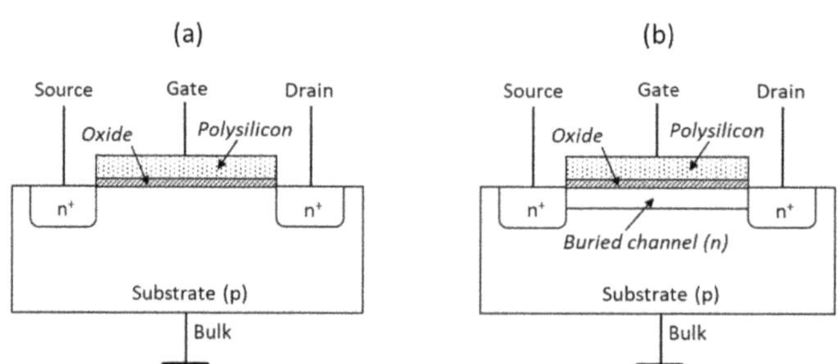

Figure 1.22. Surface (a) and buried channel (b) *n*-channel MOSFETs.

leakage diode or the gate of a JFET) has reverse current of around 1 pA at room temperature.

Both transistors in figure 1.22 can sit in *p*-wells in order to have good control over certain MOSFET parameters such as the threshold. This also allows the substrate to have different, usually lower doping concentration, which is needed to achieve the desired depletion depth and charge collection.

1.8.2 MOSFET characteristics

There are plenty of excellent books describing how MOS transistors work [26, 29]. Here we are just going to give a short summary with the most relevance to image sensors.

N-channel MOSFETs (*n*-MOSFETs) are made on *p*-type substrate and are used in the pixel of most CIS which collect electrons as photogenerated signal. The classic long channel model gives a good approximation for the characteristics of the MOSFET. In the active region, when the drain–source voltage is small, i.e. $0 < V_{DS} < (V_{GS} - V_T)$, the drain current in a transistor with channel length L and width W is given by

$$I_D = \frac{\mu_n C_{ox} W}{L}\left[(V_{GS} - V_T)V_{DS} - \frac{V_{DS}^2}{2}\right] \tag{1.69}$$

For surface channel transistors the electron mobility μ_n is much smaller than the bulk mobility used to describe charge collection because of increased electron scattering at the Si–SiO$_2$ interface ([26], p 203). The active region is called a 'linear regime' because the drain current depends linearly on V_{GS}. Also, when $V_{DS} \ll (V_{GS} - V_T)$ the second term in the brackets in (1.69) can be ignored and the drain current becomes approximately linearly dependent on the drain–source voltage:

$$I_D \approx \frac{\mu_n C_{ox} W}{L}(V_{GS} - V_T)V_{DS} \tag{1.70}$$

In the linear regime the MOSFET behaves as a resistor with voltage-controlled resistance given by:

$$R_{lin} = \frac{V_{DS}}{I_D} = \frac{L}{\mu_n C_{ox} W(V_{GS} - V_T)} \tag{1.71}$$

This is used in analogue switches which operate at very low voltage drop across the drain–source $V_{DS} \ll (V_{GS} - V_T)$ in their 'on' state, when $V_{GS} > V_T$.

For fixed V_{GS} the drain current described by (1.69) initially increases with the drain–source voltage, but as V_{DS} continues to increase, the second term in the brackets becomes larger and the drain current increase slows down. Eventually the drain current stops increasing with V_{DS}, or in other words, *saturates*. Mathematically, saturation is expressed as $\partial I_D / \partial V_{DS} = 0$, and by differentiating (1.69) we can see that this happens when $V_{DS} = V_{GS} - V_T$. For $V_{DS} \geqslant V_{GS} - V_T$ the drain current in saturation depends quadratically on the gate–source voltage as

$$I_D = \frac{\mu_n C_{ox} W}{2L}(V_{GS} - V_T)^2 \tag{1.72}$$

Equation (1.72) describes a *voltage-controlled current source* because the drain current does not depend on the drain voltage, but only on the input V_{GS}. A measure of the dependence of the drain current on the gate–source voltage is the gate transconductance g_m, a very important device parameter:

$$g_m = \frac{\partial I_D}{\partial V_{GS}} = \frac{\mu_n C_{ox} W}{L}(V_{GS} - V_T) \tag{1.73}$$

Using (1.72), the gate transconductance can be expressed also as

$$g_m = \sqrt{\frac{2\mu_n C_{ox} W}{L} I_D} \tag{1.74}$$

The transconductance describes how a small change v_{GS} of the gate–source voltage forces a drain current change $i_D = g_m v_{GS}$. This is provided that its DC gate–source voltage V_{GS} is above threshold, the MOSFET is biased in saturation, and the drain current is not limited by the supply. The changes v_{GS} and i_D can be thought of as small AC signals on top of the larger V_{GS} and I_D, correspondingly. If the drain current passes through the load resistor R_L as in figure 1.23(a), i_D will induce a change v_o of the drain voltage equal to

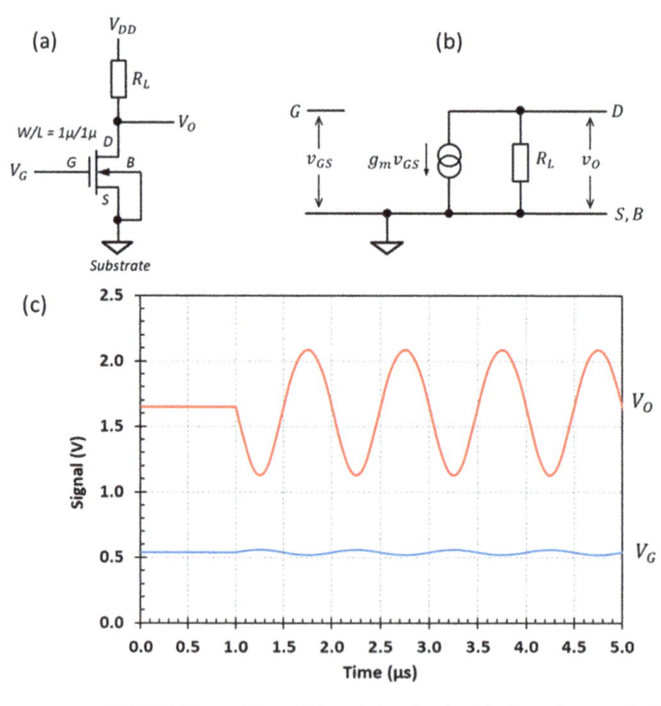

Figure 1.23. Common source MOSFET amplifier with resistive load with the substrate (bulk) connected to the source (a) and its equivalent schematic (b). SPICE simulation with $R_L = 1\,M\Omega$, $V_{DD} = 3.3$ V and 40 mV_{pp}, 1 MHz input sinewave signal starting from 1 μs onwards, showing 954 mV_{pp} at the output (c).

$$v_o = i_D R_L = -g_m v_{GS} R_L \tag{1.75}$$

The circuit is *inverting* because increasing V_G makes the drain current increase too, which in turn makes the output voltage V_O decrease due to the larger voltage across the load resistor. Since the top end of R_L is at the supply V_{DD}, which is AC ground (figure 1.23(b)), the AC gain of this circuit is the change of the output voltage divided by the change of the input voltage:

$$G = \frac{v_o}{v_{GS}} = -g_m R_L \tag{1.76}$$

The gain given by (1.76) applies only to AC signals well below the bandwidth of the circuit and should not be confused with the ratio of the DC voltages at the drain and the gate; the DC voltage ratio is a completely different matter.

The gain of this simple single-transistor circuit can be substantial, as figure 1.23(c) demonstrates. Here the input AC signal is superimposed on a DC gate voltage of 0.54 V, needed to bias the transistor in saturation. This DC bias is chosen so that the static drain voltage is at half the supply (1.65 V), thus maximising the swing at the output.

Example 1.15. Calculate the gate transconductance of an n-MOSFET with $W = L = 1\ \mu\text{m}$, $\mu_n = 400\ \text{cm}^2\ \text{V}^{-1}\ \text{s}^{-1}$, $C_{ox} = 4.93\ \text{fF}\ \mu\text{m}^{-2}$ (7 nm thick gate oxide), $V_T = 0.4$ V and $V_{GS} = 0.5$ V. Also, calculate the g_m of the transistor in figure 1.23.
Solution: Substituting the parameters in formula (1.73) we get:

$$g_m = \frac{400 \times 4.93 \times 10^{-15} \times 10^{8} \times 10^{-4}}{10^{-4}}(0.5 - 0.4) = 19.7\ \mu\text{A V}^{-1}$$

To calculate the transconductance of the transistor in figure 1.23 we can use (1.76) after calculating the gain from the given input and output AC voltages:

$$g_m = \frac{|G|}{R_L} = \frac{954/40}{10^6} = \frac{23.9}{10^6} = 23.9\ \mu\text{A V}^{-1}$$

We can also see that the DC current through the MOSFET is $I_D = (V_{DD} - V_o)/R_L = 1.65\ \mu\text{A}$, and the DC voltage ratio $V_O/V_G = 1.65/0.54 = 3.1$ has nothing to do with the AC gain, which is 945 mV/40 mV = 23.9.

The achievable gain depends not only on the product $g_m R_L$ but also on the supply voltage. Increasing g_m makes the drain current increase too, according to (1.74). With the supply fixed, the load resistance R_L cannot be made arbitrarily high because the voltage drop across it would 'eat up' into the available voltage for the MOSFET. A way to get around this limitation is to use another MOSFET as an active load instead of a resistor [29], which is the preferred method in IC design.

The formulas describing MOSFET operation assume that no current flows when the gate–source voltage is below threshold. In practice when $V_{GS} < V_T$ the drain current is very small but not zero, and the MOSFET operates in *subthreshold* (also known as weak inversion) mode. The drain current is caused by diffusion of electrons from the source because their concentration is higher than at the drain [2]. The subthreshold current can be described by [29]:

$$I_D = I_{D0}\frac{W}{L}\exp\left(\frac{V_{GS}}{n\varphi_T}\right) \tag{1.77}$$

where I_{D0} is a technology parameter of the order of 10^{-12} A called off-state leakage, $\varphi_T = kT/q$ is the thermal potential and n is the subthreshold slope factor. Formula (1.77) is valid for $V_{DS} \gg \varphi_T$.

In subthreshold mode the drain current depends exponentially on the gate–source voltage, rather than quadratically as in saturation (1.72). The gate transconductance from (1.77) is

$$g_m = \frac{\partial I_D}{\partial V_{GS}} = \frac{I_{D0}}{n\varphi_T}\frac{W}{L}\exp\left(\frac{V_{GS}}{n\varphi_T}\right) = \frac{I_D}{n\varphi_T} \tag{1.78}$$

The much higher rate of change of the drain current would appear to make the transconductance high, but the drain current in subthreshold is in the nanoamp range, so in practice g_m is much lower than in saturation. This is one reason why MOSFETs operating in subthreshold mode are rarely used as amplifiers. However, subthreshold operation occurs in many MOSFET circuits where the source is driving a high impedance load, such as the sense node connected to the reset transistor in image sensors.

TCAD models such as the one in figure 1.24 are used to extract the transistor characteristics, which after parameterisation are converted to SPICE models for more convenient and faster simulations. The simulated input characteristic in

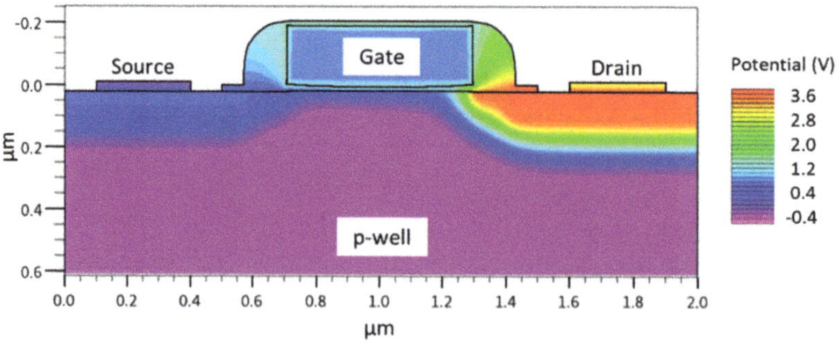

Figure 1.24. 2D TCAD model of an n-MOSFET with $L = 0.6$ µm, $W = 1.0$ µm, $t_{ox} = 12$ nm and uniform p-well with boron doping of 2×10^{17} cm^{-3}. The substrate (bulk) and V_S are at ground, and $V_G = 1.0$ V, $V_D = 3.3$ V. The structure is symmetrical, and the source and the drain are interchangeable.

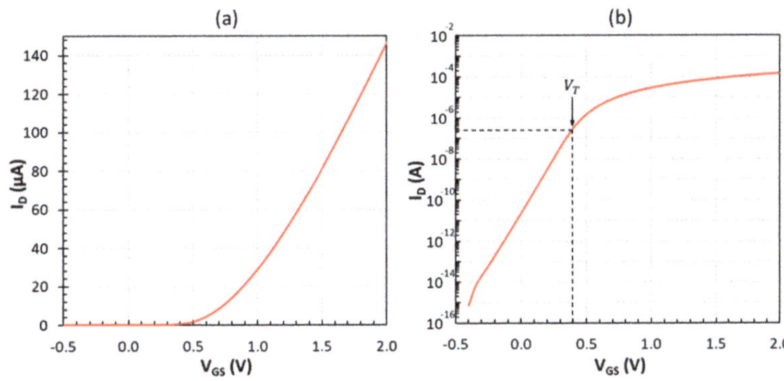

Figure 1.25. TCAD simulation of the input characteristic of the n-MOSFET in figure 1.24 for $V_{DS} = 3.3$ V plotted on a linear scale (a), and on a semi-log scale (b).

figure 1.25 shows that the threshold of this transistor, defined as the gate–source voltage at which the drain current is 100 nA $\times W/L$ [26], is approximately 0.4 V.

Below threshold, the exponential dependence (1.77) holds for nearly 8 orders of magnitude change of the drain current (figure 1.25(b)). The off-state leakage at $V_{GS} = 0$ V is around 20 pA, which may be low enough for many applications, but is enormous in the context of image sensors—125 million electrons per second. The transistor turns off completely only when its gate–source voltage is negative, and the drain current becomes too low to be reliably simulated when $V_{GS} < -0.5$ V. Such low off-state currents are indeed needed in image sensors, where the pixel dark current can be measured in few electrons per second at room temperature.

1.8.3 Output resistance and body effect

Formula (1.72) tells us that in saturation the drain current does not depend on the drain–source voltage, i.e. the MOSFET behaves as an ideal current source. This is not exactly what happens in practice; as the drain–source voltage increases, the effective channel length decreases through a mechanism known as channel modulation [26] and the drain current slightly increases. The effect is easily seen in the output transistor characteristics, such as those in figure 1.26, where the drain current continues to increase with V_{DS} after saturation is reached. In the absence of channel modulation, the drain current would be 'flat' for high V_{DS} and the output impedance would be infinity.

This effect can be approximated with a resistor connected in parallel with the channel, between the source and the drain. Mathematically this is expressed by multiplying the drain current (1.72) with a term containing the drain–source voltage and the channel modulation parameter λ:

$$I_D = \frac{\mu_n C_{ox} W}{2L}(V_{GS} - V_T)^2(1 + \lambda V_{DS}) \tag{1.79}$$

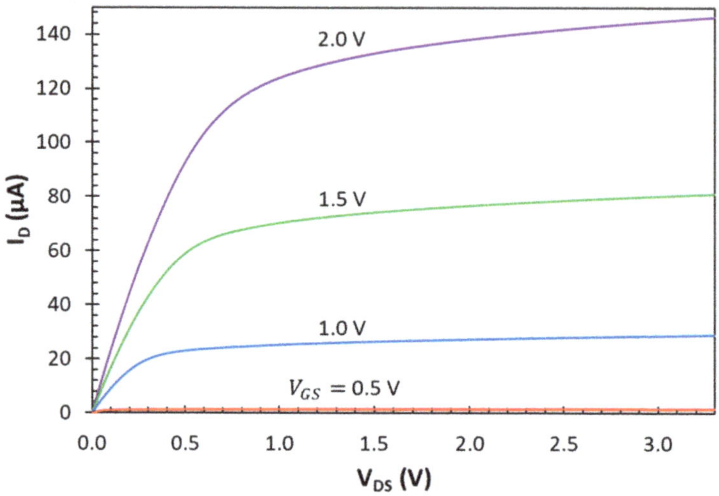

Figure 1.26. Output characteristics of the transistor in figure 1.24.

The change of the drain current caused by the drain–source voltage is called output conductance[3] g_{ds} and is defined as

$$g_{ds} = \frac{\partial I_D}{\partial V_{DS}} = \frac{\mu_n C_{ox} W}{2L}(V_{GS} - V_T)^2 \lambda = \frac{\lambda I_D}{1 + \lambda V_{DS}} \tag{1.80}$$

Normally the parameter λ is small, and as a rule of thumb g_{ds} is about a hundred times smaller than the gate transconductance g_m [29]. The output conductance (1.80) for small λ simplifies to

$$g_{ds} = \frac{\lambda I_D}{1 + \lambda V_{DS}} \cong \lambda I_D \tag{1.81}$$

The output conductance g_{ds} can be substituted by the output resistance[4] r_{ds} in the MOSFET model in figure 1.27. The output resistance is simply the inverse of the output conductance:

$$r_{ds} = \frac{1}{g_{ds}} = \frac{1}{\lambda I_D} \tag{1.82}$$

A much stronger effect on the drain current is caused by the source–substrate voltage through modulation of the transistor threshold. This is called body effect and is important for many circuits such as source followers, amplifiers and analogue switches.

[3] The drain–source voltage directly affects the drain current and therefore the term we use is conductance. The gate–source voltage influences the drain current indirectly, and the term transconductance (transfer of conductance) is used. It is also called mutual conductance, which gives the letter m in g_m.

[4] Normally we think of the output resistance as a resistor in series with a voltage source. The MOSFET is a current source, therefore its output resistance is in parallel with the channel.

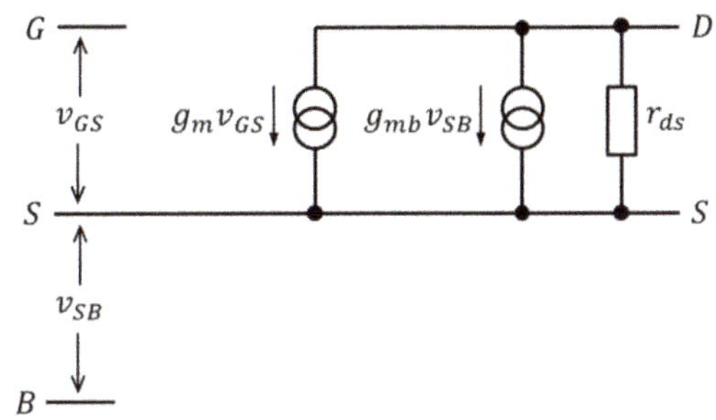

Figure 1.27. MOSFET model with output resistance and body effect.

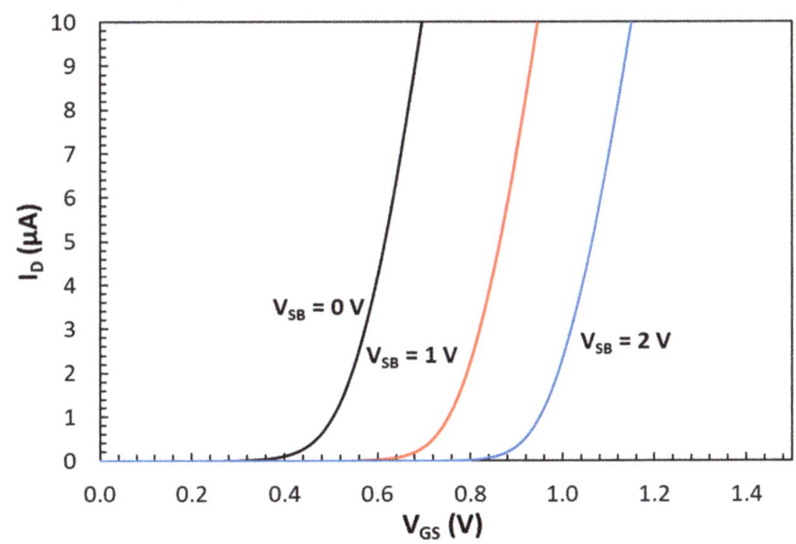

Figure 1.28. Body effect in an *n*-MOSFET with $L = 1.0$ μm and $W = 1.0$ μm, manufactured in 180 nm CMOS process.

In normal operation the source–bulk and drain–bulk *pn* junctions are reverse biased. When the source voltage is higher than the bulk ($V_{SB} > 0$), the depletion under the channel grows. This increases the amount of negative space charge, which is coupled capacitively to the inversion layer and induces in it a charge of the opposite polarity. Therefore, the gate voltage forcing channel inversion must increase to maintain the same drain current, which is equivalent to increasing the transistor threshold.

A more detailed description of the body effect is given in the following section, but here we will look at the practical aspects. Figure 1.28 shows the input characteristics of an *n*-MOSFET for three different source–bulk voltages as a parameter. The threshold increases in a sub-linear fashion by over 400 mV for a 2 V change in V_{SB} (the full dependence is given by (1.94)), and more than doubles. This is

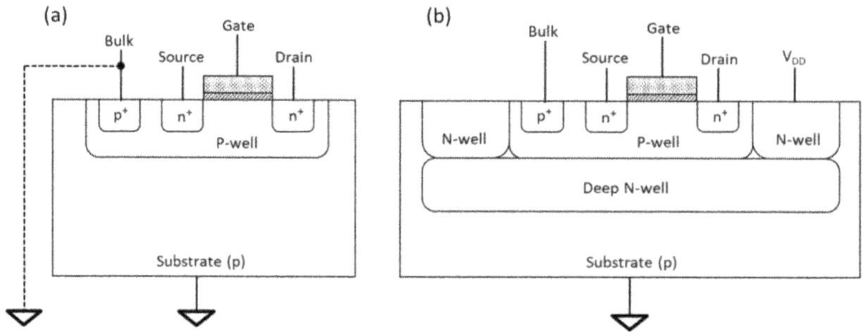

Figure 1.29. MOSFET in a *p*-well intrinsically connected to substrate (a) and in a floating *p*-well (hot *p*-well) with deep *n*-well isolation (b).

a substantial change; such large increase in V_T is usually not welcome because it reduces the available voltage range for the signal, since the supply is fixed.

Similarly to the gate transconductance, the body transconductance g_{mb} can be calculated from (1.79) as the rate of change of the drain current caused by V_{SB}:

$$g_{mb} = \frac{\partial I_D}{\partial V_{SB}} = \frac{\partial I_D}{\partial V_T}\frac{\partial V_T}{\partial V_{SB}} = -\frac{\mu_n C_{ox} W}{L}(V_{GS} - V_T)(1 + \lambda V_{DS})\frac{\partial V_T}{\partial V_{SB}} \quad (1.83)$$

and using (1.73) with added channel modulation, we can write

$$g_{mb} = -g_m \frac{\partial V_T}{\partial V_{SB}} \quad (1.84)$$

The negative sign in (1.83) is there because $\partial V_T / \partial V_{SB} > 0$ and increasing V_{SB} makes the drain current smaller by increasing the threshold. From the example in figure 1.28 and (1.84) we see that g_{mb} is roughly a factor of 10 smaller than g_m, which generally holds as a rule of thumb [29].

The body effect is not unavoidable; if the bulk and the source are at the same potential it can be eliminated. The bulk, which is the *p*-well in *n*-MOSFETs, can either be joined to the substrate intrinsically because they are both *p*-type, as in figure 1.29(a), or are floating, as in figure 1.29(b).

Unless all transistors in the circuit operate with their sources grounded, connecting to sources to the bulk is not an option for figure 1.29(a), therefore the body effect will be there. The floating *p*-well in figure 1.29(b) solves the problem because the bulk can be connected to the source of each transistor and be at a different potential, made possible by the deep *n*-well isolation.

1.8.4 Transistor threshold

Normally, transistor thresholds are chosen to be around 0.6–0.8 V. These high thresholds are necessary to reduce the subthreshold leakage in large digital circuits, where the leakage from billions of transistors can add up to unacceptable levels.

High transistor thresholds may not be optimal for analogue circuits. Supply voltages in image sensors are low and transistor thresholds often appear as undesirable voltage offsets, eating into the precious signal amplitude. Because of this, special transistors with 'low V_T' or even 'ultra-low V_T' have been developed and are increasingly being used.

Understanding how the transistor threshold depends on the oxide thickness, substrate doping concentration and bias, and its impact on performance is very important for an image sensor designer.

In an *n*-MOSFET using highly doped *n*-type poly-Si gate at zero gate bias (equilibrium state) the *p*-type body is already in depletion due to the difference in the work functions between the gate and the substrate. Starting with the energy diagram in figure 1.30, we observe that a *negative* voltage has to be applied to the gate so that the flat-band condition is achieved. The voltage V_{FB} is called flat-band voltage and is defined as the difference between the Fermi levels of the gate and the substrate, which in this condition is equal to the difference between the two work functions

$$V_{FB} = \psi_G - \psi_S \tag{1.85}$$

For the structure in figure 1.30 (highly doped *n*-type poly-Si gate over a *p*-type substrate) V_{FB} is negative, at about −0.6 to −0.9 V. From figure 1.30 we can also see that $V_{FB} = \varphi_B - \varphi_G$ and using (1.56) can write

$$V_{FB} = \varphi_B - \varphi_G = -\frac{kT}{q}\ln\left(\frac{N_A}{n_i}\right) - \frac{kT}{q}\ln\left(\frac{N_D}{n_i}\right) \tag{1.86}$$

where N_A and N_D are the dopant concentrations in the substrate and the polysilicon gate, respectively.

The threshold voltage can be calculated from the general equation (1.87) [26], which simply states that the potential drops across the oxide and the substrate balance the gate voltage.

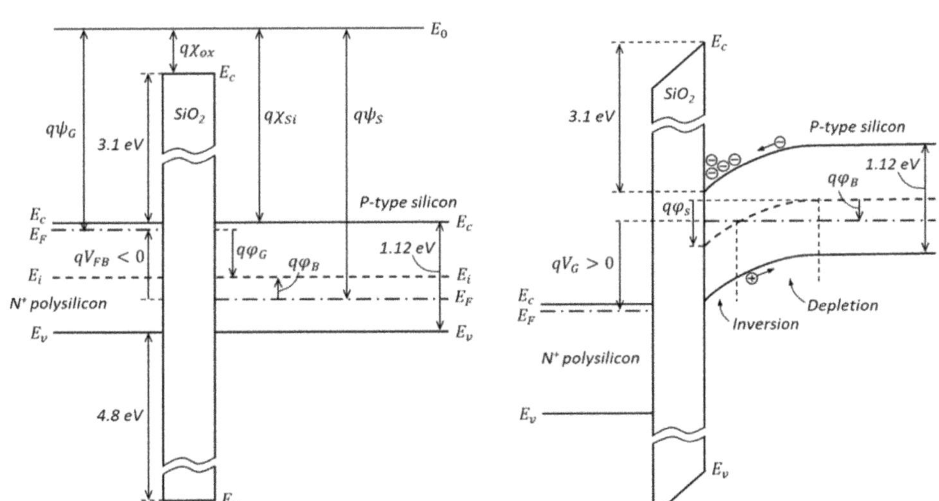

Figure 1.30. Band diagram of a MOS capacitor in flat-band condition and in strong inversion.

$$V_G - V_{FB} = \varphi_s + V_{ox} \tag{1.87}$$

In flat-band condition $V_G - V_{FB} = 0$ and $\varphi_s = V_{ox} = 0$. Starting with the flat-band condition, we then calculate the additional gate voltage change to bring the substrate into strong inversion or any other condition.

The voltage across the oxide is given by the space charge in the depleted region per unit area Q_{b0}, divided by the gate capacitance:

$$V_{ox} = \frac{Q_{b0}}{C_{ox}} \tag{1.88}$$

Using equation (1.58) for W, Q_{b0} is written as:

$$Q_{b0} = qN_A W = qN_A \sqrt{\frac{2\varepsilon_0 \varepsilon_{Si} \varphi_s}{qN_A}} = \sqrt{2\varepsilon_0 \varepsilon_{Si} q N_A \varphi_s} \tag{1.89}$$

The space charge in a p-type substrate is negative due to the negatively charged acceptor atoms. When an additional reverse voltage V_{SB} is applied between the source and the bulk the width of the depletion region expands similarly to a reverse-biased pn junction (1.45) and the space charge per unit area becomes

$$Q_b = \sqrt{2\varepsilon_0 \varepsilon_{Si} q N_A (\varphi_s + V_{SB})} \tag{1.90}$$

Strong inversion is achieved when $\varphi_s = 2\varphi_B$. In contrast with the deep depletion shown in figure 1.20, in the n-MOSFET strong inversion is achieved very quickly because the highly doped source and drain provide an abundant supply of electrons. Once inversion is reached, the surface potential and the depletion depth stop growing with the gate voltage because the semiconductor below is screened from further potential changes by the thin inversion layer.

The threshold voltage can be calculated from (1.87) using $V_T = V_G$ and $\varphi_s = 2\varphi_B$.

$$V_T = V_{FB} + 2\varphi_B + \frac{Q_b}{C_{ox}} - \frac{Q_{ss}}{C_{ox}} \tag{1.91}$$

Here Q_{ss} is the fixed surface charge density at the oxide, which is normally positive and has density Q_{ss}/q in the range 10^{10}–10^{11} cm^{-2}. Equation (1.91) can be written as

$$V_T = V_{FB} + 2\varphi_B + \frac{Q_{b0}}{C_{ox}} - \frac{Q_{ss}}{C_{ox}} + \frac{Q_b - Q_{b0}}{C_{ox}} \tag{1.92}$$

The first four terms do not depend on the substrate voltage and give the threshold voltage for $V_{SB} = 0$:

$$V_{T0} = V_{FB} + 2\varphi_B + \frac{Q_{b0}}{C_{ox}} - \frac{Q_{ss}}{C_{ox}} = V_{FB} + 2\varphi_B + \frac{\sqrt{2\varepsilon_0 \varepsilon_{Si} q N_A \, | 2\varphi_B |}}{C_{ox}} - \frac{Q_{ss}}{C_{ox}} \tag{1.93}$$

The final expression for the threshold voltage including the dependence on V_{SB} is

$$V_T = V_{T0} + \frac{Q_b - Q_{b0}}{C_{ox}} = V_{T0} + \frac{\sqrt{2\varepsilon_0\varepsilon_{Si}qN_A}}{C_{ox}}\left(\sqrt{|2\varphi_B| + V_{SB}} - \sqrt{|2\varphi_B|}\right) \quad (1.94)$$

It could be very confusing to keep track of the signs of the terms in (1.93), but the following considerations and a careful look at figure 1.30 should help. The gate voltage should bend the bands downwards to reach inversion, therefore the second term $2\varphi_B$ should be positive. Also, the gate voltage should counteract the negative space charge, therefore the third term must also be positive. The fourth term is negative because Q_{ss} reduces the needed voltage to achieve inversion, since Q_{ss} is a positive charge which acts in the same direction.

Example 1.16. Calculate the threshold voltage of an n-MOSFET with the following parameters: $N_A = 2 \times 10^{17}\,\text{cm}^{-3}$, gate doping $N_D = 10^{19}\,\text{cm}^{-3}$, $C_{ox} = 4.93\,\text{fF}\,\mu\text{m}^{-2}$ (7 nm thick gate oxide), $Q_{ss}/q = 10^{10}\,\text{cm}^{-2}$ and 300 K. Use that $n_i = 1.45 \times 10^{10}\,\text{cm}^{-3}$, $\varepsilon_0 = 8.85 \times 10^{-14}\,\text{F cm}^{-1}$ and $\varepsilon_{Si} = 11.9$. Also calculate V_T for $N_A = 1 \times 10^{16}\,\text{cm}^{-3}$.
Solution: The first step is to calculate φ_B, φ_G and the flat-band voltage from (1.86)

$$\varphi_B = -\frac{kT}{q}\ln\left(\frac{N_A}{n_i}\right) = -\frac{1.38 \times 10^{-23} \times 300}{1.6 \times 10^{-19}}\ln\left(\frac{2 \times 10^{17}}{1.45 \times 10^{10}}\right) = -0.425\text{ V}$$

$$\varphi_G = \frac{kT}{q}\ln\left(\frac{N_D}{n_i}\right) = -\frac{1.38 \times 10^{-23} \times 300}{1.6 \times 10^{-19}}\ln\left(\frac{10^{19}}{1.45 \times 10^{10}}\right) = 0.527\text{ V}$$

$$V_{FB} = \varphi_B - \varphi_G = -0.425 - 0.527 = -0.952\text{ V}$$

Next, the third term in (1.93) is:

$$\frac{\sqrt{2\varepsilon_0\varepsilon_{Si}qN_A|2\varphi_B|}}{C_{ox}}$$
$$= \frac{\sqrt{2 \times 8.85 \times 10^{-14} \times 11.9 \times 1.6 \times 10^{-19} \times 2 \times 10^{17} \times 2 \times 0.425}}{4.93 \times 10^{-15} \times 10^8} = 0.486\text{ V}$$

The last term is

$$\frac{Q_{ss}}{C_{ox}} = \frac{10^{10} \times 1.6 \times 10^{-19}}{4.93 \times 10^{-15} \times 10^8} = 0.003\text{ V}$$

Finally, the threshold is:

$$V_{T0} = -0.952 + 2 \times 0.425 + 0.486 - 0.003 = 0.381\text{ V}$$

Repeating the calculation for $N_A = 1 \times 10^{16}\,\text{cm}^{-3}$ gives $V_{T0} = -0.084$ V.

The threshold voltage can be effectively controlled by the dopant concentration of the substrate, and special 'threshold adjust' implants are used for that purpose. This example also shows that the natural threshold of n-MOSFETs in higher resistivity substrates is around 0 V. Such transistors without a threshold adjustment implant are called 'native'.

The threshold volage is temperature dependent via the thermal potential entering V_{T0}. For *n*-MOSFETs the temperature coefficient of the threshold is negative and is typically around −2 mV °C^{-1}, with a range between −0.5 and −4 mV °C^{-1}; for *p*-MOSFETs the coefficient is positive.

1.8.5 Analogue switch

MOSFETs working as analogue switches are used to route signals and connect various points together. Their other use is in sample and hold (S&H) circuits to temporarily connect a signal to a capacitor, so that its value can be stored. Figure 1.31 shows a simple S&H circuit built with a MOSFET and a capacitor. With the transistor on, the capacitor charges to the input voltage V_{in} through the resistance of the channel. After the switch turns off, the disconnected capacitor retains its voltage for a very long time (which can be seconds) and this makes it useful as an analogue memory.

To turn the switch on, the gate–source voltage must be above the transistor threshold, but the MOSFET is symmetrical and can work in both directions. The gate–source voltage can be considered as either $V_{GS} = V_G - V_{in}$ or $V_{GS} = V_G - V_{out}$; if one of them is higher than V_T the transistor will turn on. The maximum input voltage this single-transistor analogue switch can handle is $V_G - V_T$, and is limited by the highest voltage that can be applied to the gate, which is typically the supply voltage. The threshold suffers from the body effect.

The source–drain voltage in the 'off' state can be very large, but in the 'on' state it must be low because the MOSFET behaves as a resistor only when $V_{DS} \ll (V_{GS} - V_T)$. When connecting two points with very different potentials, initially the drain current can be large and the MOSFET will be operating in saturation, until the voltages equalise, and the transistor enters linear regime. This is what happens in the switch in figure 1.31—regardless of the initial capacitor voltage (within the operating limits), the end voltage is V_{in}.

Analogue switching occurs if at least one of the terminals V_{in} and V_{out} is at high impedance, so that the input and output have a chance to equalise. If both V_{in} and V_{out} are low impedance sources this will not happen, and they will maintain their own voltages—current will flow between them, but we would not call that switching.

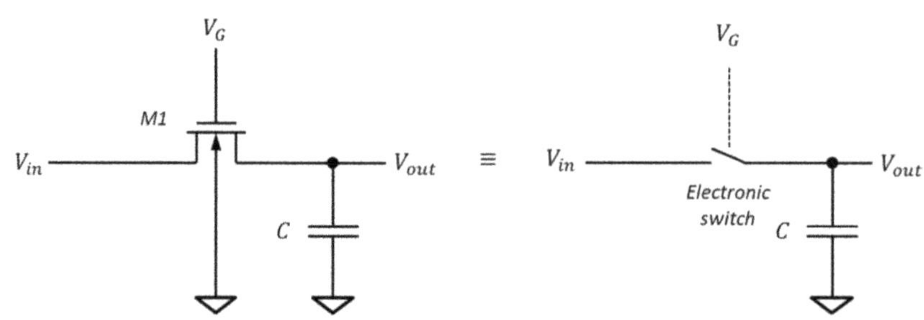

Figure 1.31. Analogue sample and hold switch built with an *n*-MOSFET (a) and its equivalent circuit.

Example 1.17. Calculate the 'on' resistance of n-MOSFET with $W = L = 1.0$ μm, $\mu_n = 400$ cm^2 V^{-1} s^{-1}, $C_{ox} = 4.93$ fF μm^{-2} (7 nm thick gate oxide), $V_T = 0.4$ V and $V_{GS} = 1.0$ V. Also, calculate the bandwidth of the circuit in figure 1.31 for $C = 1$ pF.
Solution: The drain–source voltage should be much smaller than the overdrive voltage $V_{GS} - V_T$, so that the transistor operates in the linear regime. From equation (1.71) the resistance is:

$$R_{\text{lin}} = \frac{10^{-4}}{400 \times 4.93 \times 10^{-15} \times 10^{8} \times 10^{-4} \times (1.0\text{–}0.4)} = 8.4 \text{ k}\Omega$$

The signal bandwidth is calculated from:

$$BW = \frac{1}{2\pi R_{\text{lin}} C} = \frac{1}{6.28 \times 8400 \times 10^{-12}} = 18.9 \text{ MHz}$$

Despite the MOSFET appearing to have too high a resistance for something to be called an analogue switch, for the usual small load capacitances the response of the circuits like those in figure 1.31 can be quite fast.

The simple n-MOSFET switch in figure 1.31, and its p-MOSFET counterpart, have substantial limitations on their operating voltages. For example, in a circuit supplied with 3.3 V, a typical n-MOSFET will not be able to switch voltages higher than about 2.5 V. This is because the threshold is high due to the body effect (see figure 1.28), and the maximum gate voltage is 3.3 V. Also, its 'on' resistance changes wildly as a function of the input voltage, becoming very high as the gate–source voltage approaches threshold according to (1.71).

To solve this problem, the CMOS analogue switch [29] in figure 1.32 is used. It works well for any input voltage within the supply rails and exhibits much smaller change of the 'on' resistance than the single-transistor switch. The CMOS variant has an n-MOSFET and a p-MOSFET connected in parallel; the n-channel transistor

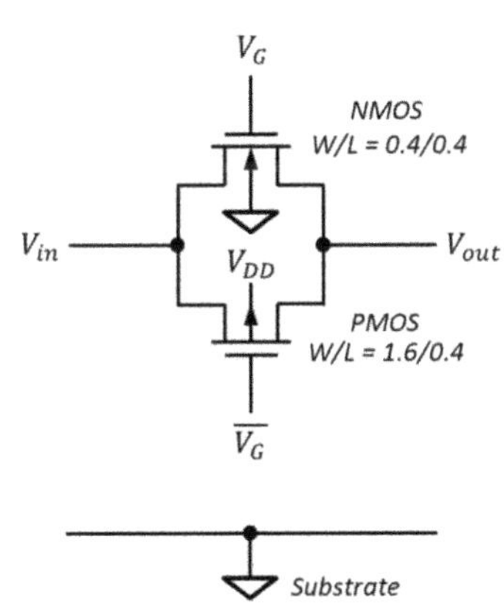

Figure 1.32. CMOS analogue switch.

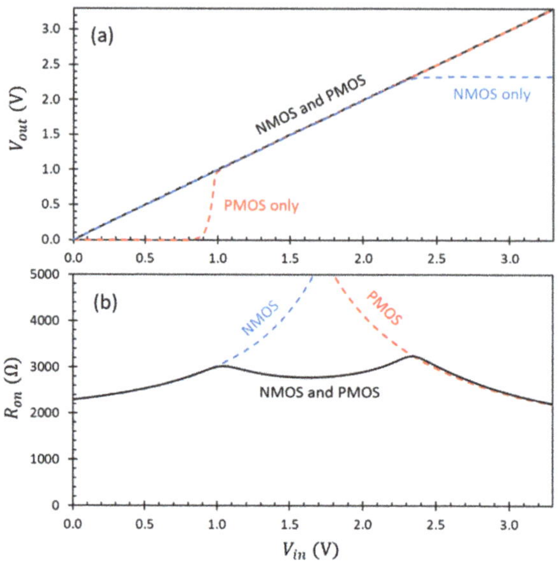

Figure 1.33. Output characteristic (a) and resistance (b) of the CMOS analogue switch in figure 1.32 for V_{DD} = 3.3 V, V_G = 3.3 V and $\overline{V_G}$ = 0 V.

takes care of the low input voltages for which V_{GS} is large. To complement this, the *p*-channel transistor works at high inputs because zero gate voltage is applied to turn it on, and its V_{GS} is maximised. The switch control requires complementary gate voltages: $V_G = V_{DD}$ and $\overline{V_G} = 0$ for the 'on' state and the opposite for the 'off' state.

Figure 1.33(a) shows a simulation of the CMOS switch in figure 1.32 for input voltage spanning from substrate potential to the supply. We see that the *n*-MOSFET transistor behaves like a resistor up to an input voltage around 2.4 V (for 3.3 V supply); above that it fails to faithfully follow the input. This is caused by its high threshold—the output can never be higher than the gate voltage minus the threshold $V_G - V_T$, and the threshold is around 0.9 V due to the strong body effect. Similarly, the *p*-MOSFET transistor begins to work only for voltages exceeding about 1 V. By connecting two complementary transistors in parallel we have an excellent switch action across the whole input range, spanning from zero (substrate potential) to the supply V_{DD}.

The resistance of the CMOS switch in figure 1.33(b) is the parallel combination of the two transistors: The *n*-MOSFET has low resistance at low input voltages and the *p*-MOSFET at high voltages, so the paralleled resistance does not suffer from large variations. The transistors can be sized appropriately so that the shape of the resistance curve is symmetrical, as done for the switch in figure 1.32.

1.8.6 MOSFET capacitor

When operating in inversion or accumulation, the concentration of free carriers in the MOSFET channel is very high. The free carriers act as one half of a parallel plate capacitor, with the other half being the gate. Normally only inversion is used; this is

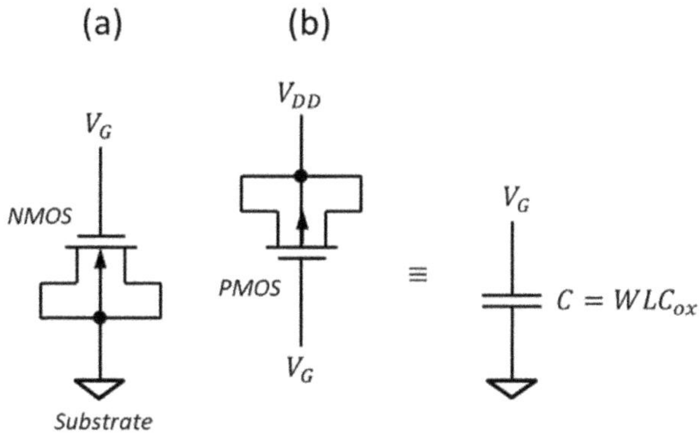

Figure 1.34. MOSFETs used as capacitors.

because to bias the channel of a *n*-MOSFET in accumulation the gate must be negative with respect to the bulk, which is possible, but rarely a practical option.

Figure 1.34 shows two ways to create MOSFET capacitors, for *n*- and *p*-channel transistors. For both the bulk, source and drain are connected, and both circuits are equivalent to a capacitor to ground, because the supply V_{DD} is AC ground too. For correct operation in inversion, we need $V_G > V_T$ for the *n*-MOSFET, and $V_G < V_{DD} - V_T$ for the *p*-MOSFET.

Example 1.18. Calculate the gate capacitance to substrate of a *n*-MOSFET with $W = L = 5.0$ μm, $C_{ox} = 4.93$ fF μm^{-2} (7 nm thick gate oxide), $V_T = 0.4$ V and $V_{GS} = 2.0$ V. Compare with the *pn* junction capacitance in example 1.13.
Solution: Because $V_{GS} \gg V_T$ the channel is in inversion, therefore the gate capacitance is

$$C = WLC_{ox} = 5 \times 5 \times 4.93 = 123.3 \text{ fF}$$

This capacitance is about 90 times higher than the *pn* junction with the same area in example 1.13.

The gate capacitance of a MOSFET offers the highest capacitance per unit area because it uses very thin oxide measured in just a few nanometres; there is simply no other way to achieve such high specific capacitance. The downside is that the transistor must always be in inversion, and failure to follow this results in much smaller capacitance, paired with nonlinearity.

1.9 Source follower

1.9.1 Gain

The source follower (SF) is among the simplest, but also the most important circuits in image sensors. The name simply means that the source voltage 'follows' the gate voltage. Because of the transistor threshold there is a DC offset between the gate and the source and the 'following' applies only to the AC component of the gate voltage.

Figure 1.35 shows two basic SF circuits using a resistor as a load; they are useful to explain how the SF works but rarely used in practice because resistors take up too much space in integrated circuits.

We can derive the voltage gain of the simple SF in figure 1.35(b). The transistor operates in saturation and the relationship between the DC voltages (marked with capital letters) for $V_{GS} > V_T$ is:

$$V_O \approx V_G - V_T \tag{1.95}$$

For small signal analysis we use the AC components of the output voltage v_O, drain current i_D and the gate–source voltage v_{GS}, marked with lower case letters.

The definition of the transconductance (1.73) applied to small signals states that a small change of the gate–source voltage v_{GS} around a DC bias point induces a small change in the drain current i_D, and we can write

$$i_D = g_m v_{GS} \tag{1.96}$$

The small signal schematic in figure 1.36 shows that the drain current flows through the load resistor R_L since the drain is connected to the supply voltage V_{DD}, which is at AC ground. For simplicity, the load resistance here includes the

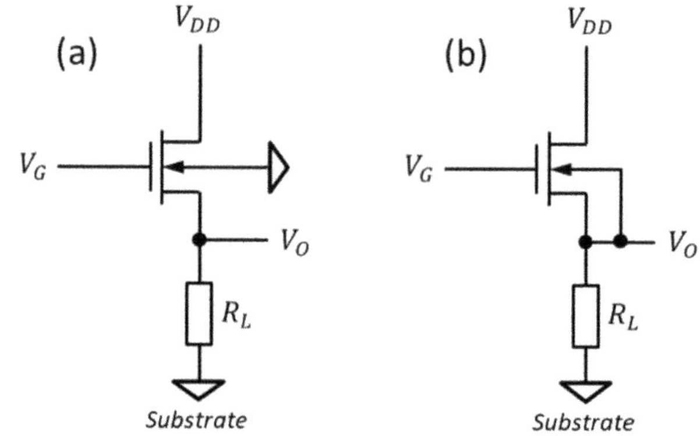

Figure 1.35. Source follower with resistor load and body connected to the substrate(a); and with the body connected to the source to avoid the body effect (b).

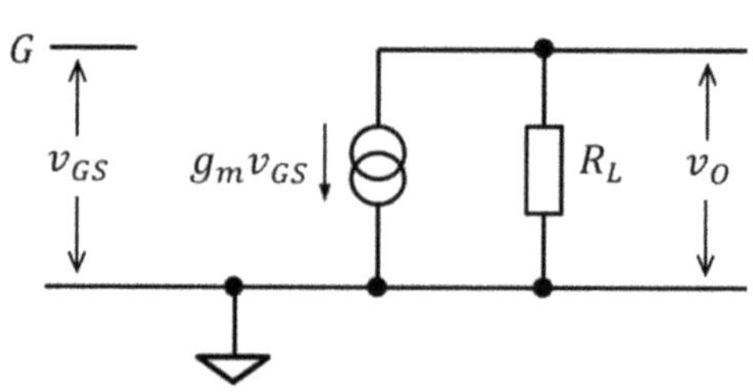

Figure 1.36. Small signal equivalent schematic of a source follower without body effect.

transistor's output resistance r_{ds}, which similarly to (1.82), can be replaced by the transconductance $g_L = 1/R_L$.

The output voltage v_O of the SF is the drain current (1.96) multiplied by the load resistance R_L:

$$v_O = i_D R_L = g_m v_{GS} R_L \tag{1.97}$$

Using that $v_{GS} = v_G - v_O$ and substituting in (1.97) we get

$$v_O = \left(\frac{g_m R_L}{1 + g_m R_L}\right) v_G \tag{1.98}$$

The term in brackets is the gain of the source follower G_{SF}:

$$G_{SF} = \frac{v_O}{v_G} = \frac{g_m R_L}{1 + g_m R_L} = \frac{g_m}{g_m + g_L} \tag{1.99}$$

In the absence of body effect, the gain G_{SF} given by (1.99) is very close to one and is typically above 0.9. With body effect the gain is

$$G_{SF} = \frac{g_m}{g_m + g_{mb} + g_L} \tag{1.100}$$

Because usually $g_{mb} \approx 0.1 g_m$ the theoretical maximum gain with body effect is around 0.9 (if R_L is very large, $g_L \approx 0$), but in practice is around 0.8.

Figure 1.37 shows the DC transfer characteristic of a source follower with resistive load, which can be used to determine the gain. Formulas (1.99) and (1.100) give the gain for small input signals when the source follower is biased in saturation and should not be confused with the ratio between the DC voltages at the source and the gate. When the input voltage is close to V_T the ratio of the DC source and gate

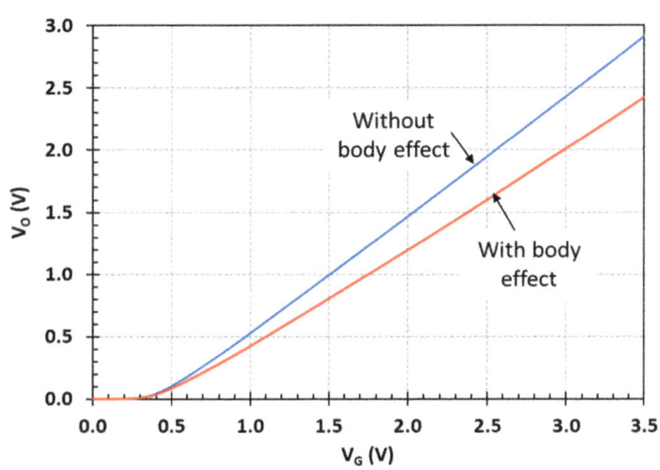

Figure 1.37. DC transfer characteristics of a source follower with $W/L = 1\mu/1\mu$, $R_L = 1$ MΩ and $V_{DD} = 3.3$ V, with and without body effect, corresponding to figure 1.35.

voltages can be very far from the AC small signal gain, but it does approach it for high input voltages. This is because the output voltage stays at zero until the input exceeds the transistor threshold.

Example 1.19. Calculate the approximate SF gain from the data in figure 1.37.
Solution: The gain is defined as the change of the output voltage over a small change in the input voltage. Since the output voltage is linear for $V_G > 0.6$ V, we can use any input voltage difference above that, for example between $V_G = 2.0$ V and $V_G = 3.0$ V. Without body effect $G_{SF} \approx (2.4 - 1.5)/1.0 = 0.9$; with body effect $G_{SF} \approx (2.0 - 1.2)/1.0 = 0.8$. A proper fit to the numerical data gives $G_{SF} = 0.94$ and $G_{SF} = 0.77$, correspondingly.

The gain of the source follower G_{SF} is always less than one. For the typical source follower operating currents of few microamps the load resistor R_L should be in the mega-ohm range. It is not practical or even possible to include such a large resistor on a chip; instead, an active load is used virtually everywhere. Figure 1.38(a) shows a source follower M1 working with an active current load. Because the current is flowing into the load it is also called a current sink.

The simplest possible current load is built with just one MOSFET, as shown in figure 1.38(b). Since in saturation the MOSFET behaves as a voltage-controlled current source according to (1.72), when an appropriate voltage V_{bias} is applied to the gate of M2 the drain current I_{bias} is nearly constant and independent of the output voltage V_O. In practice I_{bias} depends slightly on V_O due to the output resistance of M2.

The current load can be taken a step further in figure 1.38(c) with a current mirror consisting of the transistors M2 and M3. An external current I_{in} is supplied to M3, which is connected as a MOS diode. This creates the voltage drop V_{bias} across M3, which in turn supplies the gate of M2. What this does is to generate the voltage V_{bias} locally to the source follower, using a current input instead of a voltage input. If transistors M2 and M3 were the same, the same gate–source voltage V_{bias} ensures that their drain currents would be the same too. When I_{bias} is just a few microamps,

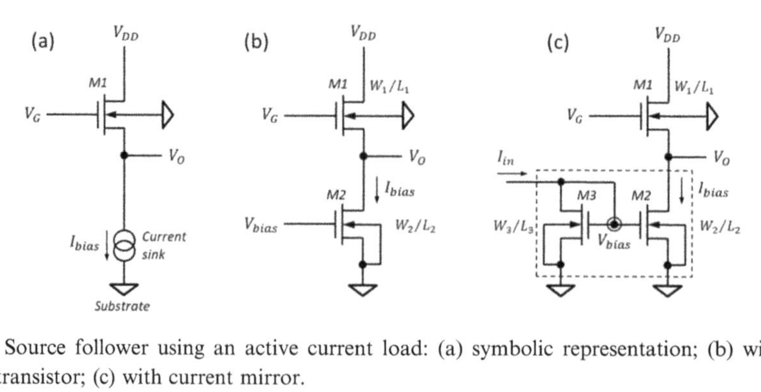

Figure 1.38. Source follower using an active current load: (a) symbolic representation; (b) with externally biased load transistor; (c) with current mirror.

the voltage V_{bias} is very close to the transistor threshold and is typically around 0.4–0.6 V.

The current I_{bias} depends on the channel width and length of M2 and M3 as [29]

$$I_{\text{bias}} = I_{\text{in}}\left(\frac{W_2}{W_3}\right)\left(\frac{L_3}{L_2}\right) \quad (1.101)$$

Equation (1.101) tells us that the bias current I_{bias} 'mirrors' the input current I_{in} through a proportionality constant, determined by the width and the length of the two transistors. Normally $L_2 = L_3$ and this allows the current ratio to be chosen simply as the width ratio of M2 over M3. The transistor M3 can generate the bias for many current load transistors and can be shared among them.

Source followers with an active load have better performance and higher gain than the resistor-loaded followers due to the high dynamic resistance of the load transistor. They are almost invariably used in all CMOS image sensors and will pop up frequently in the following chapters.

1.9.2 Input capacitance

The input capacitance of the source follower adds to the sense node capacitance, therefore, it is part of the charge-to-voltage conversion. Electrically, the input capacitance consists of two parts, as shown in figure 1.39(a):

(a) C_{d} 'detection capacitance' is all the capacitance between the gate and the substrate, with the substrate being at AC ground. The gate–drain capacitance is also connected between gate and AC ground, and for simplicity is included in C_{d}.

(b) C_{GS} is the capacitance between the gate and source.

We would like to know the effective gate capacitance to substrate C_{in}. We can find it by introducing a small charge Q at the input, which changes the gate voltage by $v_{\text{G}} = Q/C_{\text{in}}$. The change of the output voltage, including the source follower gain, is $v_{\text{o}} = QG_{\text{SF}}/C_{\text{in}}$, therefore we need to find an expression for v_{o}.

The charge Q is shared between the two capacitances, so that from conservation of change we have

$$Q = C_{\text{d}}v_{\text{G}} + C_{\text{GS}}v_{\text{GS}} \quad (1.102)$$

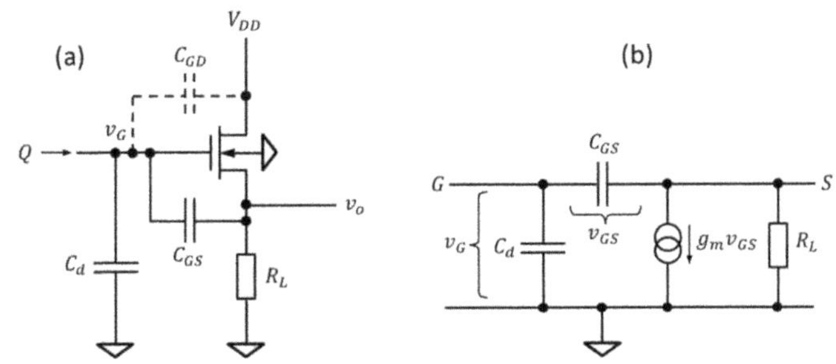

Figure 1.39. Source follower driven by a high impedance charge input (a) and its simulation schematic (b).

Using that $v_{GS} = v_G - v_o$ and $v_o = G_{SF}v_G$ from (1.99), equation (1.102) can be written as

$$Q = C_d\frac{v_o}{G_{SF}} + C_{GS}\left(\frac{v_o}{G_{SF}} - v_o\right) \tag{1.103}$$

From here, solving for v_o we get

$$v_o = \frac{QG_{SF}}{C_d + C_{GS}(1 - G_{SF})} \tag{1.104}$$

From equation (1.104) we can conclude that the effective input capacitance of the source follower is

$$C_{in} = C_d + C_{GS}(1 - G_{SF}) \tag{1.105}$$

It is not surprising that C_d appears unaltered in the input capacitance, because it is connected to the substrate. On the other hand, C_{GS} is significantly attenuated; the voltage change across C_{GS} is much smaller than the gate voltage because the source closely follows the gate. A reduced voltage change means that the capacitance is reduced too; in the extreme case when the source follows the gate exactly ($G_{SF} = 1$) there would be no voltage change across C_{GS}, therefore no current would flow through it. This is the same as if C_{GS} is not connected at all, i.e. $C_{GS} = 0$. For the typical SF gains of 0.8–0.9, C_{GS} is reduced by a substantial factor.

Figure 1.39 gives a schematic view of the capacitances as they appear electrically, but we can also consider their physical location in figure 1.40. There are three physical gate capacitances—to the source, the drain and the channel.

The gate–source and gate–drain edge capacitances are identical for the symmetrical transistor in figure 1.40 and are given by

$$C_{GSe} = C_{GDe} = C_e W \tag{1.106}$$

Here C_e is the edge capacitance per unit length of gate overlap and can be found from the process specifications for the used CMOS technology.

The gate area capacitance C_{Ga} is between the gate and the conducting channel. In a source follower configuration, the potential along the channel is superimposed on the source potential, therefore C_{Ga} is effectively connected between the gate and the source, the same as C_{GSe}. C_{Ga} is [30]

$$C_{Ga} = \frac{2}{3}C_{ox}WL \tag{1.107}$$

and is smaller than the expected $C_{ox}WL$ due to the potential distribution in the channel.

Therefore, substituting C_{GDe} for C_d and $(C_{GSe} + C_{Ga})$ for C_{GS} in (1.105) the input capacitance of the source follower becomes

$$C_{in} = C_{GDe} + (C_{GSe} + C_{Ga})(1 - G_{SF}) \tag{1.108}$$

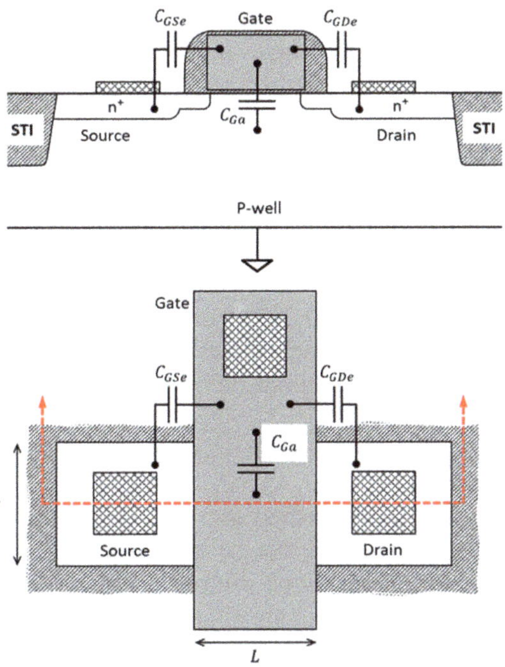

Figure 1.40. Location of the gate capacitances in a MOSFET. The shallow trench insulation (STI) oxide surrounds the MOSFET on all four sides.

Finally, using the transistor geometry, the capacitance can be written as:

$$C_{\text{in}} = C_e W + \left(C_e W + \frac{2}{3} C_{\text{ox}} W L\right)(1 - G_{\text{SF}}) \tag{1.109}$$

Usually C_{Ga} is much larger than the edge capacitances due to the large C_{ox}, but fortunately it is strongly suppressed by the factor $(1 - G_{\text{SF}})$.

Chapter summary

1. Photons generate electron–hole pairs in silicon via the internal photoeffect. Energetic charged particles generate electron–hole pairs via ionisation. Charge is generated independently of the doping concentration and the concentration of free carriers in the semiconductor.
2. A single photon with wavelength between 1100 and 369 nm can generate only one electron–hole pair in silicon. Above 1100 nm silicon is transparent and insensitive to light.
3. Photons with wavelengths shorter than 369 nm (3.36 eV) generate more than one electron–hole pair, and for photons with energies above 10 eV the electron–hole creation energy levels off to about 3.65 eV.
4. The absorption length of light in silicon depends strongly on the photon energy, and changes between few nanometres at UV to hundreds of micrometres for near-IR light and soft x-rays.

5. Carrier lifetimes in high quality silicon are in the millisecond range and decrease with increasing trap or doping concentration. Trap-assisted recombination is the dominant mechanism controlling carrier lifetime.
6. Charge is collected by diffusion and drift. Drift occurs under electric field and is the preferred way of charge collection due to its high speed.
7. Charge diffuses regardless of whether there is an electric field or not, but a short collection time under drift does not allow for significant diffusion. In field-free semiconductor the charge diffuses until it is trapped or reaches a region with electric field.
8. The rate of direct electron–hole recombination in silicon is vanishingly small except at very high carrier concentrations. This is why electrons and holes generated in a dense cloud diffuse away before they have a chance to recombine with each other.
9. Charge is converted to electrical signal, usually to voltage, at the capacitance of a sense node.
10. The depletion region established in a *pn* junction acts like an insulator and facilitates charge collection due to the internal electric field. The capacitance of the *pn* junction can be used as a sense node.
11. MOSFETs act as a voltage-controlled current sources. The drain current in saturation has a quadratic dependence on the gate–source voltage. As an amplifier, the MOSFET gain is reduced by the body effect and the output resistance.
12. Source followers are used as buffers due to their very high input impedance and well-defined gain, which is always less than one.

References

[1] Mazziotta M 2008 Electron–hole pair creation energy and Fano factor temperature dependence in silicon *Nucl. Instrum. Methods Phys. Res.* A **A584** 436–9

[2] Sze S 1981 *Physics of Semiconductor Devices* 2nd edn (New York: Wiley)

[3] Geist J and Wang C 1983 New calculations of the quantum yield of silicon in the near ultraviolet *Phys. Rev.* B **27** 4841–7

[4] Scholze F, Henneken H, Kuschnerus P, Rabus H, Richter M and Ulm G 2000 Determination of the electron–hole pair creation energy for semiconductors from the spectral responsivity of photodiodes *Nucl. Instrum. Methods Phys. Res.* A **439** 208–15

[5] Kübarsepp T, Kärhä P and Ikonen E 2000 Interpolation of the spectral responsivity of silicon photodetectors in the near ultraviolet *Appl. Optics* **39** 9–15

[6] Kuschnerus P, Rabus H, Richter M, Scholze F, Werner L and Ulm G 1998 Characterization of photodiodes as transfer detector standards in the 120 nm to 600 nm spectral range *Metrologia* **35** 355–62

[7] Green M A 2008 Self-consistent optical parameters of intrinsic silicon at 300 K including temperature coefficients *Solar Energy Mater. Solar Cells* **92** 1305–10

[8] Shevell S K (ed) 2003 *The Science of Color* 2nd edn (Oxford: Elsevier)

[9] Chantler C, Olsen K, Dragoset R, Chang J, Kishore A, Kotochigova S and Zucker D 2005 *X-Ray Form Factor, Attenuation and Scattering Tables (version 2.1)* (Gaithersburg, MD: National Institute of Standards and Technology) http://physics.nist.gov/ffast

[10] Henke B, Gullikson E and Davis J 1993 X-ray interactions: photoabsorption, scattering, transmission, and reflection at $E = 50$–30,000 eV, $Z = 1$–92 *At. Data Nucl. Data Tables* **54** 181–342

[11] Scholze F, Rabus H and Ulm G 1998 Mean energy required to produce an electron–hole pair in silicon for photons of energies between 50 and 1500 eV *J. Appl. Phys.* **84** 2926–39

[12] Lowe B and Sareen R 2007 A measurement of the electron–hole pair creation energy and the Fano factor in silicon for 5.9 keV X-rays and their temperature dependence in the range 80–270 K *Nucl. Instrum. Methods Phys. Res.* A **576** 367–70

[13] Janesick J 2001 *Scientific Charge-Coupled Devices* (Bellingham, WA: SPIE Press)

[14] Tsunemi H, Hiraga J, Yoshita K, Miyata E and Ohtani M 2000 Comparison of methods of measuring the primary charge-cloud shape produced by an x-ray photon inside the CCD *Nucl. Instrum. Methods Phys. Res.* A **439** 592–600

[15] Zula P *et al* 2020 Commonly used radioactive sources *Prog. Theor. Exp. Phys.* 083C01 https://pdg.lbl.gov/

[16] Zula P *et al* 2020 Passage of particles through matter *Prog. Theor. Exp. Phys.* 083C01 https://pdg.lbl.gov/)

[17] Bichsel H 1988 Straggling in thin silicon detectors *Rev. Modern Phys* **60** 663–99

[18] Wang F, Dong S, Nachman B, Garcia-Sciveres M and Zeng Q 2018 The impact of incorporating shell-corrections to energy loss in silicon *Nucl. Instr. Meth. Phys. Res.* A **899** 1–5

[19] Schroder D K 1997 Carrier lifetimes in silicon *IEEE Trans. Electron Devices* **44** 160–70

[20] Shockley W and Read W T 1952 Statistics of the recombinations of holes and electrons *Phys. Rev.* **87** 835–42

[21] Grove A S 1967 *Physics and Technology of Semiconductor Devices* (New York: Wiley)

[22] Richter A, Glunz S W, Werner F, Schmidt J and Cuevas A 2012 Improved quantitative description of Auger recombination in crystalline silicon *Phys. Rev.* B **86** 165202

[23] Law M, Solley E, Liang M and Burk D 1991 Self-consistent model of minority-carrier lifetime, diffusion length, and mobility *IEEE Electron Dev. Lett.* **12** 401–3

[24] Sinton R A and Swanson R M 1987 Recombination in highly injected silicon *IEEE Trans. Electron Devices* **34** 1380–9

[25] Nguyen H T, Baker-Finch S C and Macdonald D 2014 Temperature dependence of the radiative recombination coefficient in crystalline silicon from spectral photoluminescence *Appl. Phys. Lett.* **104** 112105

[26] Hu C C 2009 *Modern Semiconductor Devices for Integrated Circuits* (Upper Englewood Cliffs, NJ: Pearson)

[27] Silvaco Atlas, Silvaco Inc. https://silvaco.com/

[28] Spieler H 2005 *Semiconductor Detector Systems* (Oxford: Oxford University Press)

[29] Allen P E and Holberg D R 2012 *CMOS Analog Circuit Design* (Oxford: Oxford University Press)

[30] Boukhayma A, Peizerat A and Enz C 2016 Temporal readout noise analysis and reduction techniques for low-light CMOS image sensors *IEEE Trans Electron Devices* **63** 72–8

IOP Publishing

CMOS Image Sensors

Konstantin D Stefanov

Chapter 2

CMOS pixel architectures

2.1 History and technology

The inventor of the modern CMOS image sensor (CIS) is considered to be Eric Fossum [1, 2], who while working for NASA in the 1990s designed the first photogate-based image sensors. Earlier work by Peter Noble [3] and Gene Weckler [4, 5] in the late 1960s pioneered MOS-based passive and active image sensors, but after the invention of the charge coupled device (CCD) in 1969, it became nearly forgotten.

The performance of early CIS was poor compared to CCDs, however, due to the relentless improvements to CMOS technology since then, today's CIS demonstrate astonishing quality and usability. The vast majority of image sensors manufactured today are CIS, and billions find their way into mobile phones, cameras, toys, computers, drones and everything else. Their success is due not only to improvements in pixel performance, but also to inclusion on numerous functions on chip, such as analogue-to-digital converters (ADC) and image processing, to the point where a complete image system can be integrated.

CIS can be either passive or active pixel sensors. The passive type usually contains just one transistor per pixel, used as a switch to connect the photosensitive element to the chip periphery. Since the transistor is not used as an amplifier the sensor is called passive. Most CIS in use today are active pixel sensors (APS) because they contain at least one MOSFET per pixel to buffer or amplify the photogenerated signal, in addition to other transistors for switching and biasing. Having an amplification element per pixel, even though the gain when used as a buffer may be less than one, qualifies the image sensor to be called an APS. Virtually all CIS made today are APS, and both terms are often interchangeable.

In most sensors, the pixel uses transistors of only one type (*n*-type MOS being dominant), despite being made with CMOS process. 'Complementary' MOS means that both *n*-channel (NMOS) and *p*-channel (PMOS) transistors with similar performance can be manufactured on the same wafer in one process. True CMOS is usually used only in the chip periphery to build logic circuits, amplifiers and

doi:10.1088/978-0-7503-3235-4ch2

ADCs. Interestingly enough and perhaps a bit counterintuitively, CIS in their core are predominantly NMOS devices.

The performance breakthrough of CIS began when the pinned photodiode (PPD) was introduced in the CMOS manufacturing process in the late 1990s. Initially invented as a method of reducing image lag in interline CCDs [6, 7], the PPD proved remarkably useful and is the mainstay of CMOS imaging. The first mention of the PPD [8] in CIS describes it as 'using CMOS/CCD process technology', demonstrating the marriage between the two. It is fair to say that we would not be having high quality CMOS image sensors today without the adoption of the PPD.

The main driver of CMOS technology is the manufacture of digital integrated circuits. There is a relentless drive to pack more and more transistors per die for memories, microprocessors, field programmable gate arrays (FPGA) and graphics processing units (GPU). A density of 100 million transistors *per square millimetre* was surpassed in 2017 with the introduction of the 10 nm process node. Image sensors do not usually require very small process feature sizes because the pixels cannot be much smaller than the wavelength of light they are detecting, and 'ancient' manufacturing processes such as the 180 nm (introduced in 1998) remain popular. The trend is to build multi-die CIS with separate image sensor, processor and memory tiers [9], interconnected with 3D integration technologies. In this way the digital tiers can make use of advanced CMOS processes on a fine pitch, while the imaging tier can be made with an optimised process tailored for image sensors. Most CIS found in today's mobile phones are multi-die devices.

CIS-optimised CMOS processes feature low noise and low threshold transistors, very low dark currents, and reduced thickness of the metallisation layers. Post-processing options include backside thinning and microlenses.

In common with all small feature size processes, rapid thermal annealing (RTA) is used for dopant activation after implantation. The anneal time is measured in tens of seconds and very little diffusion takes place. Gate oxides are very thin, for example around 6–7 nm for the 180 nm process. Usually only one polysilicon layer is available. Shallow trench insulation (STI) and deep trench insulation (DTI) are used to isolate active components, in and outside the pixel.

High quality epitaxial wafers are the default starting material for CIS manufacture. In addition to the standard *p*- and *n*-wells for NMOS and PMOS transistors, most processes offer deep *p*- and *n*-wells which can be used for many purposes, such as deep *pn* junctions and charge reflective barriers, and for noise and crosstalk reduction in mixed-signal circuits.

2.2 Photodiode APS

2.2.1 Structure

Photodiode type APS, also called a 3T (from 3-transistor) APS, is one of the simplest types of CIS. It is not used widely today but is nevertheless very important as a building block for more complex pixel architectures.

As shown in figure 2.1, in the 3T pixel a reverse-biased photodiode (PD) is used as a photosensitive element, buffered by a source follower MOSFET. The reset

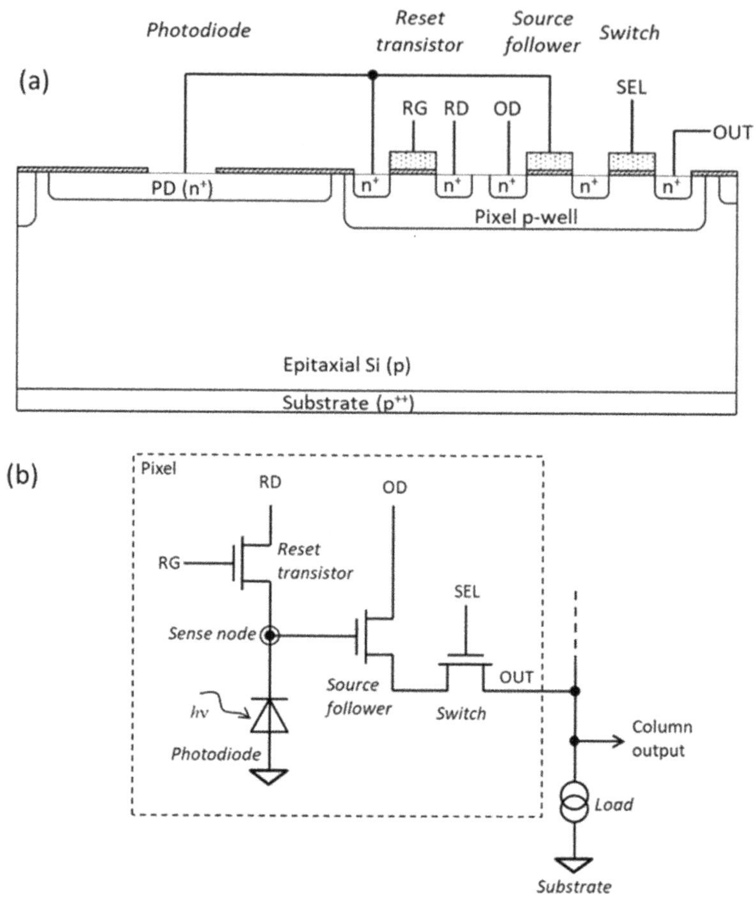

Figure 2.1. Simplified physical cross-section of a 3T pixel (a) and a schematic diagram including the column load outside the pixel (b). The bulks of all transistors are at substrate and are not shown for simplicity.

transistor is used to bring up the voltage on the photodiode's cathode to a reference value when sufficient voltage is applied to the reset gate (RG) to turn the transistor on. The cathode of the PD is the 'sense node' (SN) because it is the point where the photogenerated voltage appears. The sense node is also called floating diffusion (FD) for historical reasons, as it is floating and was formed by diffusion in the early days of image sensors. The reset transistor drain (RD) and the source follower drain (output drain, OD) are connected to stable, low noise voltage supplies.

The switch transistor (also called 'select', or 'row select' transistor) is turned on when that particular pixel is to be read out. This transistor works as an analogue switch—when 'off' (gate voltage SEL below transistor threshold) it is an open circuit, and when 'on' it behaves as a resistor. The gate voltage SEL is usually driven by a digital circuit (address decoder) on chip, and swings between logic '0', equal to the substrate voltage, and logic '1' which must be far greater than the transistor threshold. The transistor 'on' resistance is usually in the hundreds to several thousand Ohm range.

The column load is almost always a constant current sink with bias current of the order of few microamps. The output impedance of the source follower and the row select transistor are in series with the current load, but they are much smaller compared to its dynamic resistance. Because of that, the voltage appearing at the column output is nearly identical to the one at the source follower, and almost no signal amplitude is lost. More sophisticated current loads can be used [10], but the simplest one-transistor load is adequate in most cases. By controlling the gate voltage from a current mirror, the source follower bias current can be adjusted externally.

Figure 2.2 shows an example layout of a 3T pixel. The three transistors sit in the pixel *p*-well shown in figure 2.1(a), which surrounds the photodiode on all sides. The reset and the row select transistor are normally designed with the minimum possible gate length, but the source follower is larger and has optimised gate length and width for better noise performance.

You may notice that the source and drain terminals of the in-pixel transistors are *pn* junctions in their own right and can therefore have photosensitivity. Indeed, charge will collect at all transistor sources and drains which form reverse-biased *pn* junctions. Electrons reaching RD or OD will be quickly removed, as both are connected to a low impedance voltage sources. The situation at the source of the reset transistor is different—it is connected in parallel to the photodiode and takes part in charge collection. The photogenerated current flowing to RD and OD contributes to loss of 'fill factor', as most of the *p*-well does not contribute to the generation of useful signal and is effectively a dead area for light absorbed in it.

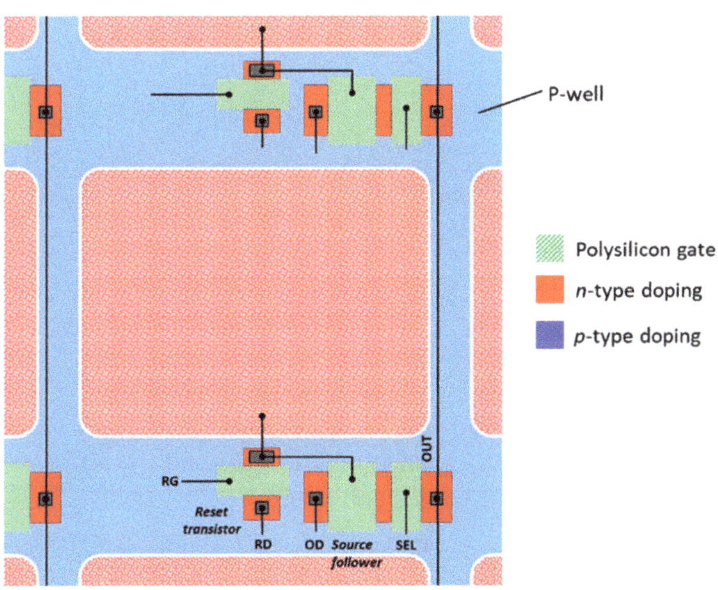

Figure 2.2. Simplified physical layout of a 3T pixel corresponding to figure 2.1. The connections to the diode, gates, RD, OD and OUT are using metal tracks on several levels and are shown only schematically.

However, for penetrating light with longer wavelengths which does not completely absorb in the *p*-well, or higher energy x-rays, some part of the collected charge on the photodiode will come from underneath the *p*-well, due to the reflective *p*/*p*+ barrier present at the backside interface.

In large image arrays there are contacts to the *p*-well in every pixel. This is necessary because the *p*-well may not be conductive enough to provide proper substrate connections to pixels far from the edge of the array. Instead, metal tracks are used to connect all pixels' *p*-wells together and to the periphery of the sensor.

2.2.2 Operation

The 3T pixel can be readily studied in simulation using SPICE as a pure electronic circuit, and the example in figure 2.3 shows the model schematic and the simulated

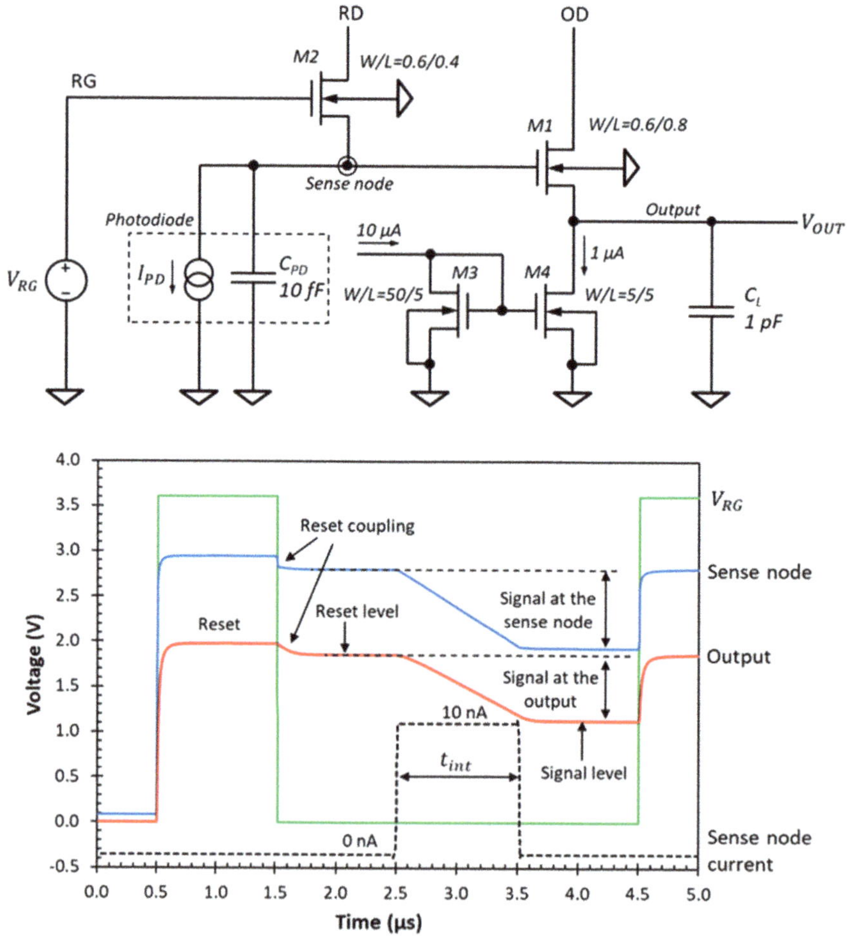

Figure 2.3. Simulation schematic and signals in a 3T pixel built with transistors on a 3.3 V, 180 nm process and with $V_{RD} = V_{OD} = 3.3$ V. The row select transistor has been omitted for simplicity. Transistor channel widths and lengths are given in micrometres.

signals for a typical 3.3 V, 180 nm process. Transistor M1 is the source follower, M2 is the reset transistor, and M3 and M4 form a current load with a scaling ratio of 10:1, supplied externally with 10 μA. The row select transistor has been omitted for clarity.

M2 is nearly the smallest transistor that can be made in this technology and is chosen like that because it is intended to work as a switch. The source follower M2 is substantially larger and has channel length of 0.8 μm; this increases its threshold, but the larger gate size is needed to reduce its noise. In general, the source follower is never a minimum size transistor. M4 is even larger because good matching is required between the column load transistors; this is easier to achieve when transistors have long channels and large area. The current mirror pair M3 and M4 resides outside the pixel and serves a whole column indicated in figure 2.1(b).

The photodiode is simulated with the capacitor C_{PD} (substituting for its depletion capacitance) and an ideal current source in parallel, representing the photogenerated current. The output capacitor C_L is intended to represent the capacitance of the entire column bus. The circuit is powered by DC voltage sources connected to RD and OD.

The voltage source V_{RG} generates a 1 μs pulse with amplitude of 3.6 V to the gate of the reset transistor. During reset, the sense node reaches around 3.0 V, lower that the 3.3 V drain supply. The reset transistor M2 is not fully on because it is in subthreshold mode, and we have a 'soft reset'. To reach the voltage at RD, M2 must be in the linear regime, therefore V_{RG} should be much higher, so that we have 'hard reset'. Approximately a volt above V_{RD} is needed to turn M2 fully on, due to the high threshold caused by the body effect.

The sense node voltage is buffered by the source follower and appears at the output, minus the threshold voltage of M1. At the falling edge of the RG pulse the voltage at the sense node drops a little due to the capacitive coupling between the gate of M2 and the sense node. This reset coupling, also called reset feedthrough offset, increases as the sense node's capacitance gets smaller. It can be much larger than in figure 2.3, and must be taken into account because it reduces the maximum voltage swing at the output. Once settled, this voltage at the source follower output is called the 'reset level' of the pixel.

From figure 2.3 we can see that threshold of the source follower V_T is around 0.8 V, calculated as the sense node voltage minus the output voltage. As M2, the source follower M1 has higher threshold due to the strong body effect. Bear in mind that the sense node voltage is not available to probe in real-life circuits; we only have the version appearing at the pixel output, buffered by the source follower.

As the photodiode collects electrons, its potential decreases. The photogenerated signal can be represented as current I_{PD} out of the PD[1], which causes its potential to decrease according to

[1] Current into the PD would cause the potential to rise. The convention in electronics is that current flows from the positive to the negative potential, despite being composed of electrons which flow the other way around.

$$\Delta V_{SN} = -\frac{1}{C_{PD} + C_{in}} \int_0^{t_{int}} I_{PD} dt \tag{2.1}$$

where C_{PD} is the capacitance of the photodiode, C_{in} is the input capacitance from the connection of M1 and M2 to the sense node, t_{int} is the time duration of the current flow, and ΔV_{SN} is the voltage change at the sense node. In figure 2.3 the current I_{PD} is a constant 10 nA during t_{int} and zero outside, therefore equation (2.1) simplifies to:

$$\Delta V_{SN} = -\frac{I_{PD} t_{int}}{C_{PD} + C_{in}} \tag{2.2}$$

The conversion gain of the pixel, measured at the output, is the voltage at the sense node created by one electron, multiplied by the gain of the source follower M1:

$$G_c = \frac{q G_{SF}}{C_{sn}} = \frac{q G_{SF}}{C_{PD} + C_{in}} \tag{2.3}$$

Example 2.1. Calculate the sense node capacitance, the gain of the source follower and the conversion gain for the pixel simulation in figure 2.3.
Solution: The sense node capacitance is $C_{sn} = C_{PD} + C_{in}$ and can be obtained from (2.2) for the voltage change at the sense node.

$$C_{sn} = -\frac{I_{PD} t_{int}}{\Delta V_{SN}} = -\frac{10 \times 10^{-9} \times 10^{-6}}{1.93 - 2.8} = 11.5 \text{ fF}$$

From this, we conclude that M1 and M2 add 1.5 fF to the sense node capacitance. We can also see that the injected charge is $I_{PD} t_{int}/q = 62.5 \text{ ke}^-$. The source follower gain can be calculated from the signal at the output, divided by the signal at the sense node:

$$G_{SF} = \frac{\Delta V_O}{\Delta V_{SN}} \approx \frac{(1.85 - 1.13)}{(2.8 - 1.93)} = 0.83$$

The conversion gain is measured at the pixel output, therefore it is smaller than what would be calculated from the sense node capacitance due to the source follower gain being less than one.

$$G_c = \frac{q G_{SF}}{C_{sn}} = \frac{1.6 \times 10^{-19} \times 0.83}{11.5 \times 10^{-15}} = 11.5 \ \mu\text{V/e}^-$$

The G_c can also be calculated without knowing the source follower gain by using equation (2.2) with the output voltage change instead; this uses the effective sense node capacitance with the SF gain included, and of course produces the same result.

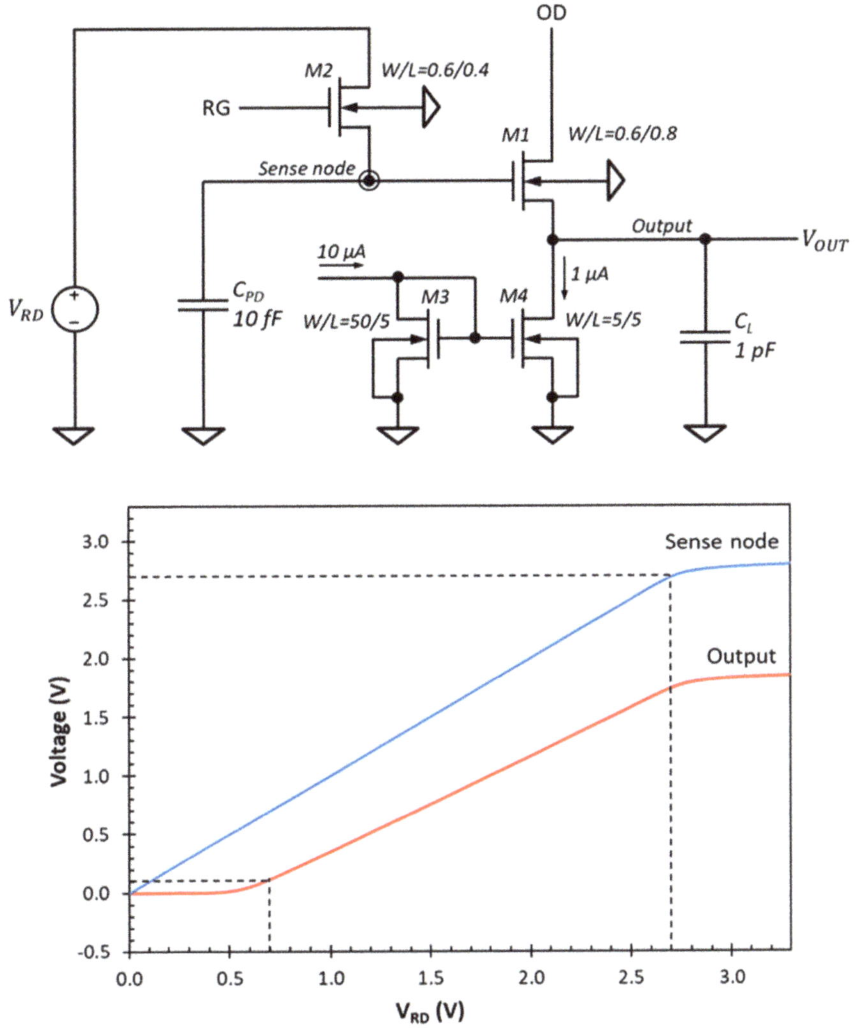

Figure 2.4. Electrical transfer function simulation for $V_{RG} = 3.6$ V and $V_{OD} = 3.3$ V.

A practical method to measure the gain and the threshold of the SF is to take the electrical transfer function (ETF) of the pixel, as shown experimentally in chapter 5. The idea behind the ETF, shown in figure 2.4, is to use the reset transistor M2 to supply the gate of the source follower M1 with the voltage on its RD terminal. For this, two things are required: the drain of M2 must be accessible as an external connection, and M2 must be operating in the linear regime by biasing its gate above $V_{RD} + V_T$.

The ETF is obtained by scanning V_{RD} from zero to the maximum possible drain voltage, while the reset gate RG is DC biased. The voltage at the sense node is equal to V_{RD} until the gate–source voltage of M2 approaches its threshold; in the simulation in figure 2.4 this occurs at $V_{RD} = 2.7$ V. The gain of the source follower is calculated as $G_{SF} = \Delta V_{RG}/\Delta V_O$ in the linear output range between 0.7 and 2.7 V. From figure 2.4 we can also see that the threshold of M1 is around 0.6 V.

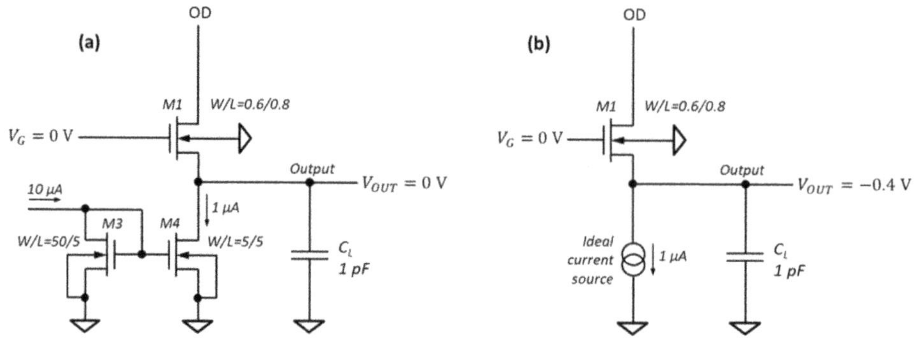

Figure 2.5. Simulation of a source follower with zero gate voltage with a transistor current load (a) and with an ideal current load (b).

In the simulation circuit diagrams in figures 2.3 and 2.4 we have used a transistor current load for the SF even though an ideal current sink, set at 1 μA, would be simpler. This is because current loads are actually used in image sensors, but there is another reason— the ideal current source, as a SPICE primitive, can cause some 'strange' behaviour, and therefore must be used with care.

Let's consider the simulation circuits in figure 2.5. When the gate voltage of the source follower is zero, the output in figure 2.5(a) shows the expected zero volts. However, for the same load current and zero gate voltage the output of the SF in figure 2.5(b) would show *minus* 0.4 V! This appears unphysical because there are no negative voltages in the circuit, but has a perfectly reasonable explanation. The ideal current load, being ideal, tries to force a current through its terminals, whatever the voltage across them. Therefore, with the SF turned off, the only way for the current to flow in figure 2.5(b) is through the *forward-biased* source-substrate *pn* junction of M1, which makes the source negative with respect to the substrate. This effect will not be seen in real circuits because ideal current sources do not exist, but can cause some head-scratching when simulating with them.

2.2.3 Performance

The 3T pixel is fairly simple to design, simulate and understand; electrically it consists of just three transistors and a diode. The maximum output signal is the difference between the reset level and the lowest output voltage, which is near zero. Knowing this, we can calculate the maximum number of signal electrons the pixel can handle. Often called full well capacity (FWC), it is the ratio of the maximum output voltage swing to the conversion gain:

$$\mathrm{FWC} = \frac{\Delta V_{\mathrm{O\,max}}}{G_{\mathrm{c}}} \tag{2.4}$$

The pixel could have higher FWC if the reset level were higher, which can be realised with higher RD and RG voltages. To achieve this, the reset drain is tied to the highest supply voltage. The RG voltage could be even higher, which requires level shifting and may not always be practical. Another way to increase the output swing is to use a lower threshold ('low V_T') process for the source follower, which is offered by most foundries. Eliminating the body effect in the SF by connecting the source to the bulk in a 'hot *p*-well' (as discussed in chapter 1) is rarely an option because the *n*-well isolation competes with the photodiode for photogenerated charge.

Example 2.2. Calculate the FWC of the pixel in figure 2.3, assuming that the output can go all the way to zero.

Solution: The reset level at the output is 1.85 V, therefore this is the maximum output swing $\Delta V_{O\max}$. With conversion gain of 11.5 μV/e⁻ the FWC is

$$\mathrm{FWC} = \frac{\Delta V_{O\max}}{G_c} = \frac{1.85}{11.5 \times 10^{-6}} = 161\ \mathrm{ke^-}$$

One of the main characteristics of the 3T pixel is that the photodiode is simultaneously the charge collection and the charge conversion element, and this is the cause if its performance deficiencies. Despite its limitations, the 3T pixel is used in applications requiring large pixel sizes, high FWC, or because of its simplicity and lower cost [11].

Resetting the sense node creates thermal noise, called reset noise, described in chapter 4. It can be eliminated by a technique called correlated double sampling (CDS) if the signal level is subtracted from the reset level preceding it, before the signal is generated. In the typical readout scheme in figure 2.6, the signal level appears on the column output once a pixel is selected, then the voltage is sampled, the pixel reset, and the reset level can be sampled too. The signal level is correlated with the reset level from the *previous readout*, therefore the previous reset level must be kept for the CDS subtraction. This is accomplished by reading the sensor twice and storing the reset samples in memory. This method is not always practical, and when not implemented, the reset-signal pair from the same readout is used. The result is readout noise measured in tens of electrons root mean square (RMS), which is not acceptable for high performance applications. Even if the reset level is stored,

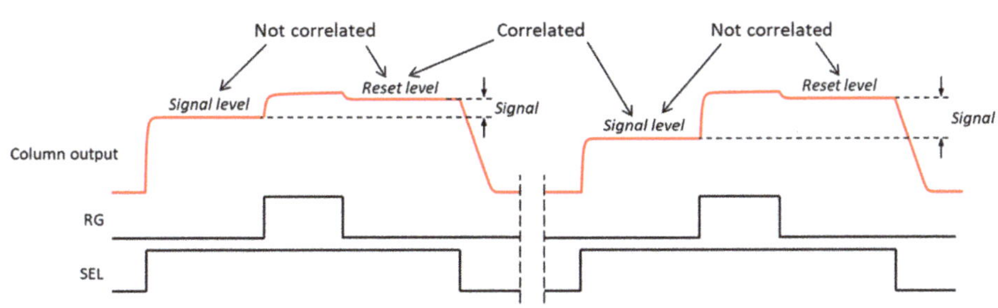

Figure 2.6. Two consecutive readouts of a 3T pixel in figure 2.1(b), corresponding to different signal levels. The column line has a valid output only when SEL is high. When a pixel is not selected, the column current load pulls the output to ground.

so that CDS is performed, it is sampled many milliseconds before its paired signal, therefore the difference can be subject to low frequency drift and other fluctuations.

In addition to the difficulties in performing effective CDS, 3T pixels suffer from fixed pattern noise (FPN)—a spatial (not temporal) noise caused by the spread of transistor thresholds. The reset level for each pixel is different because each source follower has its own threshold, and the pixel-to-pixel variations can be much larger than the voltage fluctuations from the reset. FPN can be cured by the subtraction of the reset and the signal levels during the same pixel addressing, because they have the same SF offsets. If only the signal level is sampled once, for reasons of reducing CIS complexity, the FPN can be overwhelming.

If multiple sampling is implemented, the signal level can be read out many times without resetting, also known as non-destructive readout[2]. If the signal increases linearly with time during integration, multiple sampling can be used to eliminate the SF offset and the reset noise, in addition to some noise reduction from averaging. This is possible in low frame rate applications under constant scene illumination and is known as 'sampling-up-the ramp', or Fowler sampling [12].

Another consequence from the charge collection and conversion being done in the photodiode is that the maximum signal charge, the conversion gain, and the readout noise become interconnected. The maximum charge and the depletion capacitance are approximately proportional to the diode area. Since most of the pixel is taken up by the diode, for a square pixel with pitch p we can write that FWC $\propto p^2$ and $C_{\mathrm{PD}} \propto p^2$. In 3T pixels with dominant reset noise, the noise voltage is proportional to $1/\sqrt{C_{\mathrm{PD}}}$, therefore to $1/p$. From this we can conclude that the dynamic range (DR), defined as the FWC divided by the readout noise, is proportional to the pixel pitch p [2]. For this reason, smaller 3T pixels can suffer from poor DR.

2.3 Pinned photodiode (4T)

2.3.1 Structure

The disadvantages of the 3T pixel, such as its high noise, have prompted the development of the pinned photodiode (PPD) pixel. The CMOS PPD is derived from the photodiode implemented for interline CCDs by Nobukazu Teranishi in 1982 [6, 7]. The objective of this development was to reduce the image lag due to an incomplete charge transfer between the photodiode and the charge transport channel. In interline CCDs the charge collected in each photodiode is transferred to a transport channel in the vertical direction (V-CCD), allowing the next image to be generated while the previous one is being read out.

In the conventional photodiode (figure 2.7(a)), charge is never fully transferred because the rate of transfer slows down as the signal decreases, similarly to the subthreshold current in MOSFETs. Increasing the potential of the transfer gate does not help because the photodiode's potential is not limited and continues to increase, entering subthreshold mode yet again. Thus, the image lag in the conventional

[2] Normal readout is 'destructive' because after resetting the sense node the signal is lost, and therefore 'destroyed'.

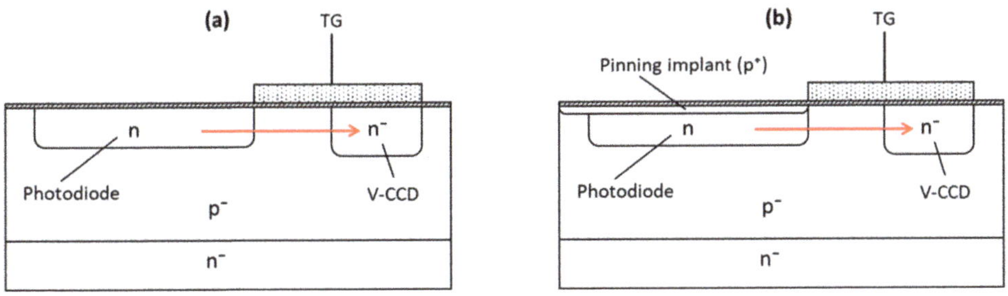

Figure 2.7. Diagram of a conventional photodiode (a) and a pinned photodiode (b) in interline CCDs (after [7]).

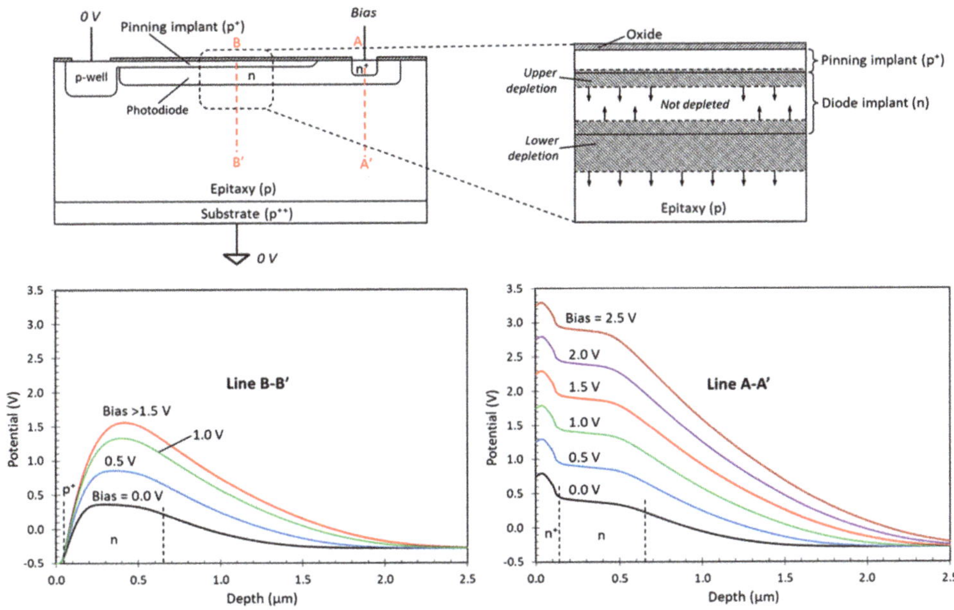

Figure 2.8. A model to illustrate why the potential under the pinning implant is limited. The potential along lines A and B in depth is shown for increasing bias voltages.

photodiode asymptotically approaches 100% as the signal approaches zero, leading to very poor performance in low light conditions.

The behaviour changes completely with the addition of a shallow p^+ layer on top of the photodiode as shown in figure 2.7(b). Now the photodiode becomes a sandwich of two back-to-back *pn* junctions with both *p*-layers at substrate potential. It turns out that in this arrangement the potential of the photodiode cannot increase indefinitely in response to the changing TG, making the charge transfer much more efficient and virtually eliminating image lag at low signals.

Why the maximum potential in the PPD is limited can be understood by considering the structure in figure 2.8. Without the pinning implant, this would simply be a *pn* junction biased from an electrode positioned in one side. With increasing bias, the whole of the *n*-type dopant would follow, and in depth the potential would show the characteristic quadratic shape. This is what we expect and see along line A–A′.

Along line B–B′ the situation is completely different because of the pinning implant, biased at substrate potential. At low bias voltages the central part of the photodiode is electrically neutral and conductive, and the two depletion regions begin to expand from both *pn* junctions upwards and downwards. The upper depletion expands mostly into the *n*-layer and barely enters the pinning implant because it is more highly doped than the photodiode. The lower depletion expands in both the epitaxial *p*-layer and in the photodiode. As the bias voltage increases, the two depletion regions grow and eventually touch and merge. At this point the potential under the pinning implant stops growing as there is no conductive path for the bias voltage; the whole of the diode is depleted and cut off from the bias terminal.

In figure 2.8 the potential along line B–B′ stops increasing when the bias voltage exceeds 1.5 V, and the peak is at depth of approximately 0.4 μm. Increasing the bias voltage further makes no difference. Higher diode doping will require higher bias potential to join the two depletions; a deeper diode implant will have similar effect because the distance to be bridged is longer, and the peak potential will be higher for both.

Combining the PPD with a transfer gate, sense node and readout transistors, as done first in [8], makes the CMOS PPD pixel shown in figure 2.9. The reset, source follower and switch transistors are identical to the 3T pixel. The crucial addition is

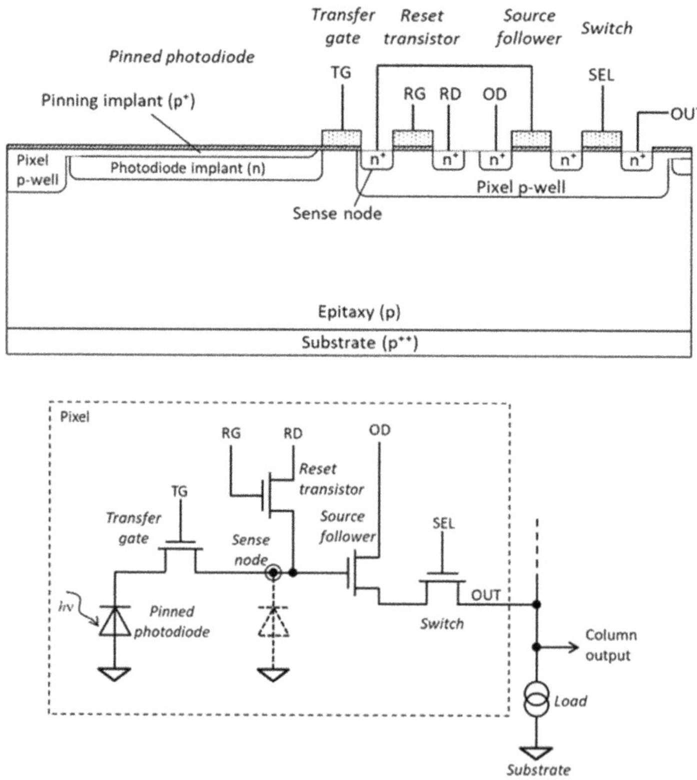

Figure 2.9. Physical diagram of a 4T (PPD) pixel (*top*) and a schematic diagram (*bottom*), including the column load outside the pixel.

the transfer gate TG between the photodiode and the sense node, allowing charge collected in the PPD to be kept isolated, and moved out for measurement. Very importantly, this separates the functions of charge collection and charge-to-voltage conversion, leading to big improvements in noise performance.

The pinning implant 'pins' the top side of the photodiode to the pixel *p*-well at substrate potential, while the *n*-type diode implant underneath is floating and free to take higher, but ultimately limited potential. The maximum potential that can be reached in the PPD does not depend on the applied voltage on the TG and the sense node, but is determined by the device construction, such as doping profiles, dimensions and dielectric thicknesses.

The layout in figure 2.10 is typical for many PPD pixels. The PPD pixel is also called 4T because there are four transistors in each pixel. Strictly speaking, there are four gates per pixel, but only three 'proper' transistors because the structure around TG does not form a traditional transistor with an n^+ implanted source—here the pinned photodiode plays this role. The 4T PPD pixel is a descendant of the photogate pixel [2, 13], which also has four gates, but is rarely used now.

In the PPD the dark current is significantly reduced in comparison with the photodiode because the Si–SiO_2 interface, normally the largest dark current contributor, is kept saturated with holes provided by the heavily doped p^+ pinning

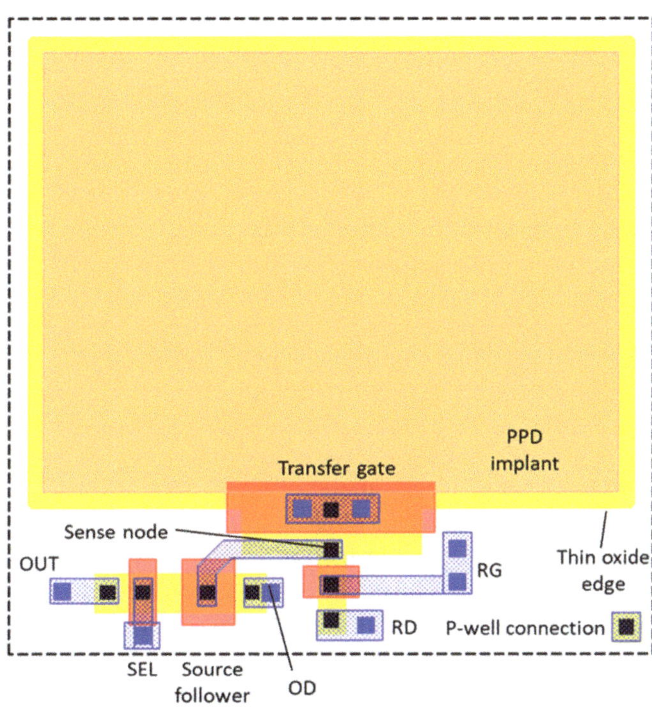

Figure 2.10. An example layout of a PPD pixel with the schematic in figure 2.9. The connections to the gates, sense node, RD, OD and OUT are displayed only in the first metal layer (in blue). The pinning implant is not shown for clarity. The *p*-well covers the area outside the PPD implant.

implant [14]. The downside is that the PPD does not have an electrode for convenient charge collection or biasing as the photodiode's cathode. Instead, the collected charge must be transferred out of the PPD to reach the sense node. Getting this charge transfer correctly is the key to the good performance of the PPD.

2.3.2 Operation

The operation of the PPD pixel is more challenging to understand and study than the 3T pixel due to addition of charge transfer. Developing a reliable SPICE model for the whole pixel is not straightforward due the subtleties of the charge transfer, therefore TCAD modelling is usually used to study the charge collection and transfer separately from the electrical response of the in-pixel transistors. Even when a simulation is possible, looking at the signals with an oscilloscope is indispensable in seeing how a pixel works. Only a small number of CIS with analogue outputs provide this valuable opportunity.

Figure 2.11 shows the simplified schematic of a PPD pixel, with the row select transistor omitted, and its response to optically generated signal. As in the 3T pixel, the reset level at the output sees the capacitive coupling from the falling edge of the RG pulse, and is lower that the level expected when RG is held permanently at 3.3 V.

Applying a TG pulse transfers the charge to the sense node, and its potential falls accordingly. At the same time the TG pulse couples capacitively to the sense node, even more than the RG because it is physically larger. TG pulse coupling and charge transfer occur simultaneously, but because of the coupling, it appears that the charge transfer happens *after* the TG has been brought low. This may look puzzling and warrants further investigation.

Everything discussed in section 2.2.2 regarding the transistors in the 3T pixel is valid here too, since they are identical. The signals in figure 2.11 are taken from a pixel and not simulated, so the sense node voltage cannot be seen. However, we can say something about the thresholds of the reset transistor M1 and the source follower M2. During reset the output voltage is around 1.4 V with the reset gate voltage $V_{RG} = 3.3$ V. Therefore, the sum of the thresholds of M1 and M2 is the difference between V_{RG} and V_{OUT}, equal to 1.9 V.

Two crucial differences in the readout of the PPD pixel, compared to the 3T pixel, contribute to its excellent performance and popularity. The charge transfer capability of the PPD is a major advantage over the 3T design because the functions of charge collection and charge-to-voltage conversion are separated. The photodiode in the PPD pixel can be large but the sense node can be very small; this allows large (or indeed small if needed) conversion gain to be achieved regardless of the size of the PPD, by selecting an appropriate sense node size and capacitance. In the 3T pixel the two functions are inseparable because the charge is converted to voltage in the diode directly, as it is collected. Increasing the diode size also increases the diode capacitance, and it grows as the square of the pixel pitch. High conversion gain plays an important role in reducing the readout noise in image sensors.

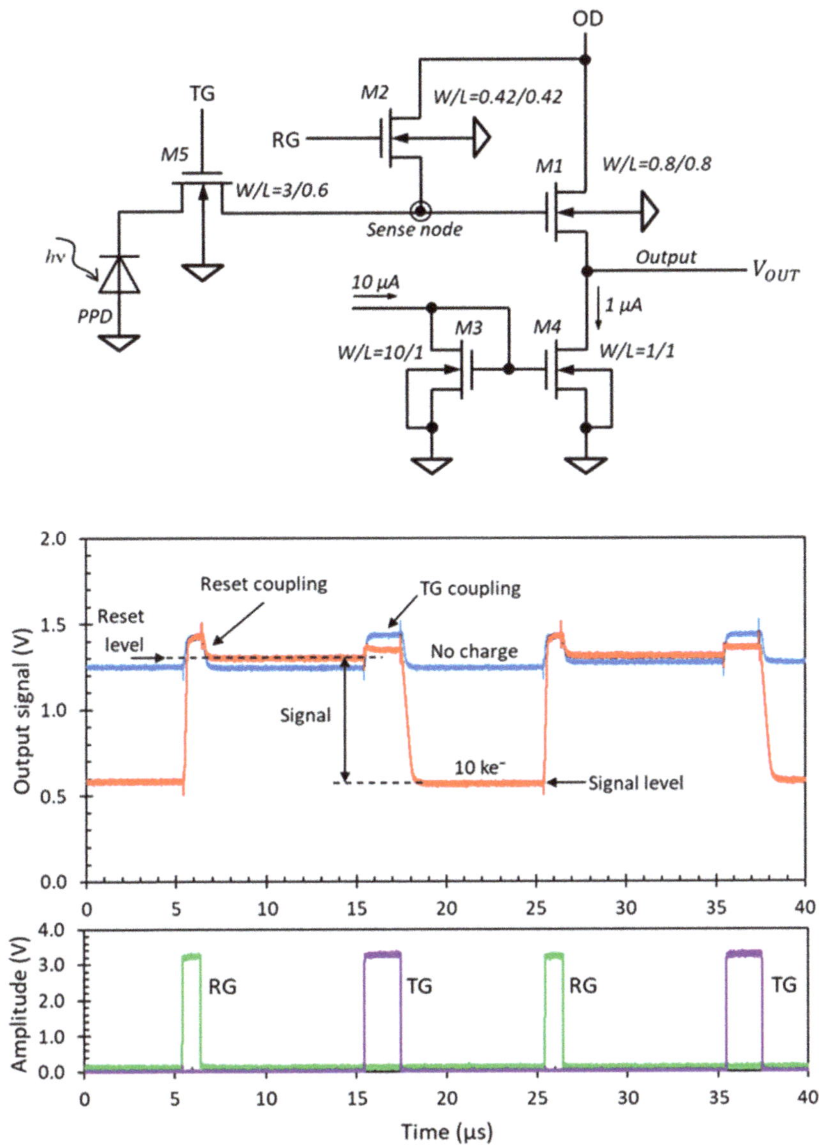

Figure 2.11. Simplified pixel schematic and oscilloscope traces of the RG, TG and the output with zero signal (blue trace) and with 10 ke$^-$ (red trace). The conversion gain of the sensor is 78 μV/e$^-$ and the OD supply is 3.3 V.

The second advantage of the 4T pixel lies in the close separation in time between the reset and the signal samples, which can be only a few microseconds. In both 3T and 4T sensors the reset noise can be eliminated by subtracting the signal level from the reset level using CDS. In 3T sensors the time between the two samples is very long and the noise performance suffers, as discussed in section 2.2.3. As we will see in chapter 4, the short time between the two samples in the 4T pixel is far better for suppressing *1/f* noise and other low frequency fluctuations. In the PPD pixel, CDS can be done properly ('true CDS'), helping to achieve very low readout noise levels.

2.3.3 Charge storage and full well capacity

Figure 2.12 shows a simplified model of a 10 μm pitch PPD pixel with only the most critical elements for its operation—the transfer gate and the sense node. The readout transistors are not needed and are omitted for clarity. The doping profile through the centre of the PPD shows the shallow p^+ pinning implant with a depth of around 100 nm, and the deeper diode implant. The doping concentrations are just an example and many other profiles are possible, for example with a shallower diode implant. The manufacturing process, described in [15], crucially relies on alignment between the transfer gate and the pinning, PPD and sense node implants to achieve good charge transfer.

Starting with an unpowered pixel, the PPD contains a large number of free electrons provided by the *n*-type diode dopant. If the sense node and the transfer gate are biased sufficiently high, the electrons begin to move out from the PPD to the

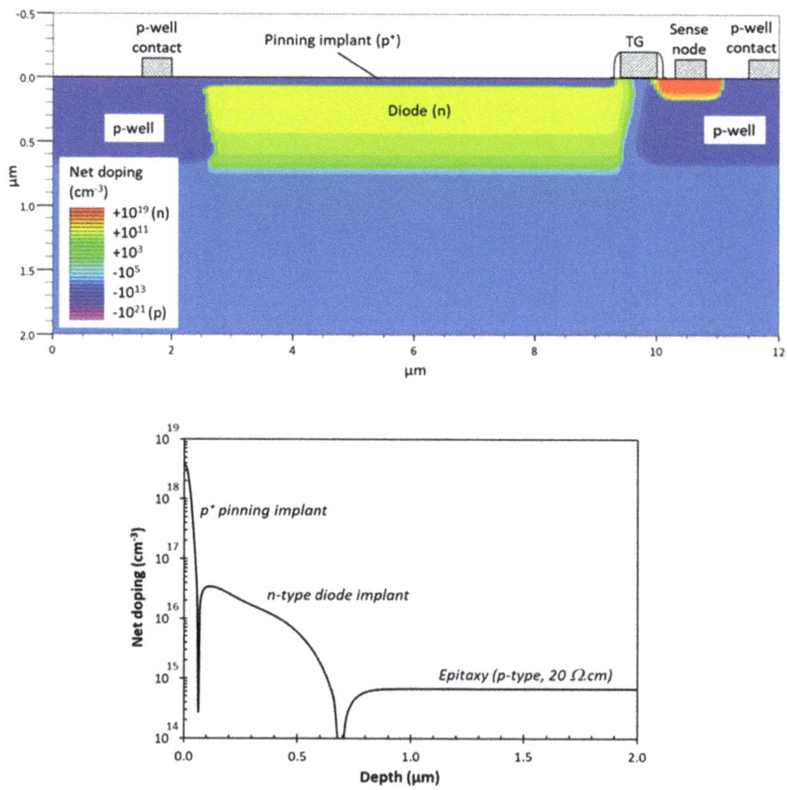

Figure 2.12. Simplified simulation model of a PPD pixel and the doping profile through the centre of the PPD. *p*-type doping is done with boron and *n*-type with phosphorous implants. Only the sense node and the TG are included for simplicity.

sense node, which should be the most positive potential. While this happens, the potential in the PPD begins to increase due to the remaining positively charged donor atoms. Eventually, all donor electrons leave the PPD, and it fully depletes. At this point its potential cannot increase any further as we saw in figure 2.8. Another way to visualise this is to think that there are no more neutral donor atoms left and able to become positively charged. At the same time, the PPD is electrically cut off from the sense node since the depletion acts as an insulator. A further increase of the sense node voltage has no effect because there is no conductive path able to influence the PPD.

The maximum potential under the PPD in full depletion is called the 'pinning voltage' V_{pin} and is one of the most important device parameters. V_{pin} does not depend on the voltages applied to the sense node and the TG.

Once the PPD is depleted, the transfer gate can be closed by biasing it to substrate potential. This establishes a potential barrier to the sense node so that any subsequent charge residing in the PPD cannot reach the sense node. Because there is no other escape path, the charge is contained in the PPD. Closing the transfer gate also prevents charge injection from the sense node back into the PPD. The sense node is a *pn* junction and is therefore photosensitive; illumination would cause its potential to decrease and free electrons can be re-introduced into the PPD if the sense node voltage falls below V_{pin} while TG is on.

Figure 2.13 shows the simulated potential and electron density in 2D for the PPD model in figure 2.12. The charge stored in the PPD is a sheet of electrons about 0.5 μm thick, and their number is calculated by numerically integrating the electron density in the box in figure 2.13(c). Because this is a 2D simulation, the charge is shown per micron of depth in the third dimension.

Figure 2.14 shows the potential and the electron concentration profiles in depth through the middle of the PPD, corresponding to the 2D plots in figure 2.13. Firstly, we can see that the pinning voltage is around 1.55 V, and the potential peak is at 0.45 μm below the top surface. Comparing with the doping profile in figure 2.12 the potential peak is near the bottom end of the diode *n*-type doping. Secondly, charge in the PPD clearly decreases its potential, and the peak of the electron concentration coincides with the potential peak. Even without comparing with the potential plot of an empty PPD and the electron concentration, we can still see that charge is present by the flattened top of the potential curve—a tell-tale indicator. With charge present, the peak of the potential moves a bit closer to the pinning layer.

Electrons generated in the epitaxial layer will collect at the potential peak in the PPD, as this is the most positive region.

A curious feature is that the potential is −0.5 V at the surface where the pinning implant resides, despite it being biased at 0 V, the same as the *p*-wells and the substrate. The reason for this is the built-in potential of the *pn* junction between the pinning and diode implants. Likewise, the *pn* junction to the epitaxy makes the potential to become negative at depths below 2 μm.

The plots in figure 2.14 let us calculate the charge handling capacity of the PPD. Knowing the change in the PPD potential with the amount of stored charge, and assuming that this relationship is linear, we can calculate the amount of charge

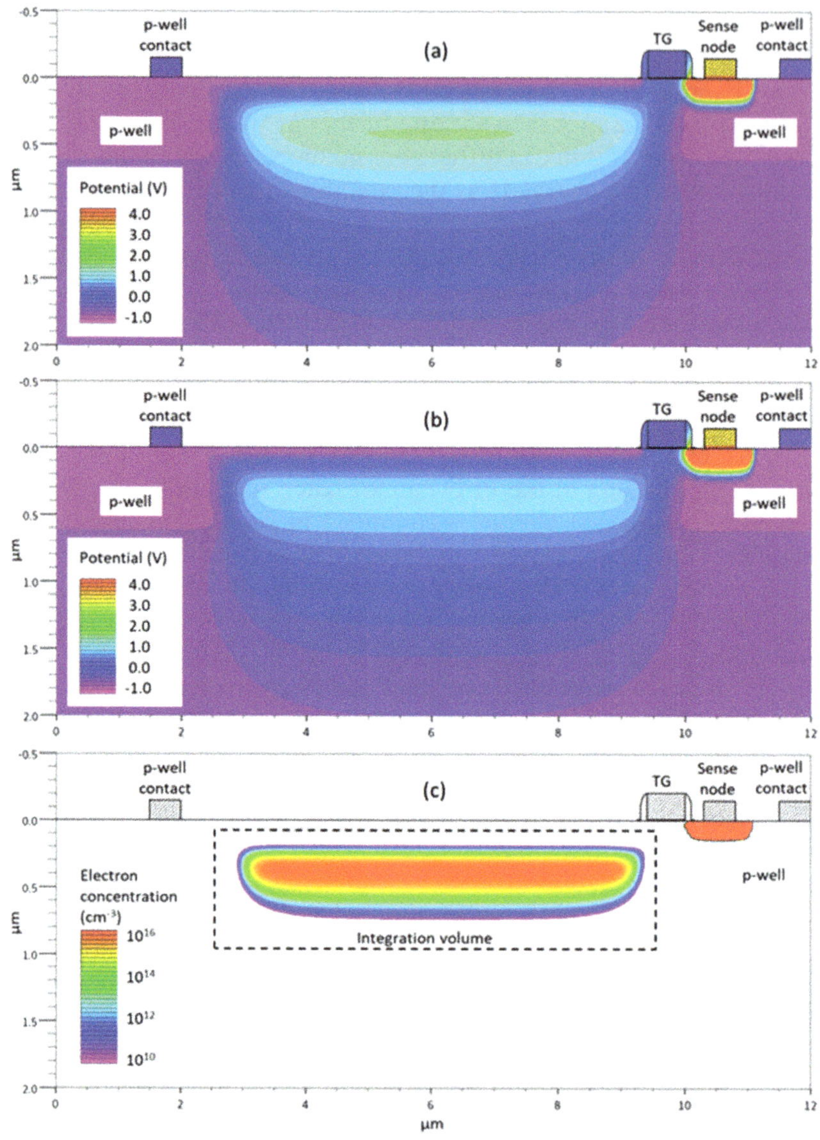

Figure 2.13. 2D potential profile of an empty PPD (a); potential (b) and electron concentration (c) with 6 ke^{-} μm^{-1} showing the volume integration box for the charge calculation.

required to reduce the potential barrier to a level where the charge begins to escape via thermionic emission [16]. A barrier of at least 20 thermal potentials ($20kT/q \approx 0.52$V at 300 K) is required to keep the charge contained in a potential well for sufficiently long time [17]. A 2D simulation can give only an approximate answer because it assumes that the PPD is infinitely long in the third dimension; a 3D simulation would be more accurate (but also much more time consuming) because it includes the edge effects on all sides of the PPD.

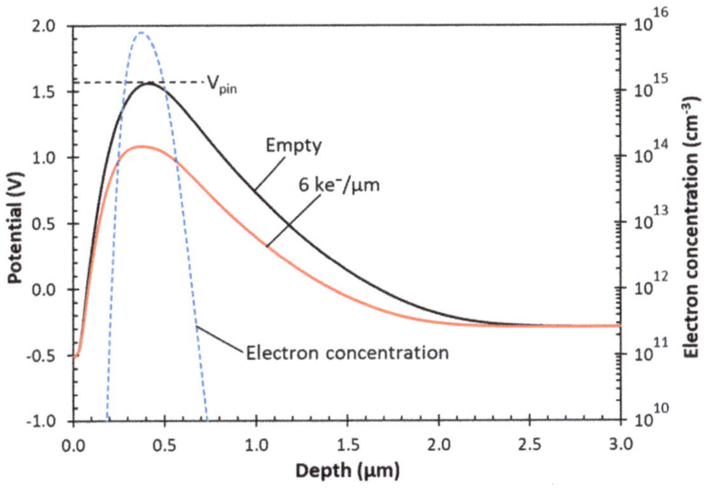

Figure 2.14. Potential profiles in depth in the middle of the PPD, with and without charge, and the free electron concentration for the 2D data in figure 2.13.

Example 2.3. Calculate the FWC of the PPD in figure 2.14, assuming that charge begins to escape via the transfer gate when the potential barrier drops to 0.52 V from the TG potential, taken as 0 V.

Solution: From the peak voltages without and with charge we can calculate the potential change in the PPD when charge of 6 ke⁻ μm⁻¹ is introduced:

$$\frac{\Delta V_{\mathrm{PPD}}}{\Delta Q} = \frac{1.55 - 1.1}{6} = 0.075\ \mathrm{V/ke^-}$$

The inverse $\Delta Q/\Delta V_{\mathrm{PPD}}$ is of course the capacitance of the PPD, equal to

$$C_{\mathrm{PPD}} = \frac{\Delta Q}{\Delta V_{\mathrm{PPD}}} = \frac{1000 \times 1.6 \times 10^{-19}}{0.075} = 2.13\ \mathrm{fF}\ \mu\mathrm{m}^{-1}$$

The amount of charge Q_{FW} required to bring down V_{pin} to 0.52 V is

$$Q_{\mathrm{FW}} = \frac{1.55 - 0.52}{\Delta V_{\mathrm{PPD}}/\Delta Q} = \frac{1.03}{0.075} = 13.7\ \mathrm{ke^-}\ \mu\mathrm{m}^{-1}$$

This result is for 1 μm of device depth in the third dimension, for a pixel pitch of 10 μm. If the depth is 9 μm in the third dimension, the full well capacity is $Q_{\mathrm{FW}} = 13.7 \times 9 = 123.3$ ke⁻ for a PPD with an area of approximately $7 \times 9 = 63\ \mu\mathrm{m}^2$. The FWC per unit area in this example is 1.96 ke⁻ $\mu\mathrm{m}^{-2}$, which is in line with the typical PPD capacities, ranging between 2 and 4 ke⁻ $\mu\mathrm{m}^{-2}$.

The FWC calculated in example 2.3 applies to long-term (hundreds of milliseconds) charge storage without loss due to emission or addition of new charge by illumination. The FWC can be measured in a different way, considering the saturation signal under strong illumination [18], when the charge overflows the potential well. The potential barrier to the TG collapses completely or even becomes negative if the PPD is forward-biased under illumination. If zero potential barrier is assumed, the FWC in the example would be nearly 50% higher at 20.7 ke$^-$ μm^{-1}.

The FWC in electrons can be calculated by integrating the PPD capacitance C_{PPD} between the lowest potential (V_{FW} under full well conditions) and the highest (V_{pin}) [19]:

$$Q_{FW} = \frac{1}{q} \int_{V_{FW}}^{V_{pin}} C_{PPD} \, dV_{PPD} \tag{2.5}$$

Equation (2.5) improves on the simple calculation in the previous example by taking into account that C_{PPD} is not constant. The PPD capacitance depends nonlinearly on V_{PPD} because it consists of two back-to-back *pn* junction capacitances under very low reverse bias [20]. Equation (2.5) also shows that the charge storage capacity increases with V_{pin}, provided that C_{PPD} and V_{FW} stay the same.

The charge storage capacity of the PPD can be understood from first principles like this: the higher the diode dopant concentration, the more donor atoms will be available to store electrons. Not all donor atoms take part in the charge storage; as we can see from figures 2.12 and 2.14 the electron density does not coincide with the PPD donor doping profile. The donor implantation dose directly correlates with the charge storage capacity. For example, implanting 5×10^{11} cm^{-2} phosphorous corresponds to 5×10^3 atoms/μm^2; if every single one could store an electron (an impossibility, but a good goal to strive for) this would mean that the full well capacity is 5 ke$^-$ μm^{-2}.

However, as the donor concentration increases the pinning voltage goes up too, because the positive donor atoms bring the potential up. In figure 2.15 the pinning voltage increases almost linearly with the implant dose, if everything else is kept the same. The pinning voltage and the position of the potential peak depend on the implant dose, energy and annealing conditions. Due to the complexity of the doping profiles this is best optimised with TCAD.

Increasing V_{pin} may not be a good way to increase the FWC because the supply voltages have to go up too. Instead, increasing the PPD capacitance could be a

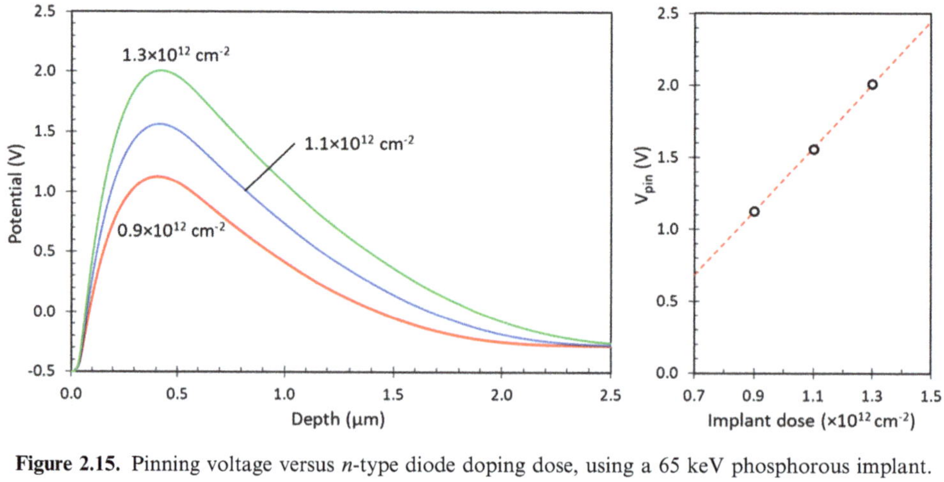

Figure 2.15. Pinning voltage versus *n*-type diode doping dose, using a 65 keV phosphorous implant.

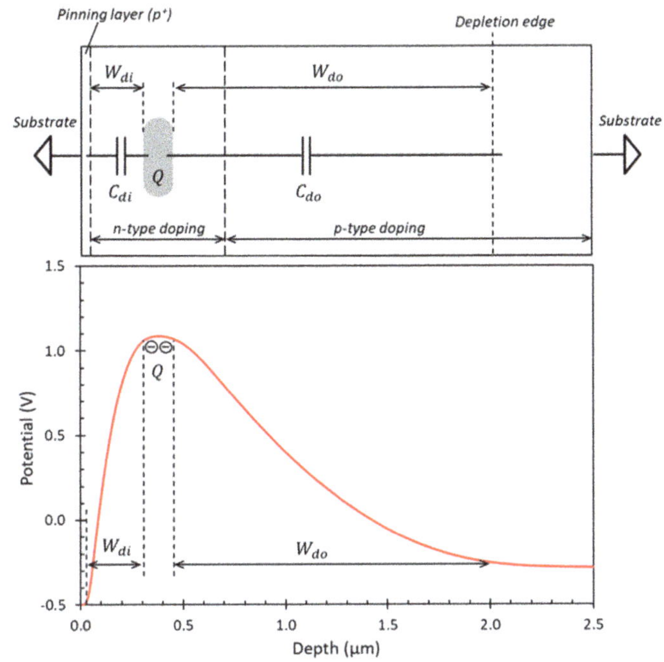

Figure 2.16. Diagram of the capacitances in the PPD.

better idea. Figure 2.16 shows how $C_{\rm PPD}$ can be visualised using the capacitances from the charge packet to the substrate. The stored charge Q is a cloud of electrons and therefore conductive; it can be thought of as the electrode of a parallel plate capacitor within the depletion region. The inner $C_{\rm di}$ and outer $C_{\rm do}$ depletion capacitances are in parallel because they both connect to substrate potential (an AC ground). Charge reaching the potential peak in the PPD will see a capacitance per unit area equal to the sum of the two depletion capacitances:

$$C_{\rm PPD} = C_{\rm di} + C_{\rm do} = \varepsilon_0\varepsilon_{\rm Si}\left(\frac{1}{W_{\rm di}} + \frac{1}{W_{\rm do}}\right) \tag{2.6}$$

Here $C_{\rm PPD}$ per unit area is written using the inner and the outer depletion widths as in a reverse-biased *pn* junction.

Since usually $W_{\rm do} \gg W_{\rm di}$, the inner capacitance is much larger and dominates the total. The PPD has large storage capacitance because $W_{\rm di}$ is small; the capacitance can be increased further by making the *n*-type diode implant shallow (to reduce $W_{\rm di}$) and by using low resistivity substrate which reduces $W_{\rm do}$ too.

2.3.4 Charge transfer

During collection the charge is confined in the PPD by potential barriers on all sides. In depth the charge is contained at the potential peak, squeezed by the pinning implant from the top and the *p*-type epitaxial layer from the bottom. The highly doped p-well surrounding the PPD provides a potential barrier for the periphery except for the area around the transfer gate. During charge collection TG is held at a low potential (typically at substrate) and creates the last barrier segment to prevent charge from escaping.

Before the charge transfer is initiated, the sense node must be biased appropriately so that it is the most positive potential for the charge to go to. This is accomplished by turning the reset transistor on for a short time, similarly to the timing diagram of the 3T pixel in figure 2.3. With the sense node biased and floating, the voltage on TG is increased, the potential barrier between the charge packet and the sense node is reduced, and the charge starts moving towards the sense node.

Figure 2.17 shows the simplified diagram of the charge transfer process using the concept of the potential well. Such diagrams are very useful in visualising the operation of devices using charge transfer, such as PPDs and CCDs. The charge behaves like a liquid which flows under the influence of the electric field from low to high potential[3]. The charge wants to find its way towards the bottom of the deepest well, just like a liquid would do under gravity.

In reality, charge transfer is far more complex than the simple diagram in figure 2.17 suggests. Under strong illumination the charge can overflow out of the PPD even if the TG is off. A potential barrier can form along the electron path which impedes the smooth transfer, and the potential at the sense node decreases during the transfer, making it less attractive to electrons.

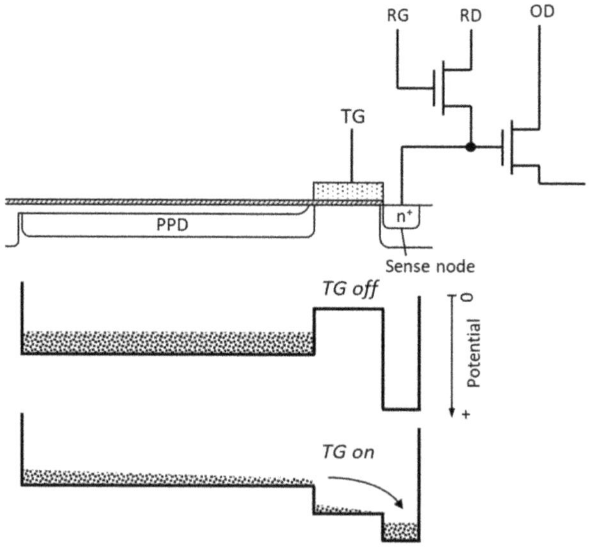

Figure 2.17. Potential diagram in the PPD with the TG off and on. Potential increases from top to bottom.

[3] This is valid for electrons. The same concept can be used for holes, which flow from high to low potential.

The interface between the PPD and TG is the most critical aspect of pixel design and operation. When TG is on, the potential along the direction of transfer should not have any potential 'bumps' or 'pockets' which could cause incomplete charge transfer due to slowing down of the charge flow or charge trapping. Similarly, when TG is off, the potential barrier should be high enough to hold the charge in the PPD for a very long time without electrons escaping thermally.

Figure 2.18 shows many of the possible scenarios than can occur. Overflow to the sense node (figure 2.18(a)) is an important mechanism to ensure that the charge is drained in a controlled manner (anti-blooming) and not spilled over the adjacent pixels. Some charge can jump thermally over the off-state TG even when the PPD is nearly full (figure 2.18(b)). This is called feedforward effect [21] and it reduces the maximum charge the PPD can hold.

Long-term, lossless charge storage is shown in figure 2.18(c), followed by a reduction of the potential barrier at the TG and a complete charge transfer. Figure 2.18(e) depicts a small potential barrier forming near the end of the charge transfer, but despite this no electrons are left in the PPD. Increasing the transfer gate voltage can reduce and even eliminate the barrier, but at high signal levels this can cause charge spillback [22], an effect resulting in some charge returning to the PPD.

Since the pinning implant is establishing a fixed potential at the top of the PPD, like a conductive plate connected to the substrate, there is almost no lateral electric field underneath. The charge moves almost entirely by diffusion until it gets near the transfer gate, where a small electric field is present; it is clear that the charge transfer is not going to be very fast.

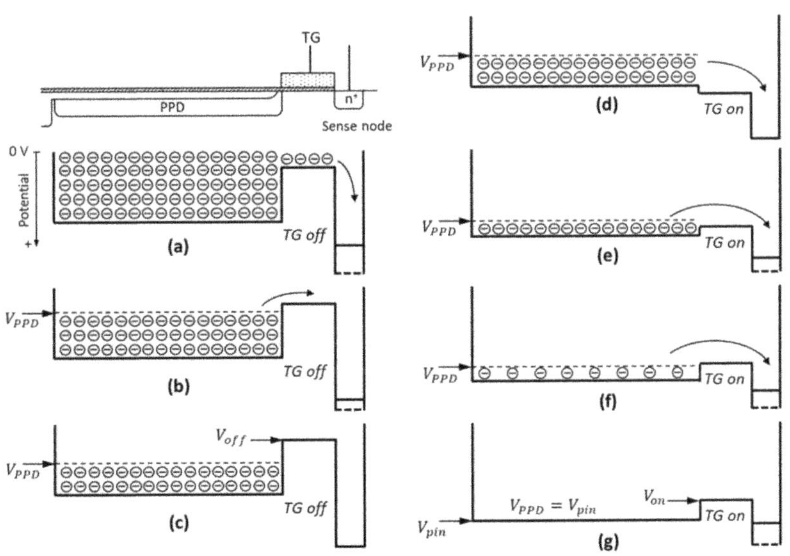

Figure 2.18. Potential diagrams of a PPD: (a) overflowing charge; (b) charge escaping thermally over the TG barrier; (c) long-term storage; (d) charge transfer immediately after TG is raised; (e) potential barrier begins to form along the transfer path; (f) most of the charge is transferred, the barrier has increased; (g) PPD empty, potential equals V_{pin}.

Charge transfer is complex because it involves charge moving from a depth of about 0.5 μm up to the surface under the TG, and then along the TG to the sense node. In figure 2.19 the charge is seen to transfer completely in less than 500 ns for the PPD model in figure 2.13. Charge leaving the PPD resembles the deflating of a thin balloon through a straw inserted in one end.

Figure 2.19 shows the electron concentration over 6 orders of magnitude which should be sufficient to spot charge transfer inefficiency down to 10^{-5} level. This simulation shows that the transfer is very efficient with virtually no charge left, which is what we wanted.

Plotting the number of electrons in the PPD over time in figure 2.20 tells us more about the efficiency of the transfer than a colourful 2D picture. As usual, the charge in the PPD is found by numerically integrating over a volume. Figure 2.20 demonstrates perfect charge transfer for TG = 2.7 and 3.0 V, with 99.9% of the charge transferred within 200 ns. Higher TG potential speeds up the transfer: with

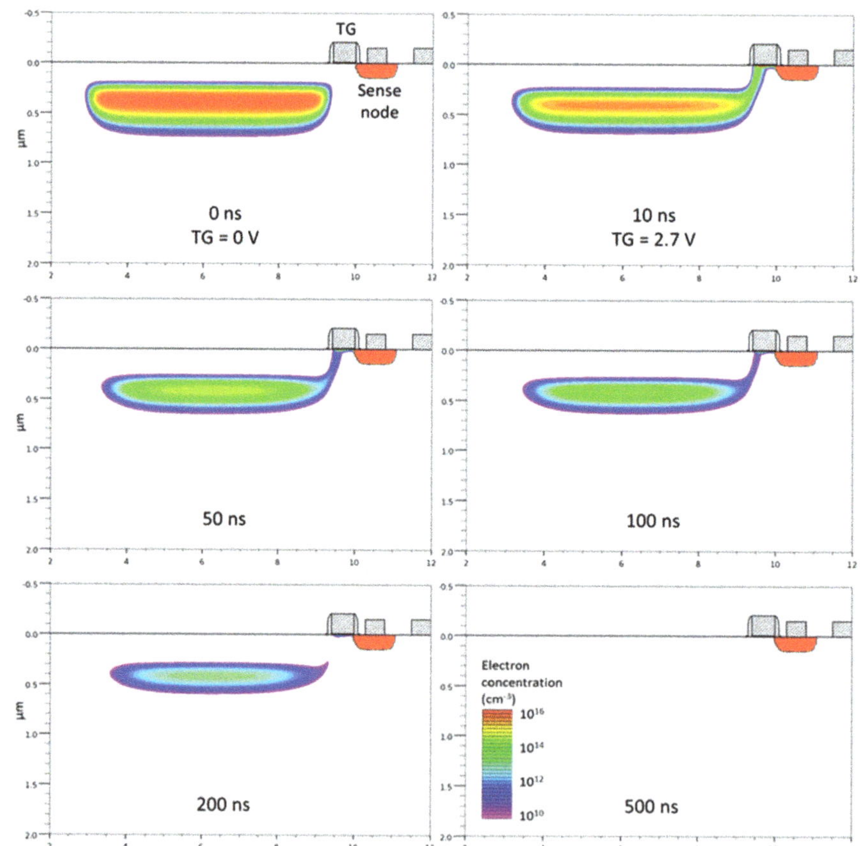

Figure 2.19. Snapshots of the electron concentration in a PPD model at different points during the charge transfer. TG rises from 0.0 to 2.7 V over 10 ns and stays constant until the end of the transfer. The sense node is biased to 3.6 V.

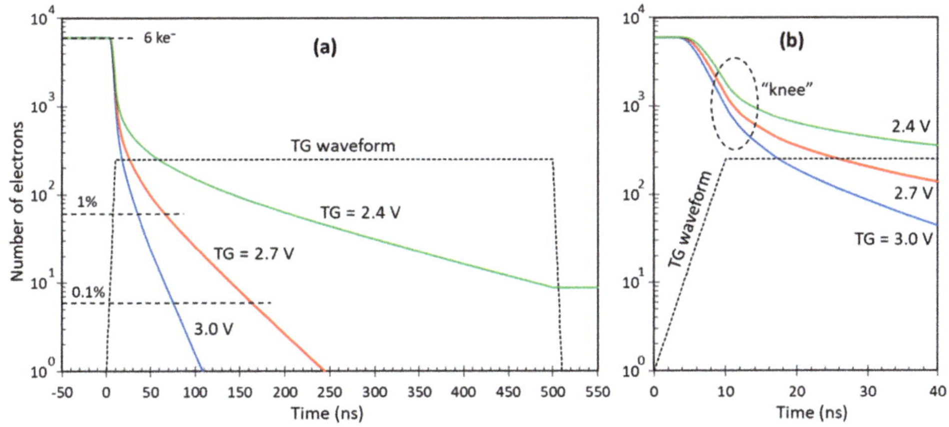

Figure 2.20. Charge transfer simulation showing the number of electrons in the PPD during transfer for three TG voltages (a) and a zoom-in of the first 40 ns (b). The rise and fall time of TG is 10 ns.

TG = 3.0 V it takes nearly 90 ns shorter to reach 99.9% efficiency than for TG = 2.7 V. Intuitively, this makes sense because higher transfer gate voltage should increase the electric field at the edge of the PPD, speeding electrons away quicker. At TG = 2.4 V the transfer is incomplete; about 10 electrons remain in the PPD after 500 ns and much longer time would be required for all the charge to clear.

Figure 2.20(b) shows that initially the transfer is quick and most of the charge moves away during the 10 ns rise time of the TG voltage. After that a slowdown begins and the number of transferred electrons gradually becomes an exponentially decreasing function of time, as the straight line (straight because it is plotted on a log scale) in figure 2.20(a) indicates. The slowing down is caused by a small potential barrier forming along the electron transfer path, which can be seen in figure 2.19 as the 'neck of the balloon'.

Initially the charge moves predominantly by diffusion since there is no potential barrier in its path, and the electron concentration is large. The time for this initial transfer can be calculated from the formula describing the diffusion length in chapter 1. Taking the maximum electron travel distance as 6 μm from figure 2.19, the characteristic diffusion time constant is:

$$\tau = \frac{d^2}{2D_n} = \frac{(6 \times 10^{-4})^2}{2 \times 36} = 5 \text{ ns} \tag{2.7}$$

The time constant in (2.7) underestimates the speed of the transfer because most of the charge does not travel the full length of the photodiode, but significantly shorter. Still, from figure 2.20(b) we can see that it takes less than 10 ns (two time constants) for 90% of the charge to transfer. This is notable by the 'knee' in the electron number, considering that not much happens in the first 5 ns during the rise of the transfer gate voltage. Despite its simplicity, formula (2.7) is a useful first-order indicator of the diffusion-dominated stage of the charge transfer.

2.3.5 Image lag

Not all charge transfers from the PPD to the sense node during the time when the TG is pulsed high. The effect is called image lag (sometimes called charge transfer inefficiency) and can be seen particularly well in a dark image following a bright one, resulting in a ghosting effect. Image lag is a characteristic of the PPD arising from the nearly field-free charge transfer and the formation of a potential barrier in the charge path. One reason for this is the pinning implant, which prevents the potential along the PPD from changing significantly, regardless of the voltage on the TG. The consequences are that the charge transfer in the PPD is driven mostly by diffusion, which is a relatively slow process.

If the barrier is small, all the charge can jump over it thermally during the transfer time, and no image lag will be observed. Severe lag can occur if there are large potential barriers or bumps, as shown in figure 2.21, so that the charge cannot clear them during the time when TG is high. The most critical areas are around the TG, and they receive special attention during design and manufacture in order to eliminate any potential irregularities. Good pixel designs do not have significant barriers or pockets, and may implement special features intended to speed up charge transfer by built-in electric field created by implants [23]. Typically, the image lag should be well below 1% for large signals.

Even if there are no potential barriers and pockets along the transfer path, charge can still be captured by traps. Reducing the number of traps and defects is an ongoing quest for the manufacturing processes, but here we are interested only in how pixel design affects the lag.

Following the initial, diffusion-dominated charge transfer for large signals, the flow of electrons slows down as a small potential barrier forms and impedes the free charge movement. If the initial charge in the PPD is small, the potential barrier is already present in the first moment after the TG is biased high. It is this barrier-dominated part of the charge transfer process that causes image lag.

Figure 2.22 shows the potential along the charge transfer path before and during the transfer. As more charge transfers to the sense node the potential in the middle of

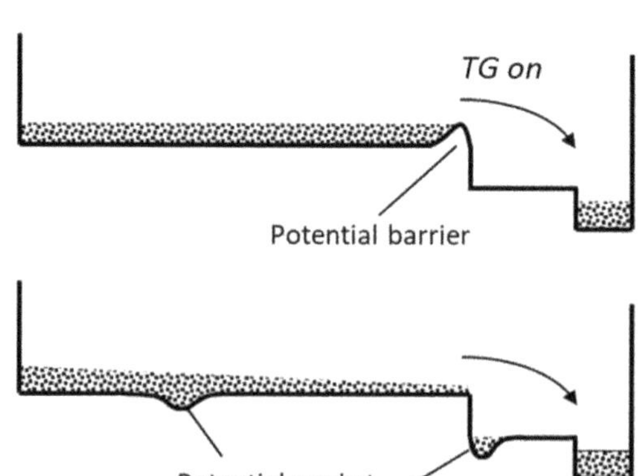

Figure 2.21. Potential barrier ('potential bump') and potential pockets in a PPD leading to image lag.

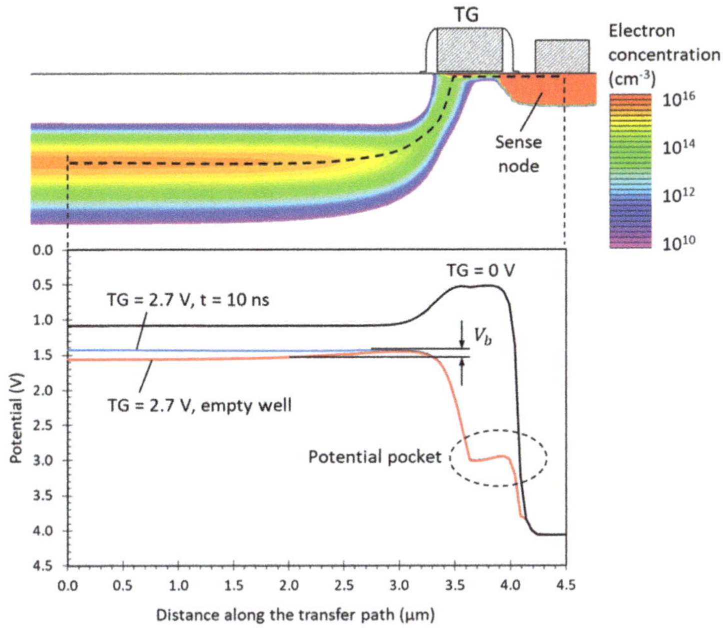

Figure 2.22. Potential along the charge transfer path for TG = 0 V, after TG has been raised to 2.7 V (blue line) and at the end of the transfer (red line).

the PPD increases, while the potential under the TG V_{on} remains the same. This is the origin of the potential barrier forming at the leading edge of the TG. If the barrier is small and the time long enough, as in this case, all the charge manages to jump over the barrier, and the PPD becomes empty. At this point the PPD is fully depleted, its potential is V_{pin} and the barrier height is $V_{\text{b}} = V_{\text{on}} - V_{\text{pin}}$. In figure 2.22 the barrier at the leading edge of the TG is $V_{\text{b}} \approx 100$ mV. We also notice a potential pocket of around 60 mV under the transfer gate; we are going to look into it later when considering various methods for image lag reduction.

The electron flow at the leading edge of TG can be modelled as thermionic emission over a barrier with height $V_{\text{b}} = (V - V_{\text{on}}) > 0$, where V is the instantaneous voltage in the middle of the PPD. The emission current I is

$$I = I_0 \exp\left(-\frac{V_{\text{b}}}{\varphi_{\text{T}}}\right) \tag{2.8}$$

where $\varphi_{\text{T}} = kT/q$ is the thermal potential, $I_0 = A^* T^2 S_A$ is the saturation current, with A^* the effective Richardson constant and S_A the cross-section of the current path. Using that the current $I = dQ/dt$, where dQ is the charge leaving the PPD and $dQ = C_{\text{PPD}} dV$, we can write:

$$C_{\text{PPD}} \frac{dV}{dt} = I_0 \exp\left(-\frac{V - V_{\text{on}}}{\varphi_{\text{T}}}\right) \tag{2.9}$$

This equation uses the simplification that C_{PPD} is constant. If the charge transfer starts when the PPD voltage is V_0 equation (2.9) can be solved by integrating both sides with the initial condition $V(t = 0) = V_0$

$$\int_{V_0}^{V} \exp\left(\frac{V - V_{\mathrm{on}}}{\varphi_{\mathrm{T}}}\right) d\left(\frac{V - V_{\mathrm{on}}}{\varphi_{\mathrm{T}}}\right) = \int_{0}^{t} \frac{I_0}{C_{\mathrm{PPD}}\varphi_{\mathrm{T}}} dt \tag{2.10}$$

$$\exp\left(\frac{V - V_{\mathrm{on}}}{\varphi_{\mathrm{T}}}\right) - \exp\left(\frac{V_0 - V_{\mathrm{on}}}{\varphi_{\mathrm{T}}}\right) = \frac{I_0 t}{C_{\mathrm{PPD}}\varphi_{\mathrm{T}}} \tag{2.11}$$

The solution for the PPD voltage is

$$V = \varphi_{\mathrm{T}} \ln\left[\exp\left(\frac{V_0 - V_{\mathrm{on}}}{\varphi_{\mathrm{T}}}\right) + \frac{I_0 t}{C_{\mathrm{PPD}}\varphi_{\mathrm{T}}}\right] + V_{\mathrm{on}} \tag{2.12}$$

Now, using that the number of transferred electrons is $N_{\mathrm{tr}} = Q/q = C_{\mathrm{PPD}}(V - V_0)/q$ we can write

$$N_{\mathrm{tr}} = \frac{C_{\mathrm{PPD}}}{q}\left(\varphi_{\mathrm{T}} \ln\left[\exp\left(\frac{V_0 - V_{\mathrm{on}}}{\varphi_{\mathrm{T}}}\right) + \frac{I_0 t}{C_{\mathrm{PPD}}\varphi_{\mathrm{T}}}\right] + V_{\mathrm{on}} - V_0\right) \tag{2.13}$$

Equation (2.13) has the problem that the number of transferred electrons increases to infinity with t, and a similar problem exists in (2.12) because V also increases uncontrollably. This happens because (2.8) does not take into account that the PPD voltage is limited by V_{pin}. Of course, infinite voltages and charges are somewhat difficult to get in practice.

This was the original treatment by Teranishi [7], who derived (2.12) for $V_{\mathrm{on}} = 0$ and non-pinned photodiode to show that the logarithmically slowing charge transfer never fully completes. Some charge remains in the photodiode and shows itself as remnant signal in the dark images following a bright one, which we call trailing edge image lag.

It is obvious that equations (2.12) and (2.13) do not describe PPD operation with complete charge transfer. What we do know is that given long enough time the PPD voltage will reach the pinning voltage, or $V(t \to \infty) = V_{\mathrm{pin}}$, and will not increase to infinity as (2.12) suggests. After most of the charge has transferred, the PPD voltage is close to V_{pin} and the potential barrier is approximately $V_{\mathrm{b}} = V_{\mathrm{pin}} - V_{\mathrm{on}}$ and barely increasing.

When the PPD holds only a small number of electrons, equation (2.8) stops being valid because it relies on unlimited supply of charge able to jump over the barrier, so that a steady state current can be maintained. Therefore, a different model is needed to describe the tail end of the charge transfer when the electron supply is limited. If we make the reasonable assumption that the number of electrons jumping over the barrier ΔN_{e} in time Δt is proportional to the number of electrons left in the PPD N_{e}, we can write the following:

$$\Delta N_e \propto N_e \exp\left(-\frac{V_b}{\varphi_T}\right)\Delta t \tag{2.14}$$

Developing equation (2.14) into a differential equation with the proportionality constant λ we can write:

$$\frac{dN_e}{dt} = -\lambda N_e \exp\left(-\frac{V_b}{\varphi_T}\right) \tag{2.15}$$

The negative sign is to reflect that the number of electrons in the PPD is decreasing, i.e. $dN_e < 0$. If the number of electrons in the PPD at time $t = 0$ is N_{e0}, the number remaining after time t can be calculated by solving

$$\int_{N_{e0}}^{N_e} \frac{dN_e}{N_e} = -\lambda \exp\left(-\frac{V_b}{\varphi_T}\right)\int_0^t dt \tag{2.16}$$

$$N_e = N_{e0} \exp\left[-\lambda \exp\left(-\frac{V_b}{\varphi_T}\right)t\right] \tag{2.17}$$

Equation (2.17) shows an exponential time dependence of the number of electrons remaining in the PPD, something we saw in the simulation in figure 2.20. Similar dependence is obtained in [24] but using different considerations.

The double exponential in equation (2.17) makes the speed of charge transfer strongly dependant on the barrier height V_b. We can consider the term multiplying the time as the inverse time constant τ of the charge transfer, so that $1/\tau = \lambda \exp(-V_b/\varphi_T)$. For prompt transfer the time constant should be short, for example if $\tau = 100$ ns, in 500 ns (five time constants) 1% of the charge will remain, which becomes only 0.005% after 1 μs. A mere 60 mV increase of V_b would increase τ by a factor of 10, and this could seriously limit how the sensor is used. Taking this further, a barrier increase of 360 mV would make the charge transfer time increase a million times, from 1 μs to 1 s!

For cooled sensors, such as those used in scientific applications requiring very low dark current, the transfer time increases due to the reduced thermal potential φ_T. This should be taken into account when choosing the transfer time duration because the increase can be large, as the following example demonstrates.

Example 2.4. Calculate the increase of the transfer time caused by cooling the sensor from 20 °C to −50 °C, assuming that the barrier height is $V_b = 100$ mV and the coefficient λ in equation (2.17) is temperature-independent.

Solution: We need to calculate the ratio of the time constants at 20 °C ($T_1 = 293$ K) and at −50 °C ($T_2 = 223$ K):

$$\frac{\tau_1}{\tau_2} = \frac{\exp\left(V_b/\varphi_{T_1}\right)}{\exp\left(V_b/\varphi_{T_2}\right)} = \exp\left(\frac{V_b}{\varphi_{T_1}}\frac{\varphi_{T_2} - \varphi_{T_1}}{\varphi_{T_2}}\right) = \exp\left(\frac{V_b}{\varphi_{T_1}}\frac{T_2 - T_1}{T_2}\right)$$

$$\frac{\tau_1}{\tau_2} = \exp\left(\frac{100}{26} \times \frac{223 - 293}{223}\right) = 0.3$$

Therefore, the time constant at −50 °C is 1/0.3 = 3.3 times larger than at 20 °C, and the transfer time must be 3.3 times longer to achieve the same transfer efficiency.

After looking into the charge transfer, one important question to answer is: does a potential barrier always form along the transfer path? There is no simple answer, but unless the TG voltage, and therefore the sense node voltages is very high, it is difficult to prevent a barrier from forming. Naturally, the potential in the middle of the PPD is higher than at its periphery, as can be seen in figure 2.13(a), so the tendency is there.

Increasing the transfer gate voltage is effective in reducing the image lag for small signals as the potential barrier at the leading edge of the TG is reduced or eliminated. Unfortunately, increasing the TG further causes a different lag mechanism, called charge spillback [22], to kick in for large signals. Spillback occurs when some of the charge stored under the TG returns to the PPD instead of going to the sense node, as intended after the TG potential is lowered. This can happen when the charge is large and the sense node potential has decreased so much that the charge is shared with the TG, as shown in figure 2.23.

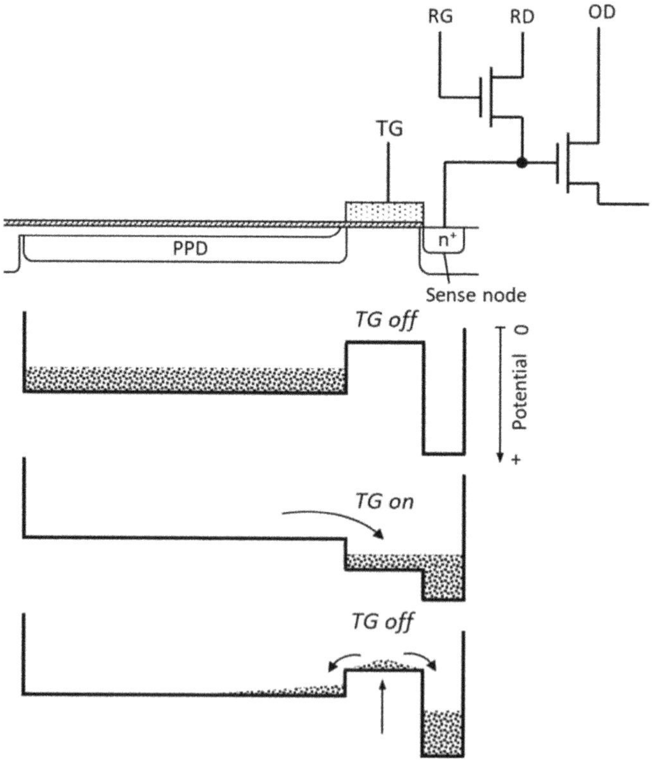

Figure 2.23. Mechanism of charge spillback in PPDs.

Increasing the sense node voltage reduces the spillback, and so does reducing the TG amplitude [25]. Slowing down the falling edge of the TG is also used in some image sensors. A different method, described in [15] and [23], uses a threshold adjustment *p*-type implant under the TG (figure 2.24) to perform two functions. Firstly, the threshold adjust increases the TG threshold to the desired value as the epitaxial layer usually has high resistivity and therefore the native TG threshold would be low or even negative. Secondly, it creates a potential gradient from the photodiode towards the sense node and removes the potential pocket, as seen in figure 2.25. Having a potential gradient under the TG counteracts the charge spillback because the electrons move preferentially towards the sense node when the TG voltage is lowered [26].

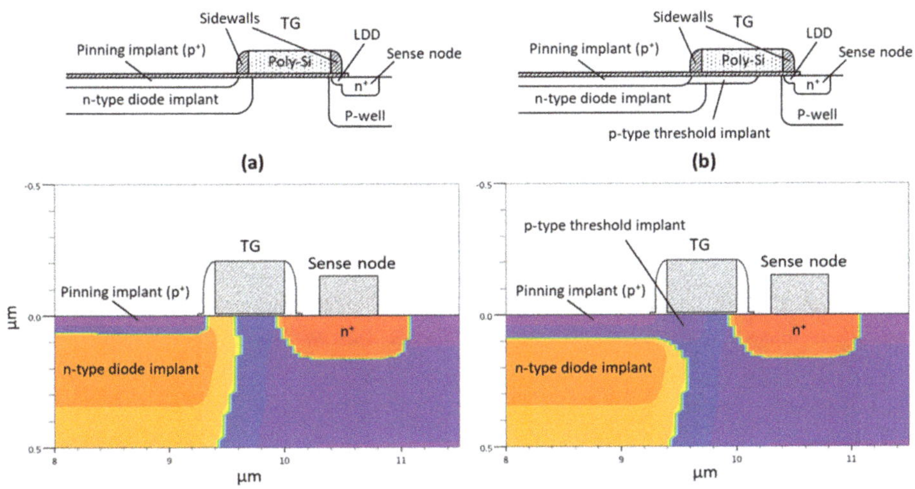

Figure 2.24. PPD pixel without (a) and with a threshold adjustment implant under the TG (b).

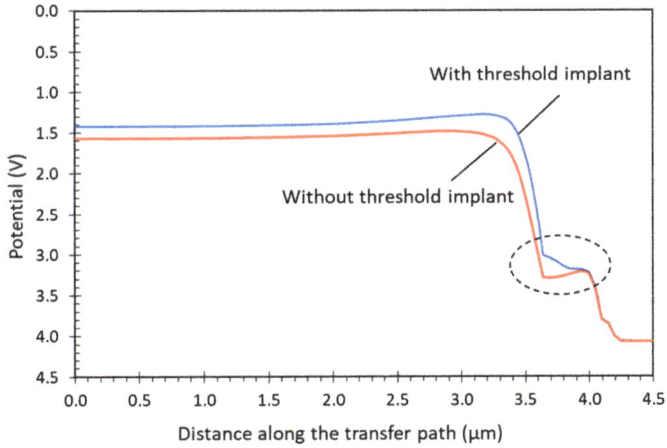

Figure 2.25. Potential along the charge transfer path with and without a threshold implant under the TG.

The structures in figure 2.24 use typical manufacturing processes relying on self-alignment between the gate, the sidewall spacers and the implants making the PPD. Without this self-alignment it is not possible to make reliable transistors and to control precisely the position of the implants at the leading edge of the TG, leading to poor charge transfer performance. A PPD CIS process might employ the following steps [15]:

1. *p*-well and STI formation
2. Threshold adjustment implant
3. Gate oxide growth, poly-Si gate deposition and patterning
4. *n*-type photodiode implant, self-aligned to the edge of the TG
5. Lightly doped drain (LDD) implantation, self-aligned to the TG edge
6. Sidewall spacer on the poly-Si gates
7. Pinning implant, self-aligned to the sidewall spacer of the TG
8. Transistor drain and sense node n^+ implants, self-aligned to the sidewall spacer, and covering the transfer gate.

The last step makes the transfer gate *n*-type doped, the same as the gates of the in-pixel transistors.

Another method for reducing image lag involves storing the charge near the TG, so that it travels shorter distances. For example, [27] and [28] describe methods of introducing potential gradients within the PPD for fast transfer. These methods are especially useful and even indispensable for large pixels above 10–20 μm pitch [29]. Image lag is less of a problem in smaller pixels, below about 5 μm pitch.

2.3.6 Transistor sharing

Charge transfer opens the possibility for the sense node to be shared between two or more PPD pixels. Each sense node needs three transistors for readout; therefore three transistors can be eliminated for each shared sense node. For example, if two PPDs share their readout, the total number of gates would be five: two transfer gates and three for the readout transistors, or 2.5 gates per pixel. For four pixels the number is 7/4 = 1.75, and for eight pixels it is 11/8 = 1.375 gates per pixel, as shown in figure 2.26.

As the pixels get smaller, maintaining large PPD area relative to the readout transistors, which are not photosensitive, becomes increasingly important. Shared readout is crucial for reducing the pixel size of PPD-based CIS and is one of the factors allowing imagers with sub-micron pixels to exist.

2.4 Other PPD-based pixels

2.4.1 Global reset (5T)

Charge can be transferred out of the PPD pixel by more than one transfer gate. Any side edge of the PPD can in principle be used for this, if it is equipped with a gate.

The 5T pixel adds one more transfer gate to the 4T design to create a *global reset* functionality. This is used in global shutter (GS) readout to reset all the pixels in an image sensor, so that exposure starts simultaneously for all. The term '5T' is

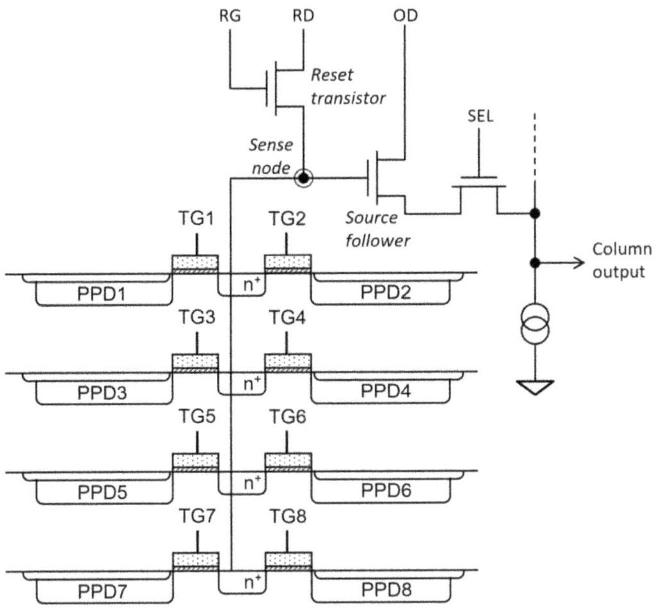

Figure 2.26. Transistor sharing between 8 PPD pixels, based on [15].

generally reserved to describe a PPD pixel with global reset, but other architectures can also have five transistors with different functionality, for example the LOFIC pixel.

The 5T pixel in figure 2.27 [30] is simply created by adding one more transfer gate, called global reset (GRST), and an n^+ implant on the opposite side of the TG. The additional n^+ contact is biased to a fixed voltage and acts as a charge drain. The bias on this drain is normally the RD voltage, but OD can be used as well. When the voltage on GRST is raised sufficiently high in a short pulse just like in a normal transfer using the TG, the charge in the PPD is transferred and drained away to the RD supply.

Global shutter operation is realised by simultaneously transferring the photo-generated signal from the PPD to the sense node in all pixels in the device. GRST is activated before the next exposure to clear the charge collected during the readout of the previous image, or it can be kept on throughout the readout. A downside of this technique is that effective CDS cannot be performed, so the reset noise is not removed. Therefore, the readout noise of the 5T pixel operating in GS mode is very similar to the 3T pixel for the reasons described in section 2.2.3, despite using a PPD.

A design with one globally controlled GRST, positioned as in figure 2.27, is the simplest possible 5T pixel. Often, the pixel layout allows two GRST gates to be used instead of one—instead of opposite the TG, they can be on the two remaining sides of a rectangular PPD. In principle, global reset can also be achieved by using the reset transistor and the transfer gate simultaneously, however, this clears the signal at the sense node too, and requires global control of both TG and RG.

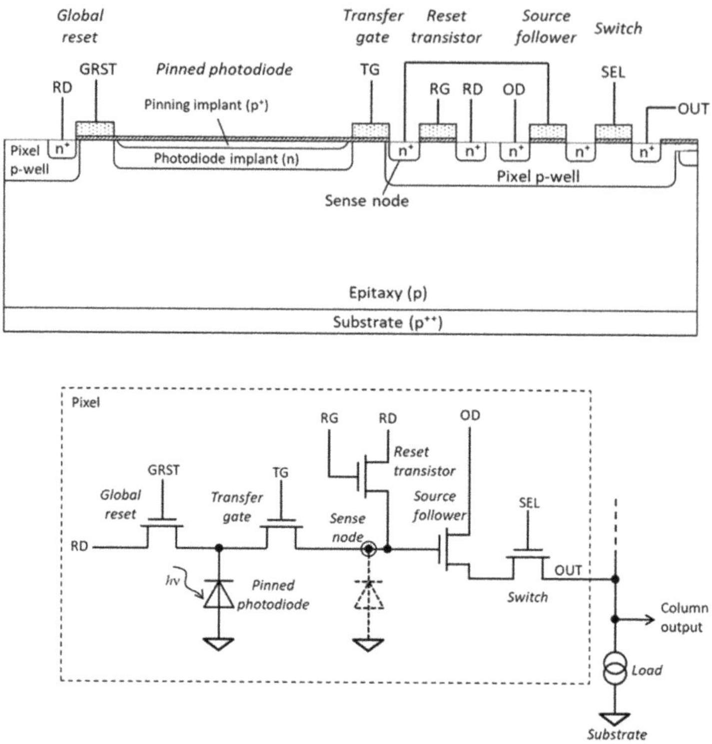

Figure 2.27. Global reset (5T) pixel structure and schematic.

The 5T global reset pixel has the capability to be operated in global shutter mode if the signal charge is simultaneously transferred to the sense node in all pixels. However, global reset can be used independently of the GS functionality, and the pixel can be read out like a normal 4T pixel without using GRST, so 5T does not always mean a global shutter.

2.4.2 In-pixel signal storage

High performance PPD-based GS imagers use two-stage readout to allow removal of the reset noise on the sense node via CDS. Once the signal charge is transferred out of the PPD, it can be stored by converting it to voltage while the reset level on the sense node is sampled. This is called 'voltage domain GS' and relies on several capacitors and switches per pixel for the conversion [31].

The other method for intermediate signal storage is to keep it as charge under a *storage gate* (SG) or in a *storage diode* (SD), without converting it. This is known as 'charge domain GS'. Both voltage and charge domain GS pixels have their pros and cons [32], but operation in the charge domain is more closely coupled to PPD operation and offers unique functionality and challenges.

The operating principle of charge domain GS pixels can be understood from figure 2.28. Here two PPDs are used—'pinned PD' for signal collection, and 'pinned

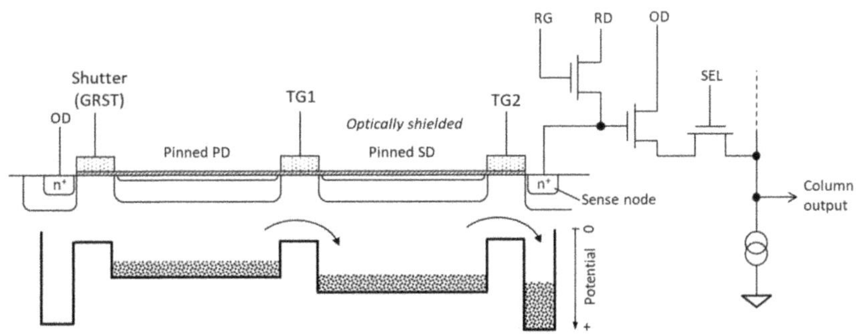

Figure 2.28. Charge domain global shutter pixel using pinned storage diode, based on [33].

SD' for signal storage. After integration, photogenerated charge is transferred from the PD to the SD, which has a higher pinning voltage to allow this to happen. The sense node (FD) is reset, its level sampled, and then the charge is transferred from the SD. The reset noise is removed due to the CDS and the readout noise can be very low, similarly to the 4T pixel. The shutter gate (GRST) is used to clear the photogenerated signal as usual.

Even more advanced functionality has been demonstrated in [34] by using a shaped PPD and dual readout, combining a PD–SD storage (path A–A′ in figure 2.29) with a 5T structure (path B–B′). Higher pinning voltage in SD is achieved by higher dopant concentration (n_2) relative to the PD (n_1).

In-pixel signal storage almost inevitably increases the number of gates per pixel and reduces the available area for the photodiode. The pixel in figure 2.28 has six gates and the one in figure 2.29 has eight; it is also clear that the PD occupies only about 40% of the pixel area in figure 2.29, reducing the fill factor. Despite this, the added complexity is justified for many applications, such as machine vision and automotive.

An important requirement for all GS pixels is for the storage node to be protected from light, so that the stored signal does not charge once the integration is complete. This is characterised by a parameter called parasitic light sensitivity (PLS). The PLS, defined as the ratio between the light-induced signal in SD and the signal in PD, should be as low as possible. In front-side illuminated (FSI) sensors, the storage elements are protected by metal shields and assisted by lightguides and microlenses to direct the light towards the PD [33]. This is, however, not enough for charge domain GS pixels because charge can enter both the PD and the SD, and the SD has higher potential. This is why the additional *p*-type implant p_2 is used in the design in figure 2.29—it creates a reflective barrier for electrons coming from below and prevents most of them from entering the SD directly.

Despite all the measures taken, charge domain GS is more susceptible to PLS, especially in backside illuminated (BSI) sensors where metal shielding becomes ineffective [32]. Voltage domain GS pixels are better in this respect because charge is stored on capacitors, but are not immune. This is because the source and drain *pn* junctions of all transistors are photosensitive and the storage capacitors can be

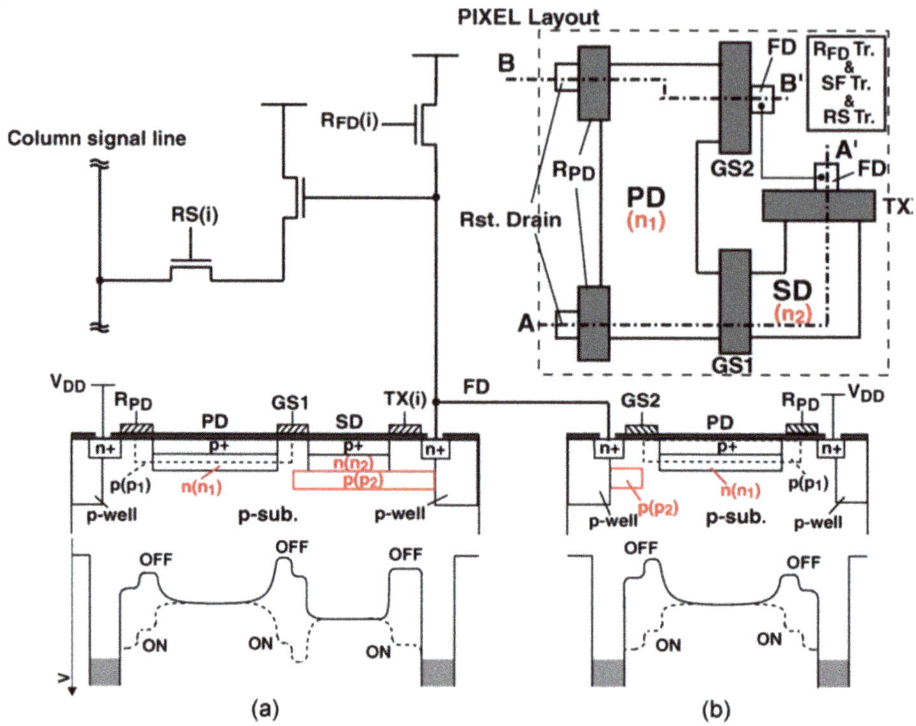

Figure 2.29. PPD pixel with charge domain signal storage and dual readout. Copyright IEEE (2011). Reprinted with permission from [34]).

discharged through them. 3D integration with separate photosensitive and storage tiers has been shown to have the best PLS [32]. An additional advantage of the 3D integration is that the photodiode can occupy the maximum possible area, resulting in the highest fill factor.

2.4.3 High dynamic range

Many applications, such as automotive imaging, surveillance, and science encounter scenes with very large difference between the brightest and the dimmest parts of the image. The dynamic range (DR) of the sensor, defined as the ratio of its FWC to the readout noise, must exceed 10^5 or even 10^6 to be able to faithfully capture those challenging scenes. A huge variety of CIS with high dynamic range (HDR), normally taken to mean above 10^5 (100 dB), have been developed to deal with this.

The typical PPD pixel achieves DR between 1000 (60 dB) and 10 000 (80 dB), and that depends on the pixel size, conversion gain and the system noise. A quick calculation can show why DR above 80 dB is difficult to achieve. A 5 μm pixel may have 60% fill factor and an area capacity of 2 ke$^-$ μm^{-2}, so it can store 30 ke$^-$. For a maximum output signal of 1.5 V the conversion gain must not exceed 50 μV/e$^-$ and the noise is likely to be around 4 e$^-$ RMS, therefore the maximum DR is 30 000/4 = 7500 (77.5 dB).

Reducing the readout noise seems like a good way to increase the DR, and PPD imagers with very low, even sub-electron noise have been developed [35, 36]. However, reducing the noise comes at the expense of increasing the conversion gain much above 100 μV/e$^-$, while the maximum output signal still cannot exceed about 1.5 V. As a result, the FWC becomes limited by the output signal and not by the PPD's FWC, and the DR does not take a great leap forward. For example, the pixel in [35] achieves 0.5 e$^-$ RMS readout noise but 6400 e$^-$ FWC for a DR of 82 dB, and the 0.27 e$^-$ RMS pixel in [36] manages only 75 dB.

The other 'simple' way to increase the DR—boosting the FWC and the output signal span, is also not very productive. The operating voltages in newer CMOS processes tend to be lower and the pixels shrink too, making the PPD area and the FWC lower. Additionally, HDR requires on-chip ADC resolution above 16-bit, which is a big challenge.

Since the maximum saturation signal in low-noise pixels is limited by the readout path and not the FWC of the PPD, the solution to HDR often lies in the column readout circuit, and not in new pixel architectures. A very successful solution is the dual gain column amplifier with dual ADC conversion [37], shown in figure 2.30. Here, a low noise 5T pixel image array is served by two column amplifiers: one with a gain of 1, the other with 30, and their outputs are digitised by separate 11-bit ADCs. Each signal processing channel achieves DR of only 64 dB (high gain) and 74 dB (low gain), but the combined result has a DR of 92 dB with noise of only 1.2 e$^-$ RMS. CIS using similar architecture and further improvements have spurred the very successful product line of 'scientific CMOS' (sCMOS) image sensors [38], used in many high-performance applications.

Subjecting the sensor to multiple exposures is one of the most powerful methods for achieving very high DR and does not require novel pixels. Multiple exposure is also the oldest HDR method, used since the days of photographic film. Many techniques based on this method have been invented and implemented in 4T and 5T pixels, such as partial charge transfer ('charge skimming'), multiple sub-frame

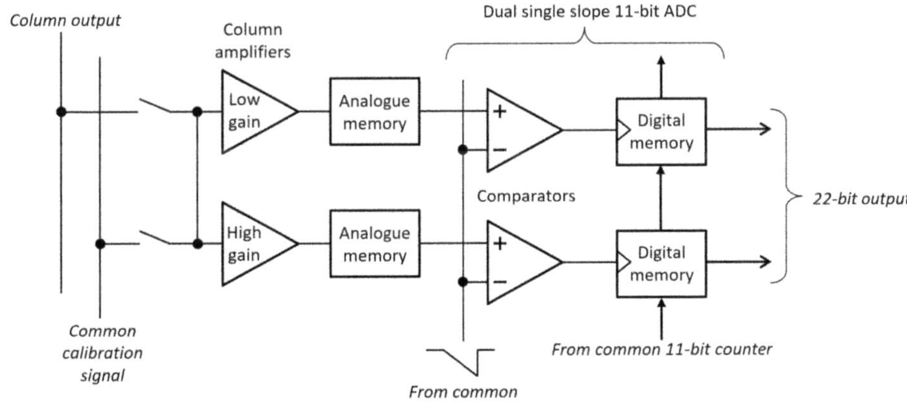

Figure 2.30. Dual gain column amplifier and ADC, based on [37].

readout, and interleaved exposure [39, 40]. A common feature of these techniques is that they are susceptible to motion artifacts due to the exposures happening at different times; this is detrimental in fast moving scenes, in particular in automotive imaging.

Considering pixel architectures capable of HDR, a single exposure is much preferred because the motion artifacts are minimised. Linear response is also preferred; logarithmic pixels [2] can cover light levels in excess of 120 dB, but suffer from slow reaction time and poor temporal and fixed pattern noise.

Further considerations are given to linear pixels only, falling into three main categories:

1. Dual photodiode;
2. Multiple in-pixel gain;
3. Overflow signal storage.

The pixel in figure 2.31 [41] uses two pinned photodiodes with sensitivity ratio of 6.5:1. A transistor switch (DFD) connects the two sense nodes FDL and FDS, and when activated, the conversion gain is reduced due to their combined capacitance. The photodiode SPD is read out with low conversion gain, while LPD can use both gains. The 3-way readout, coupled with a sophisticated ADC and digital periphery results in a DR exceeding 120 dB and readout noise of 0.94 e^- RMS.

HDR methods with dual and triple in-pixel gain, and only one PPD, have demonstrated very good performance too. The pixel in figure 2.32 uses the transistor HDR to add the capacitance of point B to the capacitance of the sense node (point A) for a reduction of the conversion gain. Measuring the signal twice as per the timing diagram with low gain (LG) and high gain (HG), a dynamic range of 87 dB and 2.0 e^- RMS noise have been achieved [42]. With triple gain the DR has been increased to 91 dB [43], and a combination of dual gain and dual exposure has raised it to an impressive 120 dB [44].

Lateral overflow integration capacitor (LOFIC) is one of the most elegant solutions for HDR imaging. Invented in 2005 [45] and further refined [46], the LOFIC pixel works by allowing charge to overflow from the PPD over the transfer

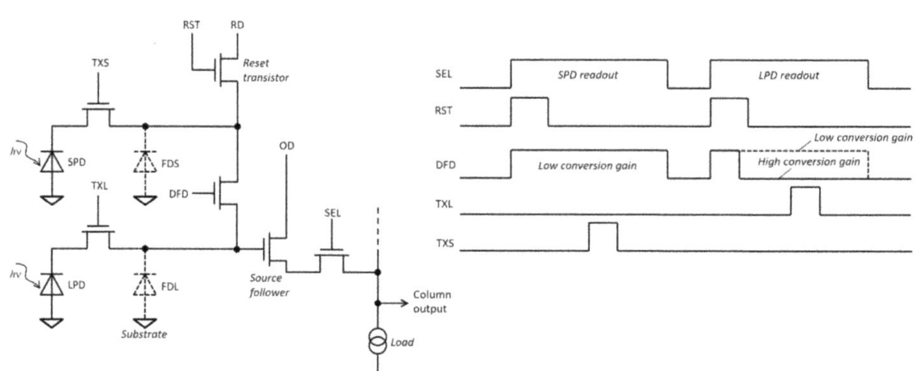

Figure 2.31. Pixel with dual photodiode and multiple gain, based on [41].

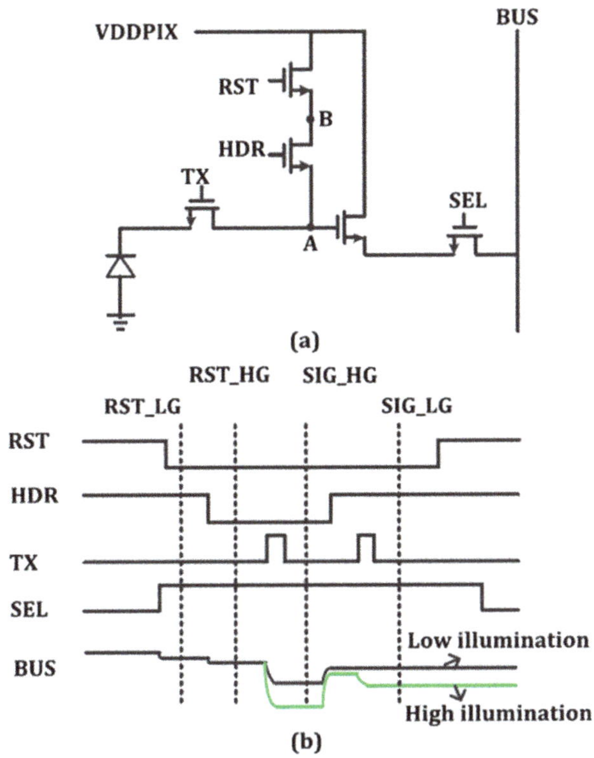

Figure 2.32. HDR pixel using dual gain, realised via a switchable conversion gain. Copyright (2017) IEEE. Reprinted with permission from [42].

gate into a high value signal storage capacitor. In this way, the FWC of the PPD is not a limiting factor, and the DR is greatly increased due to the storage capacitor using thin gate oxide for the highest possible density. The LOFIC takes the design in figure 2.32 one step further and preserves the low sense node capacitance for very high conversion gain, while allowing large signals to be measured too, all in a single exposure.

As shown in figure 2.33, during integration M3 is turned on. Any charge overflowing the barrier created by the transfer gate M1 ends up in the combined capacitance of the sense node FD and the storage capacitor CS at time t_3. For high gain readout M3 is turned off and the remaining PPD charge is transferred to the sense node by pulsing ϕT in t_5. Low gain signal readout is done by turning M3 on again, so that all the charge on FD and CS is sensed.

Small signals do not exceed the FWC of the PPD and are read out similarly to a normal 4T pixel. CDS is accomplished by reading the reset level of the sense node in step t_2. The LOFIC method achieves HDR beyond 100 dB with very low noise, and a two-stage design has been demonstrated with DR of 120 dB [47].

Many more HDR concepts exist, and there is no apparent limit to the ingenuity of the designers. One of the most creative ones uses *both* electron and hole collection, applying different conversion gain to each, and achieving DR of 115 dB in a tiny 3.2 μm pixel [48].

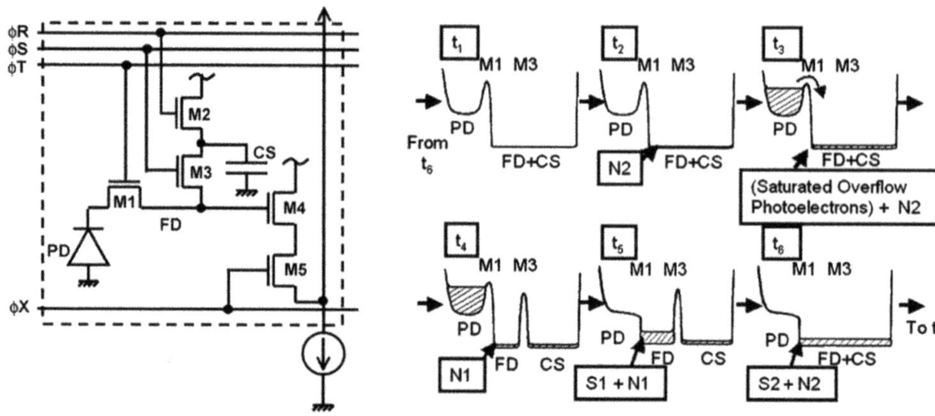

Figure 2.33. LOFIC pixel design and its timing diagram. Copyright (2005) IEEE. Reprinted with permission from [46].

2.5 Hybrid and 3D image sensors

Hybrid imagers are assemblies between a passive pixel array and a CMOS readout integrated circuit (ROIC). 3D imagers take this a step further, integrating several layers of ICs (called tiers) through wafer-scale bonding, with the pixel array tier normally being an active sensor.

The origins of hybrid sensors can be traced to IR imagers using CCDs, long before CMOS technology was invented. The most common hybrid sensor has a 2D array of photodiodes, bump-bonded to an ROIC as shown in figure 2.34 [49]. In this example, every photodiode in the sensor pixel array is connected to a CMOS readout circuit with an indium bump bond on 10 μm pitch.

The two dies are made with different, optimised manufacturing technologies to achieve the desired performance—for example high resistivity bulk silicon for the sensor array and a ROIC made on a 65 nm CMOS process. The sensor array can achieve 100% fill factor because it does not contain any transistors, can also be made very thick (>100 μm) for high QE and be fully depleted by applying appropriate bias. Most importantly, the pixel array can be made from a material other than silicon and be tailored for imaging in the IR [50], visible, UV and x-ray bands [51].

Because the pixel readout is separate, it is no longer limited to using only NMOS transistors as in the standard monolithic imagers. Any CMOS circuit can be employed—for example a PMOS reset transistor for higher sense node voltage, a PMOS source follower for reduced noise, and in-pixel CDS with multiple signal storage for high speed, burst image capture [52]. The most widely used CMOS readout is based on a source follower as the first stage [49], which makes it functionally identical to the monolithic 3T pixel. Source follower readout is preferred because the pixel does not consume power during integration and has low input capacitance. Transimpedance amplifiers are also used and offer increased dynamic range due to the ability to dynamically change the gain depending on the input signal, but at the expense of greater complexity and power dissipation.

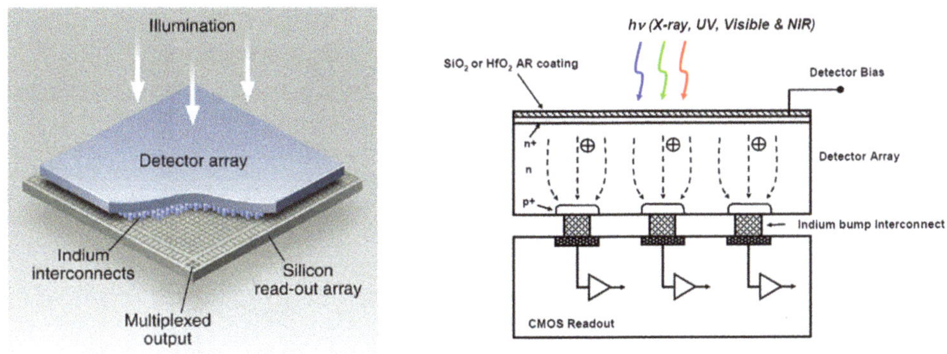

Figure 2.34. Hybrid image sensor architecture. Reprinted with permission from [49].

Among the disadvantages of hybrid pixel sensors are their higher noise and lower conversion gain than monolithic sensors, which is a consequence of the increased capacitance at the sense node. Also, it is difficult to manufacture devices with pitch below 10 μm, and they are more expensive than monolithic sensors due to the complex bump bonding and lower yield.

In the last decade, 3D image sensor integration has made rapid progress [53] to the point that today almost all high-end mobile phones contain stacked image sensors. The typical 3D stacked sensor integrates a pixel tier based on PPDs and at least one digital tier for signal processing. 3D integration between a passive pixel array and readout, similar to hybrid sensors has been shown to work [54], but the main strength of this technology is that the pixel tier can contain the source follower, the reset and the row select transistors associated with the PPD. This ensures high sensitivity and low noise readout because the small sense node capacitance of the PPD is preserved. The connections between the tiers are made with through silicon vias (TSV) at their periphery. Using a 65 nm PPD CMOS process for the image sensor tier and a 14 nm FinFET process for the digital tier, a 12 million pixel, 1.4 μm pitch sensor has been fabricated [55].

In another development [9], three tiers—image sensor, DRAM and logic are stacked using TSVs (figure 2.35). This 3D integration between pixels and digital circuits creates image sensors with powerful processing capabilities, far beyond what can be achieved in single die imagers.

TSV connections can only be used in the periphery of the chips because they cannot go through active circuits, and their pitch is far larger than the pixel's. Interconnects over the pixel area allow many more connections to be made, including to individual pixels, and offer yet another step-up in integration and performance. Copper connections, named Cu–Cu direct bonding, have been demonstrated on 3 μm pitch [9]. Using this technology, two connections per pixel between the PPD-based image sensor and logic tiers have been made [56]. This has allowed a differential amplifier instead of a source follower buffering the sense node, and an ADC in each pixel.

Besides their complexities, the economies realised from wafer-scale bonding and mass production have made 3D stacked sensors much more widespread and affordable than traditional hybrid sensors.

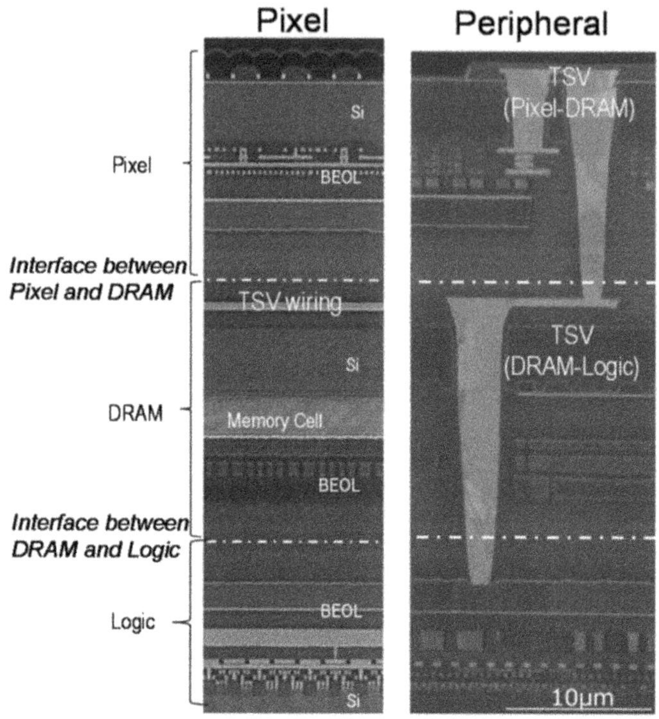

Figure 2.35. 3D stacked image sensor with 3 tiers. Copyright (2019) IEEE. Reprinted with permission from [9].

Chapter summary

1. Photodiode (3T) pixel is the basis for most APS. It uses a source follower to buffer the photodiode, a reset transistor to periodically connect the photodiode to a fixed voltage, and row select transistor to enable the output. It is rarely used today due to difficulties in suppressing the reset noise.
2. In the pinned photodiode the charge is stored in a potential well away from any Si–SiO_2 interface. The pinning voltage depends only on the doping profiles used.
3. The PPD has large charge storage capacity due to the high effective capacitance of the *pn* junction between the diode and the pinning implant.
4. Due to the separation of the functions of charge collection and charge-to-voltage conversion, the sense node in the PPD can be very small, and high conversion gain can be obtained.
5. Effective CDS can be employed to suppress the reset noise in PPDs. When used together with a high conversion gain, the readout noise can be very low.
6. Charge transfer in PPDs is relatively slow due to the lack of lateral electric field. Image lag can arise due to potential barriers and pockets in the charge transfer path.
7. 5T pixels add another transfer gate to the PPD to provide global reset, used in global shutter mode readout.

8. The PPD can be used as an intermediate signal storage to allow global image capture and global shutter operation with CDS.
9. HDR pixels use multiple conversion gains, multiple photodiodes, or charge overflow storage to increase the dynamic range of the PPD.
10. Hybrid image sensors combine a passive pixel array with a CMOS readout chip using a bump-bonded assembly.
11. 3D stacked image sensors typically integrate a PPD-based active pixel array with one or more CMOS 'tiers' for much greater functionality and performance.

References

[1] Fossum E 1995 CMOS image sensors: electronic camera on a chip *Proc. Int. Electron Devices Meeting* 17–25

[2] Fossum E 1997 CMOS image sensors: electronic camera-on-a-chip *IEEE Trans. Electron Devices* **44** 1689–98

[3] Noble P 1968 Self-scanned silicon image detector arrays *IEEE Trans. Electron Devices* **15** 202–9

[4] Weckler G 1967 Operation of p-n junction photodetectors in a photon flux integrating mode *IEEE J. Solid-State Circuits* **2** 65–73

[5] Dyck R and Weckler G 1968 Integrated arrays of silicon photodetectors for image sensing *IEEE Trans. Electron Devices* **15** 196–201

[6] Teranishi N, Kohno A, Ishihara Y, Oda E and Arai K 1982 No image lag photodiode structure in the interline CCD image sensor *1982 Int. Electron Devices Meeting (San Francisco, CA, USA)* 324–7

[7] Teranishi N, Kohno A, Ishihara Y, Oda E and Arai K 1984 An interline CCD image sensor with reduced image lag *IEEE Trans. Electron Devices* **31** 1829–33

[8] Lee P, Gee R, Guidash R, Lee T-H and Fossum E 1995 An active pixel sensor fabricated using CMOS/CCD process technology *1995 IEEE Workshop on Charge-Coupled Devices (Dana Point, CA, USA)*

[9] Kagawa Y and Iwamoto H 2019 3D integration technologies for the stacked CMOS image sensors *Int. 3D Systems Integration Conf. (3DIC) (Sendai)*

[10] Allen P E and Holberg D R 2012 *CMOS Analog Circuit Design* (Oxford: Oxford University Press)

[11] Teranishi N 2016 Analysis of subthreshold current reset noise in image sensors *Sensors* **16** 663

[12] Fowler A M and Gatley I 1991 Noise reduction strategy for hybrid IR focal-plane arrays *Proc. of SPIE 1541 (San Diego)*

[13] Mendis S K *et al* 1997 CMOS active pixel image sensors for highly integrated imaging systems *IEEE J. Solid-State Circuits* **32** 187–97

[14] Teranishi N 2016 Effect and limitation of pinned photodiode *IEEE Trans. Electron Devices* **63** 10–5

[15] Fossum E R and Hondongwa D B 2014 A review of the pinned photodiode for CCD and CMOS image sensors *IEEE J. Electron Devices Soc.* **2** 33–43

[16] Sze S 1981 *Physics of Semiconductor Devices* 2nd edn (New York: Wiley)

[17] Kawai S, Mutoh N and Teranishi N 1997 Thermionic-emission-based barrier height analysis for precise estimation of charge handling capacity in CCD registers *IEEE Trans. Electron Devices* **44** 1588–92

[18] Pelamatti A, Goiffon V, Estribeau M, Cervantes P and Magnan P 2013 Estimation and modeling of the full well capacity in pinned photodiode CMOS image sensors *IEEE Electron Device Lett.* **34** 900–2

[19] Pelamatti A *et al* 2015 Temperature dependence and dynamic behavior of full well capacity in pinned photodiode CMOS image sensors *IEEE Trans. Electron Devices* **62** 1200–7

[20] Khan U and Sarkar M 2018 Dynamic capacitance model of a pinned photodiode in CMOS image sensors *IEEE Trans. Electron Devices* **65** 2892–8

[21] Sarkar M, Buttgen B and Theuwissen A J P 2013 Feedforward effect in standard CMOS pinned photodiodes *IEEE Trans. Electron Devices* **60** 1154–61

[22] Bonjour L, Blanc N and Kayal M 2012 Experimental analysis of lag sources in pinned photodiodes *IEEE Electron Device Lett.* **33** 1735–7

[23] Rizzolo S, Goiffon V, Estribeau M, Marcelot O, Martin-Gonthier P and Magnan P 2018 Influence of pixel design on charge transfer performances in CMOS image sensors *IEEE Trans. Electron Devices* **65** 1048–55

[24] Han L, Yao S and Theuwissen A J P 2016 A charge transfer model for CMOS image sensors *IEEE Trans. Electron Devices* **63** 32–41

[25] Xu J, Wang R, Han L and Gao Z 2020 Analysis and modeling of spill back effect in high illumination CMOS image sensors *IEEE Sensors J.* **20** 3024–31

[26] Cao Z, Zhou Y, Li Q, Qin Q, Liu L and Wu N 2013 Design of pixel for high speed CMOS image sensors *Int. Image Sensor Workshop* 7.11

[27] Shin B, Park S and Shin H 2010 The effect of photodiode shape on charge transfer in CMOS image sensors *Solid-State Electron.* **54** 1416–20

[28] Miyauchi K *et al* 2014 Pixel structure with 10 nsec fully charge transfer time for the 20 m frame per second burst CMOS image sensor *Proc. SPIE* 9022

[29] Cao X *et al* 2015 Design and optimisation of large 4T pixel *Int. Image Sensor Workshop*

[30] Janesick J, Andrews J, Tower J, Grygon M, Elliott T, Cheng J, Lesser M and Pinter J 2007 Fundamental performance differences between CMOS and CCD imagers: part II *Proc. of SPIE (San Diego, CA)*

[31] Wang X *et al* 2010 A 2.2M CMOS image sensor for high-speed machine vision applications *Proc. of SPIE 7536 (San Jose)*

[32] Miyauchi K, Mori K, Otaka T, Isozaki T, Yasuda N, Tsai A, Sawai Y, Owada H, Takayanagi I and Nakamura J 2020 A stacked back side-illuminated voltage domain global shutter CMOS image sensor with a 4.0 µm multiple gain readout pixel *MDPI Sensors* **20** 486

[33] Velichko S *et al* 2016 CMOS global shutter charge storage pixels with improved performance *IEEE Trans. Electron Devices* **63** 106–12

[34] Yasutomi K, Itoh S and Kawahito S 2011 A two-stage charge transfer active pixel CMOS image sensor with low-noise global shuttering and a dual-shuttering mode *IEEE Trans. Electron Devices* **58** 740–7

[35] Boukhayma A, Peizerat A and Enz C 2016 A sub-0.5 electron read noise VGA image sensor in a standard CMOS process *IEEE J. Solid-State Circuits* **51** 2180–91

[36] Seo M, Kawahito S, Kagawa K and Yasutomi K 2015 A 0.27e-RMS read noise 220-µV/e-conversion gain reset-gate-less CMOS image sensor with 0.11-µm CIS process *IEEE Electron Device Lett.* **36** 1344–7

[37] Vu P *et al* 2011 Low noise high dynamic range 2.3Mpixel CMOS image sensor capable of 100 Hz frame rate at full HD resolution *Int. Image Sensor Workshop (Hokkaido)*

[38] CMOS sensors (https://fairchildimaging.com/products/scmos-sensors)

[39] Solhusvik J *et al* 2013 A comparison of high dynamic range CIS technologies for automotive applications *Int. Image Sensor Workshop (Snowbird, UT)*

[40] Kabir S, Guidash M, Vogelsang T, Smith C, Schneider A and Endsley J 2017 A small pixel high performance full frame HDR sensor *Int. Image Sensor Workshop (Hiroshima)*

[41] Willassen T *et al* 2015 A 1280×1080 4.2 μm split-diode pixel HDR sensor in 110 nm BSI CMOS process *Int. Image Sensor Workshop (Vaals, The Netherlands)*

[42] Ma C, Liu Y, Li Y, Zhou Q, Wang X and Chang Y 2017 A 4-M pixel high dynamic range, low-noise CMOS image sensor with low-power counting ADC *IEEE Trans. Electron Devices* **64** 3199–205

[43] Tanaka S *et al* 2018 Single exposure type wide dynamic range CMOS image sensor with enhanced NIR sensitivity *ITE Trans. MTA* **6** 195–201

[44] Solhusvik J *et al* 2017 A 1392x976 2.8 μm 120 dB CIS with per-pixel controlled conversion gain *Int. Image Sensor Workshop (Hiroshima)*

[45] Sugawa S, Akahane N, Adachi S, Mori K, Ishiuchi T and Mizobuchi K 2005 A 100 dB dynamic range CMOS image sensor using a lateral overflow integration capacitor *IEEE Int. Solid-State Circuits Conf.*

[46] Akahane N, Sugawa S, Adachi S, Ishiuchi M K T and Mizobuchi K 2006 A sensitivity and linearity improvement of a 100-db dynamic range CMOS image sensor using a lateral overflow integration capacitor *IEEE J. Solid-State Circuits* **41** 851–8

[47] Fujihara Y, Murata M, Nakayama S, Kuroda R and Sugawa S 2021 An over 120 dB single exposure wide dynamic range CMOS image sensor with two-stage lateral overflow integration capacitor *IEEE Trans. Electron Devices* **68** 152–7

[48] Lalanne F, Malinge P, Hérault D and Jamin-Mornet C 2017 A native HDR 115 dB 3.2 μm BSI pixel using electron and hole collection *Int. Image Sensor Workshop (Hiroshima)*

[49] Bai Y *et al* 2012 4K×4K format 10-micron pixel pitch H4RG-10 hybrid CMOS silicon visible focal plane array for space astronomy *Proc. of SPIE 84530M (Amsterdam)*

[50] Garnett J *et al* 2004 2K×2K molecular beam epitaxy HgCdTe detectors for the James Webb Space Telescope NIRCam instrument *Proc. of SPIE 5499 (Glasgow)*

[51] Bai Y *et al* 2008 Teledyne imaging sensors: silicon CMOS imaging technologies for x-ray, UV, visible, and near infrared *Proc. of SPIE 702 102 (Marseille)*

[52] Douence V *et al* 2005 Hybrid image sensor with multiple on-chip frame storage for ultrahigh-speed imaging *Proc. of SPIE 5580 (Alexandria, VA)*

[53] Gove R J 2020 CMOS image sensor technology advances for mobile devices *High Performance Silicon Imaging* ed D Durini 2nd edn (Cambridge: Woodhead Publishing) pp 185–240

[54] Suntharalingam V *et al* 2005 Megapixel CMOS image sensor fabricated in three-dimensional integrated circuit technology *2005 IEEE Int. Digest of Technical Papers. Solid-State Circuits Conf.*

[55] Kwon M *et al* 2020 A low-power 65/14 nm stacked CMOS image sensor *2020 IEEE Int. Symp. on Circuits and Systems (ISCAS)*

[56] Miura T *et al* 2019 A 6.9 μm pixel-pitch 3D stacked global shutter CMOS image sensor with 3M Cu-Cu connections *2019 Int. 3D Systems Integration Conf. (3DIC) (Sendai)*

IOP Publishing

CMOS Image Sensors

Konstantin D Stefanov

Chapter 3

Advanced image sensor topics

3.1 Photocurrent

In chapter 1 we looked into the current caused by the electron–hole pairs generated by light. Assuming that the light is fully absorbed without reflections and all electrons are collected, the photocurrent can be simply determined by the number of e–h pairs generated per second. In practical photosensitive devices such as *pn* junctions, PPDs, and MOS capacitors full light absorption is not guaranteed because of their finite thickness and light reflections off the front surface. Complete charge collection is also not certain because some carriers may not reach the electrodes or could recombine.

In this section we will consider the photocurrent in a *pn* junction used as a photodiode because this is applicable to many other devices and configurations. In image sensors photodiodes are usually operated in integrating mode, after they have been reverse-biased and disconnected. The generated electrons do not leave the photodiode[1], but discharge its capacitance instead, and reduce the voltage across it. This is the case for 3T pixels with the reset transistor off, and for also 4T pixels, which do not have any electrodes connected to the PPD. The *pn* junction model in figure 3.1 is shown with a bias electrode to the *n*-side, through which a constant voltage is applied. This is just for illustration; even if the bias electrode is floating and the voltage on it decreases under illumination, we will show that the photocurrent has negligible bias dependence.

Electron–hole pairs are generated everywhere in silicon, regardless of the concentration of the dopants or the free carriers. Pairs created in the depleted region with depth W are separated and collected quickly as drift current, while those in neutral regions must diffuse out before collection, and give rise to the diffusion current.

[1] But the holes do, via the substrate and the ground connection.

doi:10.1088/978-0-7503-3235-4ch3

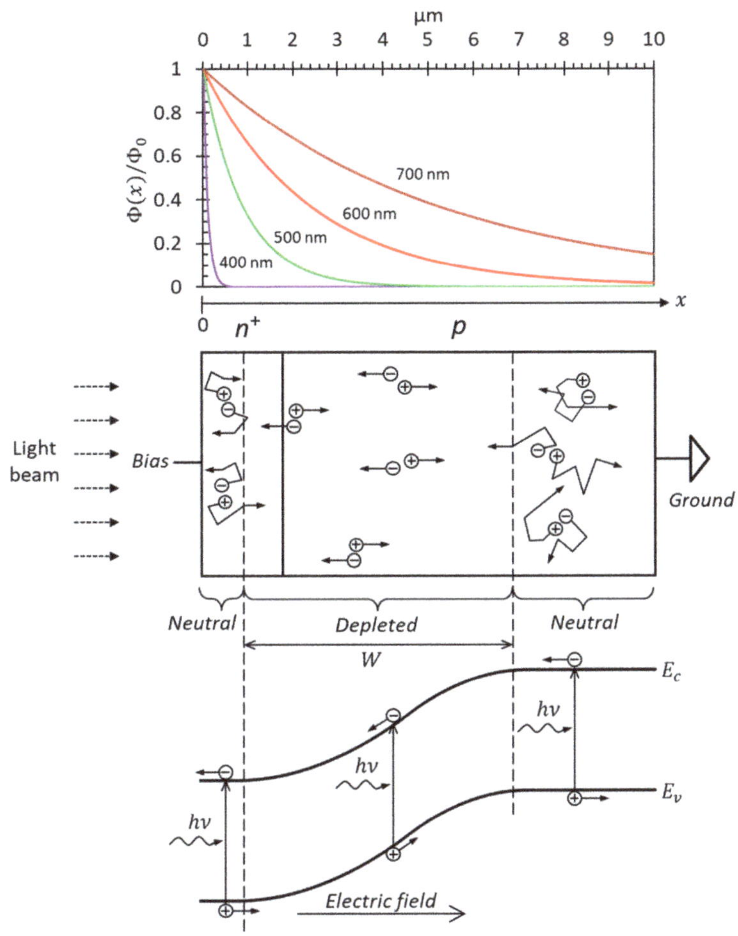

Figure 3.1. Absorption length of light and photocurrent in a *pn* junction.

As stated in chapter 1, the incoming light flux Φ_0 (number of photons per unit area per second) hitting the surface of the photodiode in figure 3.1 decreases according to the Beer–Lambert law $\Phi(x) = \Phi_0 e^{-\alpha x}$, where α is the absorption coefficient and x is the depth. The decrease of the photon flux is due to absorption and generation of e–h pairs. Since every absorbed photon generates one pair, we can say that the rate of *decrease* of the photon flux equals the rate of *increase* of the e–h pairs' concentration. The rate of change of the photon flux at depth x is

$$\frac{d\Phi(x)}{dx} = -\alpha\Phi_0 e^{-\alpha x} \tag{3.1}$$

Therefore, the e–h generation rate $G(x)$ (the number of generated carriers per cm^3, per second) is

$$G(x) = -\frac{d\Phi(x)}{dx} = -\alpha\Phi_0 e^{-\alpha x} \tag{3.2}$$

The drift current density J_{dr} can be found by integrating (3.2) over the depletion depth W. The pn junction in figure 3.1 is asymmetric with much more heavily doped n-side, therefore the depletion extends predominantly on the p-side. For simplicity, we can take that the n-side is very thin, the depletion starts at $x = 0$, and can obtain

$$J_{dr} = q\int_0^W G(x)dx = q\Phi_0(1 - e^{-\alpha W}) \tag{3.3}$$

If W is much larger than the absorption depth $1/\alpha$, the exponential term in (3.3) becomes very small and can be ignored, therefore the drift current would depend only on the photon flux.

In figure 3.1 the thickness of the n-side of the junction is grossly exaggerated to illustrate how the position of e–h pair generation strongly depends on the wavelength. Violet light ($\lambda = 400$ nm) creates almost all carriers in the first 0.2 μm, which is entirely contained in the neutral n-type silicon. On the other hand, deep red ($\lambda = 700$ nm) is not fully absorbed in the 10 μm depth and creates e–h pairs everywhere.

Electron–hole pairs generated in the neutral regions diffuse out until some reach the depletion and are collected. This creates a diffusion current with density proportional to the gradient of the electron concentration n_p at the edge of the depletion

$$J_{\mathrm{diff}} = qD_n\left(\frac{\partial n_p}{\partial x}\right)_{x=W} \tag{3.4}$$

As with the depletion current, we consider the n-side to be very thin and contributing negligibly. The gradient of n_p can be found from the diffusion equation (3.5) in steady-state condition for the neutral p-type semiconductor, balancing three terms: diffusion, recombination, and optical generation from (3.2):

$$D_n\frac{\partial^2 n_p}{\partial x^2} - \frac{n_p - n_{p0}}{\tau_n} + G(x) = 0 \tag{3.5}$$

Here D_n is the diffusion coefficient, τ_n is the electron lifetime and n_{p0} is the electron concentration in the p-type neutral silicon. Using the solution in [1] (p 756) with conditions $n_p = n_{p0}$ for $x \to \infty$ (thick neutral semiconductor), and $n_p = 0$ for $x = W$ (zero electron concentration at the edge of the depletion), and substituting in (3.4), the diffusion current becomes

$$J_{\mathrm{diff}} = q\Phi_0\left(\frac{\alpha L_n}{1 + \alpha L_n}\right)e^{-\alpha W} + qn_{p0}\frac{D_n}{L_n} \tag{3.6}$$

where $L_n = \sqrt{D_n\tau_n}$ is the electron diffusion length. The total photocurrent density is the sum of (3.3) and (3.6)

$$J_{\mathrm{ph}} = J_{\mathrm{dr}} + J_{\mathrm{diff}} = q\Phi_0\left(1 - \frac{e^{-\alpha W}}{1 + \alpha L_n}\right) + qn_{p0}\frac{D_n}{L_n} \tag{3.7}$$

The second term in (3.7) does not depend on the illumination. We will see in section 3.2.3 that this is the diffusion dark current (3.26), which is usually negligibly small. Therefore, the photocurrent density becomes

$$J_{ph} = q\Phi_0\left(1 - \frac{e^{-\alpha W}}{1 + \alpha L_n}\right) \tag{3.8}$$

The term in brackets multiplies the maximum possible current $J_{ph} = q\Phi_0$ and is a measure of the photodiode's efficiency η, called quantum efficiency (QE) [1]. So far, we have not included light losses due to reflections, but if the fraction of reflected light R is added to (3.8), the QE becomes [1]

$$\eta = (1 - R)\left(1 - \frac{e^{-\alpha W}}{1 + \alpha L_n}\right) \tag{3.9}$$

For the longest wavelength in figure 3.1 (λ = 700 nm) the absorption coefficient is 1900 cm^{-1}, corresponding to an absorption length $1/\alpha$ = 5.26 μm (see table 1.1). With $W \approx$ 6 μm, the exponential term in (3.8) is 0.34, which on its own would significantly reduce the photocurrent. However, in p-type silicon with doping $N_A < 10^{15}$ cm^{-3}, typical for epitaxial material, the diffusion length is hundreds of micrometres because the electron lifetime is in the millisecond region [2]. This makes the denominator $1 + \alpha L_n$ sufficiently large, so that the photocurrent is very close to the maximum $q\Phi_0$. Therefore, in thick photodiodes all of the charge can be collected by diffusion even if the depletion is small, provided that L_n is large. Despite this, it is a good idea to minimize the field-free regions so that most of the charge is collected by drift. Diffusion is slow and reduces the spatial resolution of a sensor because charge generated in adjacent pixels would mix, which is best avoided if we want a sharp image.

This conclusion is the consequence of the derivation of (3.8), which assumed that the p-side of the junction is very thick ($x \to \infty$). In practice, this is not true because image sensors are built on epitaxial silicon wafers with thickness t_{epi} in the range between 5 and 20 μm. The epi layer is grown on a heavily doped p^{++} type substrate with thickness around 700 μm. Figure 3.2 shows a photodiode built on such an epitaxial wafer. The n^+ implant for the cathode is very shallow to minimise the field-free region inside.

The substrate has a resistivity around 0.01 Ω.cm, corresponding to a doping concentration of $\approx 10^{19}$ cm^{-3}. At such high doping concentrations the electron lifetime is much shorter (below 10^{-7} s [2]), and the diffusion length is barely around a micron. This means that most of the photoelectrons generated in the substrate are lost. The situation in the n^+ cathode is much better because it is usually very shallow (0.1 μm in figure 3.2) and much smaller than the diffusion length for electrons, so little charge is lost due to recombination.

In the photodiode in figure 3.2 the electrons generated in the epitaxial layer are fully collected due to the combination of drift and efficient diffusion from the neutral silicon. Most of the electrons generated in the substrate are lost, and this reduces the

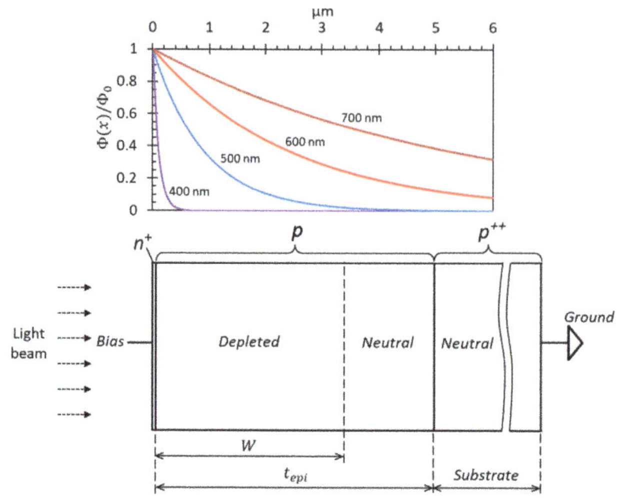

Figure 3.2. Light absorption in a *pn* junction built on an epitaxial silicon with highly doped p^{++} substrate and thin n^{+} cathode.

QE at wavelengths longer than about 600 nm. This rather simple model allows us to calculate the QE directly from the Beer–Lambert law. The QE is reduced by the fraction of light not absorbed at distance t_{epi}, or mathematically

$$\eta = (1 - R)\left[1 - \exp(-\alpha t_{\mathrm{epi}})\right] \tag{3.10}$$

Example 3.1. Calculate the QE of the photodiode in figure 3.2 for $\lambda = 600$ nm ($\alpha = 4140\ \mathrm{cm}^{-1}$), $\lambda = 700$ nm ($\alpha = 1900\ \mathrm{cm}^{-1}$) and $t_{\mathrm{epi}} = 5\ \mu\mathrm{m}$, assuming no light reflections ($R = 0$).
Solution: From (3.10) we obtain for $\lambda = 600$ nm

$$\eta = 1 - \exp(-4140 \times 5 \times 10^{-4}) = 0.874$$

and for $\lambda = 700$ nm

$$\eta = 1 - \exp(-1900 \times 5 \times 10^{-4}) = 0.613$$

The structure in figure 3.2 is very similar to the PPD despite not using the same implant types. Instead of the shallow n^{+} implant for the cathode, the PPD has the p^{+} pinning implant of similar depth. The depletion includes the n-type PPD implant and a part of the epitaxial layer. Since it is depleted, it does not matter what the doping type inside is. The neutral part of the epitaxial layer and the substrate are the same. Therefore, the same considerations for the photocurrent apply for both the *pn* junction-based photodiode and the PPD.

3.2 Dark current

3.2.1 Sources of dark current

Dark current, as the name implies, creates signal in the sensor in total darkness. The term 'dark current' is normally reserved for the fundamental causes which are difficult to circumvent, such as the defects in the bulk semiconductor and at interfaces, or the intrinsic carrier concentration. More peculiar but largely preventable causes of dark signal include things like transistor glow [3], light leaks (e.g. from an imperfect dark enclosure), or embarrassing ones such as a forgotten LED shining on the sensor—we are not going to consider those.

Dark current is measured most often in units of electrons per pixel per second (e^- pixel^{-1} s^{-1}), which is very convenient for most uses. To be able to compare sensors with different pixel sizes the dark current is often expressed as area density in units of current per square centimetre (A cm^{-2}). Since the pixel size is eliminated, the current density can be used as a characteristic of the process technology and not a given sensor, provided that the same thickness of active silicon is used. The current density can also be expressed in units of e^- cm^{-2} s^{-1}, but this is not normally used because the numbers become a bit cumbersome.

Dark current depends exponentially on the temperature, therefore stating a number without the temperature at which it was measured can be meaningless. The term 'room temperature' should also be used with care because in different parts of the world the normal room temperature can be quite different. If the range for 'room temperature' is within 18 °C to 27 °C, the dark current can be different by more than a factor of two. Most often either 20 °C (293 K) or 27 °C (300 K) are used instead.

Dark current is inescapable but can be made very small by cooling. This is routinely done in astronomy for long observations, or in life sciences where the signal is tiny and even small dark signal would interfere with the results. Astronomical sensors are typically CCDs and can be cryogenically cooled to achieve extremely low dark current. For example, the CCDs in the WFC3 in the Hubble Space Telescope have dark current of only 2 electrons per pixel *per hour* at −83 °C [4], before any radiation damage kicks in. CMOS sensors should be able to deliver similarly low dark current too, but not much data is available below −45 °C and the performance at lower temperatures is yet to be convincingly demonstrated. A recent example has 0.01 e^- pixel^{-1} s^{-1} (36 e^- pixel^{-1} h^{-1}) at −40 °C [5].

Bulk and surface defects in silicon are considered the main source of dark current. Impurities, point defects [6], defect clusters and the crystal mismatch at Si–SiO_2 interface create electrically active[2] centres in the bandgap, called generation–recombination (GR) centres (centres for short). They are also called 'traps' due to their ability to capture and emit free electrons or holes. Their behaviour is described by the Shockley–Read–Hall (SRH) theory, as already used in chapter 1 for the carrier recombination.

Trap-assisted thermal generation and recombination, shown in figure 3.3, creates electron–hole pairs in a two-step process. First, an electron must transition from the

[2] Not every impurity or defect is electrically active.

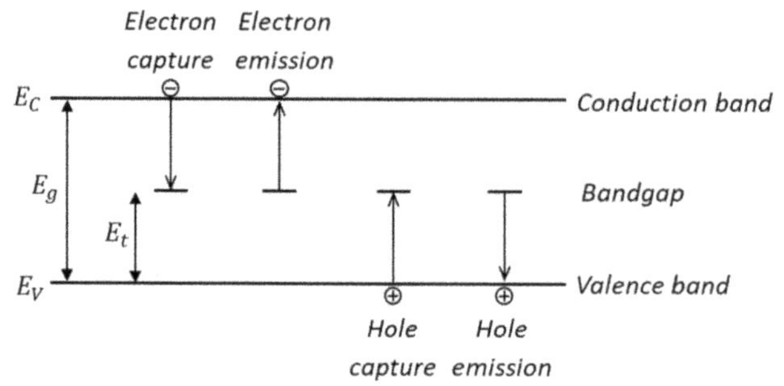

Figure 3.3. Carrier recombination and generation through inter-bandgap centres.

valence band to the trap, which is the same as hole emission from the trap to the valence band. Next, the trapped electron is emitted to the conduction band. Hole capture means that the valence band has lost a hole; this is equivalent to an electron jumping from the trap to the valence band and recombining. Hole emission is the same as electron transitioning from the valence band to the trap; the valence band gains a hole and the trap becomes negatively charged. Traps can capture only one electron or hole; therefore the trap must emit the carrier it is holding in order to capture another.

If the newly generated electron–hole pair is not re-captured and the electron reaches a charge collection element it will show as dark current. Dark current generated by traps and described by the SRH theory is called GR dark current. The GR mechanism operates at near zero free electron and hole concentrations which is a characteristic of a depleted semiconductor; therefore, it is also called *depletion dark current*.

Traps at the Si–SiO_2 interface are called interface states, or 'states' for short, and cause dark by the same GR mechanism as the depletion dark current. However, this *surface dark current* is treated separately for two reasons: (a) its generation is proportional to the surface area and not the volume of the semiconductor; and (b) the interface states have continuous distribution in the bandgap, unlike bulk traps which have discrete levels. Very often, the terms states, traps and centres are used interchangeably.

Traps are not the only way to generate dark current; it can exist even in perfect silicon, free of any bulk or surface traps. This is because electron–hole pairs are continually generated thermally in any semiconductor; in ultra-pure, undoped silicon the natural concentration of electrons and holes is $n_i = 1.45 \times 10^{10}$ cm^{-3} at 300 K. The carriers constantly appear and recombine in a state of thermal equilibrium, without any traps being present. If some of these carriers diffuse out to a region where there is an electric field, they can be collected as *diffusion dark current*.

The total dark current I_d is the sum of the depletion, diffusion and surface components:

$$I_d = I_{dep} + I_{diff} + I_s \tag{3.11}$$

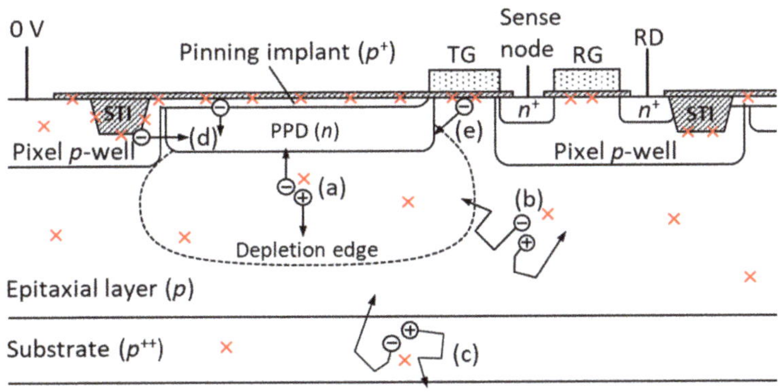

Figure 3.4. Dark current sources in a PPD (after [7]): (a) depletion; (b) diffusion in the epitaxial layer; (c) diffusion from the substrate; (d) surface in pinning; (e) non-pinned surface. Interface and bulk traps are indicated by red crosses.

Figure 3.4 shows a diagram of a PPD and the location of the different sources of dark current. Bulk and surface traps are assumed to be uniformly distributed over the epitaxial layer or at the Si–SiO_2 interface, correspondingly. Bulk dark current is generated within the PPD depletion (a); in the neutral regions of the epitaxial layer (b); and in the substrate (c) by diffusion. Surface dark current is generated at the Si–SiO_2 interfaces above the pinning implant and the shallow trench insulation (STI) (d) and under the transfer gate (e).

The surface dark current depends strongly on the free carrier concentration at the Si–SiO_2 interface and this is the reason why the dark current from the oxide under the transfer gate is considered separately from the oxides at the heavily doped p-well and the pinning implant. The surface dark current density is typically in the range of several nA cm^{-2} at room temperature for depleted interfaces, but as we will see in section 3.2.5, in the PPD it is heavily suppressed. Combined with the very low bulk dark current in high quality epitaxial layers, the dark current in PPDs is regularly in the pA cm^{-2} range at room temperature. The dark current is also often specified at 60 °C due to applications at elevated temperatures, such as automotive and industrial.

Without additional measures, the surface dark current in image sensors can be dominant. This is why special attention is paid on improving the quality and minimising the area of Si–SiO_2 interfaces. For example, the STI area can be reduced [8] or the STI eliminated altogether to achieve dark current density below 30 pA cm^{-2} at 60 °C [9], which is below 1 pA cm^{-2} at 20 °C.

Example 3.2. A PPD image sensor has dark current density J_d of 1 pA cm^{-2} and 5 μm square pixels. What is the dark current per pixel in electrons per second?
Solution: To get the answer we multiply the dark current density by the pixel area and divide by the elementary charge:

$$I_d = \frac{10^{-12} \times 25 \times 10^{-8}}{1.6 \times 10^{-19}} = 1.56\ \mathrm{e^-/s}$$

Presented in a different way, one electron comes on average every 640 milliseconds. Such low, discrete currents make one wonder how semiconductor theory can explain the observed phenomena despite that it deals with carrier densities and not with discrete charges. It is not difficult to see that the same current density in a 1 μm square pixel would give 0.0625 e^-/ pixel/s, or one electron every 16 seconds!

The dark current is very often non-uniform, with some pixels exhibiting much higher dark current than the average for the array. They are called 'hot pixels' and can be seen as a 'star field' in a dark image.

3.2.2 Depletion dark current

In the depletion region the concentration of electrons and holes is vanishingly small. Only carrier emission is of relevance here and the carrier capture process is negligible because there is almost nothing to capture.

Under the approximation that the trap cross-sections are the same for electrons and holes ($\sigma_p = \sigma_n = \sigma$) the SRH recombination rate (from chapter 1) can be written as

$$U = \frac{\sigma v_{\mathrm{th}} N_t (pn - n_i^2)}{p + n + 2n_i \cosh\left(\frac{E_t - E_i}{kT}\right)} \tag{3.12}$$

Here we assume a single trap type with concentration N_t at energy position E_t above the valence band. In depletion $p \approx 0$ and $n \approx 0$ and therefore

$$U = -\frac{\sigma v_{\mathrm{th}} N_t n_i}{2\cosh\left(\frac{E_t - E_i}{kT}\right)} = -\frac{n_i}{\tau_g} \tag{3.13}$$

In (3.13) the recombination rate is negative, meaning that the carrier concentration is increasing. We have here *generation* of electron–hole pairs instead of recombination.

The term dividing n_i has dimension of time and is called *generation lifetime* τ_g, similarly to the recombination lifetime described in chapter 1. The generation lifetime (3.14) is the effective carrier lifetime for both electrons and holes in a depletion region. Because their concentrations are nearly zero, the intrinsic carrier concentration n_i is in the numerator of (3.13) instead of the excess concentration.

$$\tau_g = \frac{2}{\sigma v_{\mathrm{th}} N_t} \cosh\left(\frac{E_t - E_i}{kT}\right) \tag{3.14}$$

Mid-band traps are the most effective in generating dark current due to the denominator in (3.13) reaching minimum for $E_t = E_i$. The effect of traps located away from the mid-band exponentially decreases as a function of their energy position. For traps exactly at the mid-band the recombination rate is at its maximum U_{max}:

$$U_{max} = -\frac{\sigma v_{th} N_t n_i}{2} \tag{3.15}$$

The maximum rate corresponds to the minimum generation lifetime, equal to twice the recombination time:

$$\tau_g^{min} = \frac{2}{\sigma v_{th} N_t} = 2\tau_n \tag{3.16}$$

Once generated, the electron–hole pairs are quickly swept away by the electric field in the depletion region. The result of that is dark current.

Often, the dark current is expressed as current density (current per photodiode area, or per pixel area) to eliminate the dependence on the size, and to allow easier comparison between designs and technologies. The GR current through a depletion region with depth W and area A is the charge generated in the volume per unit time, and is equal to the generation rate (3.13) multiplied by the region's volume AW. Since all the carriers are collected as dark current, the current density J_d is then this product divided by the area A, which for a single mid-band trap type becomes

$$J_{dep} = q\,|U_{max}|\,W = \frac{q n_i W}{\tau_g} = \frac{q\sigma v_{th} N_t n_i W}{2} \tag{3.17}$$

If there is more than one trap type, the current densities for each are simply added together. It is useful to put some numbers in (3.17) to see what dark currents can be expected.

Example 3.3. Calculate the dark current in a photodiode with area $A = 25\ \mu m^2$ and depletion depth $W = 5\ \mu m$. The dark current is caused by a single mid-band trap type with $N_t = 10^{10}\ cm^{-3}$ and $\sigma = 10^{-15}\ cm^2$. Use $v_{th} = 10^7\ cm\ s^{-1}$ and $n_i = 1.45 \times 10^{10}\ cm^{-3}$ at 300 K.

Solution: First we use (3.17) to calculate the current density and then we are going to scale it by the diode area.

$$J_{dep} = \frac{1.6 \times 10^{-19} \times 10^{-15} \times 10^7 \times 10^{10} \times 1.45 \times 10^{10} \times 5 \times 10^{-4}}{2} = 58\ pAcm^{-2}$$

This current density is high compared to most PPDs, which can have dark current density below 1 pA cm^{-2} at 300 K. The current in the photodiode is

$$I_d = J_d A = 58 \times 10^{-12} \times 25 \times 10^{-8} = 0.015\ fA$$

or 90.6 electrons per second. Interestingly, in the whole of the depletion region (a cube with 5 μm side) there are only $AWN_t = 25 \times 10^{-8} \times 5 \times 10^{-4} \times 10^{10} = 1.25$ traps!

Since $v_{\rm th} \propto T^{1/2}$ and $n_i \propto T^{3/2}\exp\left(-\frac{E_{\rm g}}{2kT}\right)$, the temperature dependence of the depletion dark current (3.17) is

$$J_{\rm dep} \propto T^2 \exp\left(-\frac{E_{\rm g}}{2kT}\right) \tag{3.18}$$

The temperature dependence is dominated by the exponential term in (3.18). The pre-exponential term T^2 has negligible influence except at very low temperatures.

If the energy position of the trap is away from the mid-band so that $|E_t - E_i| \gg kT$, the generation lifetime (3.14) becomes

$$\tau_{\rm g} = \frac{2}{\sigma v_{\rm th} N_t} \cosh\left(\frac{E_t - E_i}{kT}\right) \approx \frac{1}{\sigma v_{\rm th} N_t} \exp\left(\frac{|E_t - E_i|}{kT}\right) \tag{3.19}$$

The dark current density is then

$$J_{\rm dep} = \frac{q n_i W}{\tau_g} = q \sigma v_{\rm th} N_t n_i W \exp\left(-\frac{|E_t - E_i|}{kT}\right) \tag{3.20}$$

And the overall temperature dependence is

$$J_{\rm dep} \propto T^2 \exp\left(-\frac{E_{\rm g}}{2kT}\right) \exp\left(-\frac{|E_t - E_i|}{kT}\right) \tag{3.21}$$

This formula clearly demonstrates that traps away from the mid-band energy position are exponentially less effective in generating dark current.

Equation (3.21) can also be written as

$$J_{\rm dep} \propto T^2 \exp\left(-\frac{E_{\rm a}}{kT}\right) \tag{3.22}$$

where $E_{\rm a} = E_{\rm g}/2 + |E_t - E_i|$ is called *activation energy* of the trap. The traps with $E_{\rm a} = E_{\rm g}/2 \approx 0.56$ eV contribute the most to the dark current.

The dark current (3.17) in proportional to the depletion depth, which in turn depends as $(V_{\rm bi} + V_{\rm r})^{1/2}$ on the reverse bias $V_{\rm r}$ for abrupt *pn* junctions. This dependence can be used as a diagnostic tool.

Normally we assume that the bulk traps are uniformly distributed in the device, i.e. the density N_t is constant. As the reverse bias increases the depletion covers more traps, while fewer traps remain in the neutral semiconductor as shown in figure 3.5. Once the *pn* junction is fully depleted, all the traps become active and the dark current stops increasing because W cannot grow any further.

The key to this dependence on the depletion depth is that traps inside neutral semiconductor generate much less dark current. Electron–hole pairs are still continuously generated, but in neutral semiconductor the number of created pairs

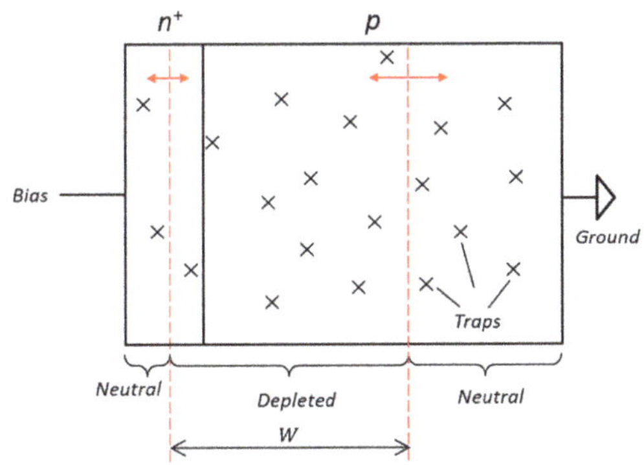

Figure 3.5. Depletion dark current in a *pn* junction under increasing reverse bias.

is equal to the number recombining, and that keeps the carriers to their equilibrium concentrations. In neutral semiconductor diffusion is the only mechanism for charge movement. There is no electric field to quickly separate the electrons and the holes and to prevent their recombination. The zero electric field brings about a different type of dark current driven by diffusion. A more mathematical way to say this is that in equilibrium we have $pn = n_i^2$, therefore the rate of change of the carrier concentration (3.12) is zero.

3.2.3 Diffusion dark current

The equilibrium hole concentration p_{p0} in neutral *p*-type silicon is approximately equal to the acceptor concentration N_A. At the same time, the equilibrium electron concentration n_{p0} (minority carriers in *p*-type silicon) is

$$n_{p0} = \frac{n_i^2}{p_{p0}} = \frac{n_i^2}{N_A} \tag{3.23}$$

and can be many orders of magnitude smaller. In a depleted semiconductor the concentration of free carriers (both electrons and holes) is practically zero, and so is the carrier concentration at the edge between the depleted region and the field-free region. In the *p*-side of the junction in figure 3.6, any electron reaching this edge will be swept away by the electric field and the holes pushed back to the field-free region. Exactly the opposite would happen on the *n*-side of the junction.

Therefore, the diffusion of the minority carriers out of the field-free region of the junction is causing a current to flow, called *diffusion dark current*. Its density is given by the product of the diffusion coefficient and the concentration gradient. For the electron current at $x = 0$ (the depletion edge on the *p*-side)

$$J_{\text{diff}} = qD_n \frac{\partial n_p}{\partial x} \tag{3.24}$$

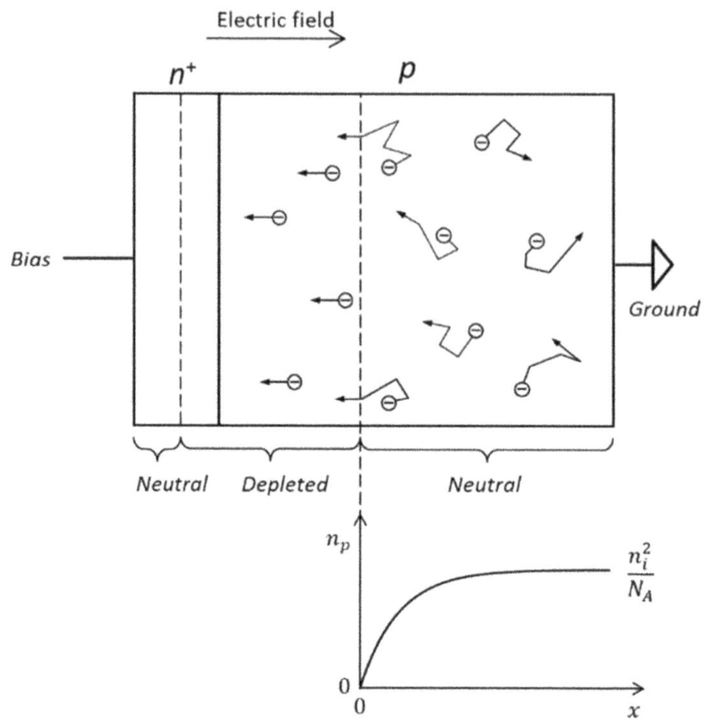

Figure 3.6. Diffusion of electrons from a neutral p-type semiconductor into the depleted region. Holes in the neutral p-type semiconductor, which are much more abundant than the electrons, are not shown.

and a similar expression can be written for the holes on the n-side. Only carriers generated within approximately one diffusion length[3] from the depletion edge can make it out of the neutral semiconductor and create diffusion current.

Deep into the p-side, far away from the depletion edge the electron concentration is given by (3.23), while at the edge it is zero. By solving the diffusion equation with these two boundary conditions, the electron density in depth is given by [10]

$$n_p = n_{p0}\left[1 - \exp\left(-\frac{x}{L_n}\right)\right] \tag{3.25}$$

Substituting (3.23) and (3.25) in (3.24) for $x = 0$ gives

$$J_{\text{diff}} = \frac{qD_n n_i^2}{L_n N_A} \tag{3.26}$$

Using that the diffusion length $L_n = \sqrt{D_n \tau_n}$, (3.26) can also be written as

$$J_{\text{diff}} = \frac{qn_i^2}{N_A}\sqrt{\frac{D_n}{\tau_n}} \tag{3.27}$$

[3] The diffusion length L_n can be thought of as the distance over which the carrier concentration decreases to $1/e$ (37%) of the original.

The classic formula (3.26) assumes that the neutral region is much longer than the diffusion length because in (3.25) the minority electron concentration reaches n_i^2/N_A only for $x \gg L_n$. We have seen in chapter 1 that the diffusion lengths can be hundreds of micrometres, but in most devices the neutral region is much shorter. If there is no neutral region, as in fully depleted *pn* junctions, there would be *no diffusion current.*

The equilibrium electron density in the case of a short neutral region L_{ff} is given by [11]

$$n_{p0} = \frac{n_i^2}{N_A}\left[1 - \exp\left(-\frac{L_{ff}}{L_n}\right)\right] \tag{3.28}$$

and can be much smaller than n_i^2/N_A. When $L_{ff} \ll L_n$ we can use that $\exp(x) \approx 1 + x$ for $x \to 0$ and the expression (3.26) for the diffusion current becomes

$$J_{diff} = \frac{qD_n n_i^2 L_{ff}}{L_n^2 N_A} = \frac{qn_i^2 L_{ff}}{\tau_n N_A} \tag{3.29}$$

A similar expression can be written for the hole component of the dark current by using τ_p and N_D. For the asymmetric n^+p junction in figure 3.6 $N_D \gg N_A$ and therefore the dark current by hole diffusion can be safely ignored.

Since $n_i \propto T^{3/2}\exp\left(-\frac{E_g}{2kT}\right)$ and $\tau_n \propto T^{-1/2}$, the temperature dependence of the diffusion dark current is

$$J_{diff} \propto T^{7/2}\exp\left(-\frac{E_g}{kT}\right) \tag{3.30}$$

As with the depletion dark current the temperature dependence is dominated by the exponent and not the pre-exponential term. An important difference is that the exponent for the diffusion current is $\exp(-E_g/kT)$ while for the depletion current it is $\exp(-E_g/2kT)$. This indicates that the diffusion current becomes more important at higher temperatures.

Example 3.4. Calculate the diffusion dark current at 300 K (27 °C) for $\tau_n = 1$ ms, $L_{ff} = 5$ μm, $N_A = 6.7 \times 10^{14}$ cm^{-3} (20 Ω cm) and $n_i = 1.45 \times 10^{10}$ cm^{-3}. Compare with the current at 60 °C.

Solution: First, we see that the electron diffusion length $L_n = \sqrt{D_n \tau_n} = \sqrt{36 \times 0.001} = 0.19$ cm is much longer than the field-free region, therefore (3.29) should be used:

$$J_{diff} = \frac{1.6 \times 10^{-19} \times (1.45 \times 10^{10})^2 \times 5 \times 10^{-4}}{10^{-3} \times 6.7 \times 10^{14}} = 25\ \text{fAcm}^{-2}$$

Compared with the depletion dark current in example 3.3 this is three orders of magnitude lower. At 60 °C we can scale the current using (3.30), ignoring the pre-exponential term:

$$J_{\text{diff}}(T_2) = J_{\text{diff}}(T_1)\exp\left[\frac{E_g}{k}\left(\frac{1}{T_1} - \frac{1}{T_2}\right)\right]$$

The dark current increases by a factor of

$$\exp\left[\frac{1.12 \times 1.6 \times 10^{-19}}{1.38 \times 10^{-23}}\left(\frac{1}{300} - \frac{1}{333}\right)\right] = 73,$$

and has grown to $25 \times 73 = 1800$ fAcm^{-2}. For comparison, the pre-exponential term increases the dark current only by $(333/300)^{3.5} = 1.44$ times.

Diffusion dark current originates from the p^{++} substrate too, which is normally doped to around 10^{19} cm^{-3} and is hundreds of micrometres thick. At such high doping concentrations, the electron lifetime is limited by Auger recombination [12] and is below 10^{-7} s. This would increase the diffusion dark current, but the electron mobility is also much reduced, and so is the diffusion coefficient $D_n = (kT/q)\mu_n$. Limited by impurity ion scattering, at acceptor doping above 10^{19} cm^{-3} the electron mobility is barely reaching μ_n = 100 cm^2 V^{-1} s^{-1} [1]. Substituting these numbers in (3.27) gives $J_{\text{diff}} = 17$ fA cm^{-2}, comparable to the diffusion current from the epitaxial layer calculated in example 3.4.

3.2.4 Surface dark current

The surface is a major source of dark current which is usually dominant. Due to the lattice mismatch between Si and SiO_2, the interface between them has many electrically active traps, called interface states. The energy positions of the states follow a continuous distribution with a characteristic U-shape with a minimum at mid-gap [1], unlike bulk traps which have discrete energy positions. The typical surface energy density N_{ss} of the states is between 10^9–10^{11} cm^{-2} eV^{-1}.

To calculate the recombination rate we can use the SRH theory (3.12) with the difference that the surface recombination rate is measured in units of carriers per area per second (cm^{-2} s^{-1}), while the bulk one has units of cm^{-3} s^{-1}. We need to account for all states in the bandgap by including the distribution of N_{ss}.

The surface recombination rate in depletion ($p \approx 0$ and $n \approx 0$), assuming that N_{ss} is constant[4], can be found from (3.13) by integrating over the bandgap [11]

$$U_s = -\int_{E_V}^{E_C} \frac{\sigma v_{\text{th}} N_{ss} n_i}{2\cosh\left(\frac{E_t - E_i}{kT}\right)} dE_t \qquad (3.31)$$

[4] This is acceptable despite N_{ss} rising steeply towards the edges of the bandgap because states near the mid-gap have the greatest contribution to the dark current.

Using that $\int \frac{dx}{\cosh x} = 2\arctan(e^x)$ ([12] 2.423–9) and $E_g/2 \gg kT$ the solution is

$$U_s = -\frac{\pi\sigma v_{th} N_{ss} n_i kT}{2} \tag{3.32}$$

Therefore, the surface dark current density is

$$J_s = q\,|U_s| = \frac{q\pi\sigma v_{th} N_{ss} n_i kT}{2} \tag{3.33}$$

Comparing to the depletion dark current in the bulk (3.15) there is an additional term of πkT which makes the pre-exponential term of the temperature dependence T^3 instead of T^2.

$$J_s \propto T^3 \exp\left(-\frac{E_g}{2kT}\right) \tag{3.34}$$

The current from even a good quality Si–SiO_2 interface can easily be in the nA cm^{-2} region and several orders of magnitude higher that the bulk dark current in depletion. The following example illustrates this.

Example 3.5. Calculate the surface dark current density for $N_{ss} = 10^9$ cm^{-2} eV^{-1}, $\sigma = 10^{-15}$ cm^2. Use $v_{th} = 10^7$ cm s^{-1} and $n_i = 1.45 \times 10^{10}$ cm^{-3} at 300 K.
Solution: We can use (3.33) with some care to convert the kT term to electron-volts (because N_{ss} is in units of cm^{-2} eV^{-1}) by dividing by the elementary charge, and then q cancels from the numerator.

$$J_s = \frac{3.14 \times 10^{-15} \times 10^7 \times 10^9 \times 1.45 \times 10^{10} \times 1.38 \times 10^{-23} \times 300}{2} = 0.94\ \text{nAcm}^{-2}$$

Pixel designs employ several measures by eliminate depleted Si–SiO_2 interfaces, such as *p*-wells to enclose the STI (as in figure 3.4) and most importantly for the PPD—the pinning p^+ implant.

3.2.5 Dark current suppression by pinning

It is clear that if the Si–SiO_2 interface is not depleted, the dark current from it can be much reduced. If the interface borders neutral semiconductor we have $pn = n_i^2$, the recombination rate (3.12) becomes zero and the dark current is eliminated entirely. However, providing a substantial thickness of neutral semiconductor at every interface is usually not possible or practical.

Another way to prevent depletion is to saturate the interface with high concentration of holes. Provided that the interface holes and the signal electrons never meet (because they will recombine, and the signal will be destroyed) this is a very good method for suppressing the dark current.

Saturation with holes can be created by inversion of the surface and is widely used in certain types of CCDs (called 'inverted' or 'multi-pinned' mode) to supress the dark current by more than three orders of magnitude [11].

In the PPD the Si–SiO_2 interface over the diode's surface is populated by a high concentration of holes provided by the doping of the p^+ pinning layer. The pinning by an implant is permanent, unlike the hole concentration provided by surface inversion. Because the pinning implant is very shallow (of the order of 100 nm) the electrons at the surface diffuse out and their concentration is much lower than in equilibrium. Therefore, at the interface in pinning condition we have $p \gg n_i \gg n$. Adapting (3.12) for interface traps, we can write the recombination rate in pinning similarly to (3.31) as an upper limit determined by $| pn - n_i^2 | \leqslant n_i^2$ [13]

$$|U_{\text{sp}}| = \int_{E_V}^{E_C} \frac{\sigma v_{\text{th}} N_{\text{ss}} \, | pn - n_i^2 |}{p + n + 2n_i \cosh\left(\frac{E_t - E_i}{kT}\right)} dE_t \leqslant \sigma v_{\text{th}} N_{\text{ss}} n_i \int_{E_V}^{E_C} \frac{dE_t}{\frac{p}{n_i} + 2\cosh\left(\frac{E_t - E_i}{kT}\right)} \tag{3.35}$$

We know that the recombination rate is negative because $pn \leqslant n_i^2$. Using that $p \gg n_i$ the solution of (3.35) ([12] 2.443–3) is approximately

$$U_{\text{sp}} \cong -\frac{\sigma v_{\text{th}} N_{\text{ss}} n_i^2 E_{\text{g}}}{p} \tag{3.36}$$

Comparing (3.36) with the recombination rate in depletion (3.32) we see that the dark current is suppressed by a factor of

$$\frac{U_{\text{s}}}{U_{\text{sp}}} = \frac{p}{2n_i} \frac{\pi kT}{E_{\text{g}}} \tag{3.37}$$

This result is similar to [13] but here we have taken into account the energy distribution of the interface states in the bandgap. Since $n_i \approx 10^{10}$ cm^{-3}, $p \approx 10^{17}$ cm^{-3} and $\pi kT/E_{\text{g}} = 0.073$ at 300 K the surface dark current is supressed by a factor of 10^6, thanks to the pinning implant. This brings it down from a few nA cm^{-2} to the fA cm^{-2} range and it becomes comparable to the diffusion dark current.

It is reasonable to ask the question: what is the minimum achievable dark current, and what is limiting it? In PPDs the surface dark current is much reduced, and the bulk trap concentration in high quality silicon can be negligible. The dominant mechanism for dark current generation then becomes the diffusion. The carrier lifetime is limited by the Auger recombination, which does not need any traps, and occurs even in 'perfect' silicon. At 300 K the diffusion-limited dark current limit appears be around 0.1 pA cm^{-2} [7], close to what is achieved today. The diffusion current is almost eliminated in fully depleted sensors with the p^{++} substrate removed. This should bring the expected dark current down to about 0.1 fA cm^{-2} at 300 K [7], a value yet to be seen in practice.

3.2.6 Temperature for dark current doubling

Very often the temperature change for the doubling of the dark current is used to gain insight into its origins and to predict the current at different temperatures. The usual rule of thumb is that in silicon the dark current doubles for every 7 °C increase in temperature.

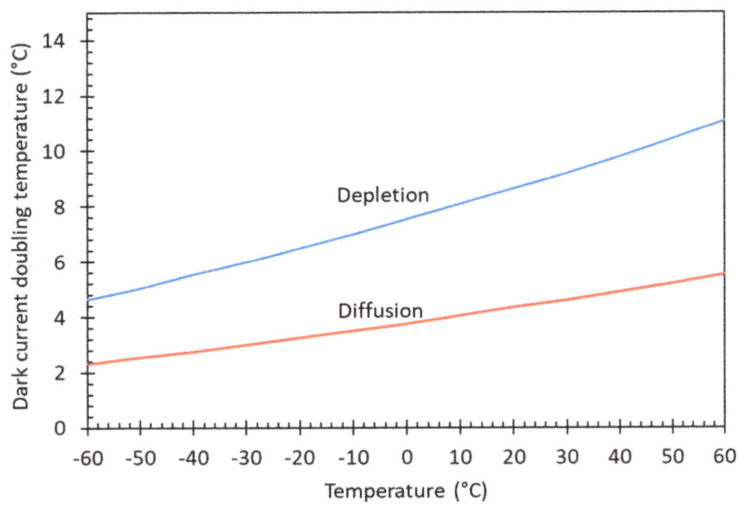

Figure 3.7. Temperature for dark current doubling for the depletion and diffusion dark currents.

This can be useful for rough estimations but doubling at a certain temperature increase would imply a power law, not an exponential dependence, and is therefore only approximate. Using the expressions for the depletion (3.18) and the diffusion (3.30) dark currents, the doubling temperature can be calculated according to the theoretical dependences.

As shown in figure 3.7, the doubling temperature changes significantly, and is 7 °C only for the depletion current around 0 °C. For the diffusion current the doubling temperature is about half as much, due to the steeper exponential. At 60 °C the doubling temperatures are 11 °C and 6 °C, respectively [7].

3.3 Reflective barrier

Most image sensors are built on *p*-type epitaxial wafers, where the lightly doped epitaxial layer is grown on a very low resistivity ($\ll$1 Ω.cm), heavily doped *p*-type substrates. The doping of the pixel *p*-wells is also much higher than the epitaxial layer. This difference in doping concentration creates a built-in potential at the boundaries between the epitaxial layer and the heavily doped regions. The built-in potential acts as a reflective barrier to electrons diffusing in the epitaxial layer and makes it harder for them to enter the *p*-wells and the substrate.

Let us consider the band diagram along the bottom red dashed line in the 4T pixel in figure 3.8. The heavily doped substrate has acceptor concentration N_{A1} and the lightly doped epitaxial layer N_{A2}. No external electric is applied across the *p*-regions[5], the *p*-well and the substrate are connected to ground, the silicon is not

[5] Applying voltage between the *p*-well and the substrate would cause large hole current to flow through the structure because it is essentially a silicon resistor.

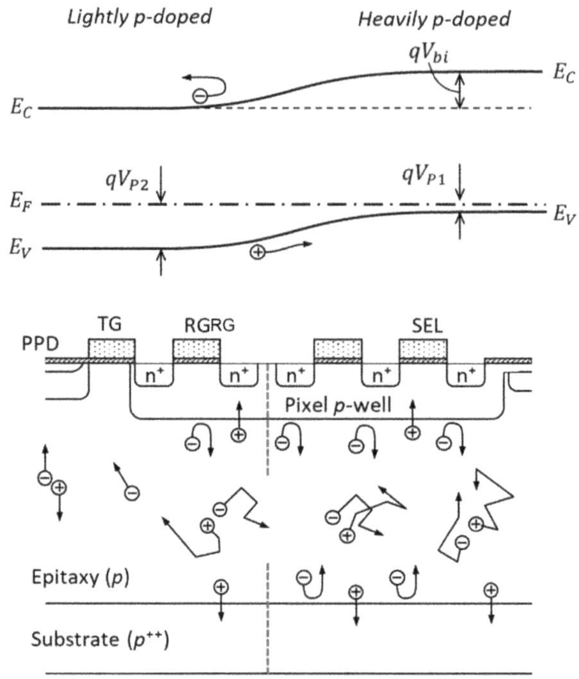

Figure 3.8. Reflective barriers in a 4T pixel on a *p*-type epitaxial layer and the band diagram along the red dashed lines.

depleted and is in thermal equilibrium. On the heavily doped side the difference between the Fermi level E_F and the valence band E_V is given by

$$qV_{P1} = kT\ln\left(\frac{N_V}{N_{A1}}\right) \tag{3.38}$$

where N_V is the effective density of states in the valence band [1]. Similarly, on the lightly doped side we have

$$qV_{P2} = kT\ln\left(\frac{N_V}{N_{A2}}\right) \tag{3.39}$$

The semiconductor is in thermal equilibrium and the Fermi level is the same everywhere, therefore the conduction and the valence bands must bend. This creates a potential difference given by

$$V_{bi} = V_{P2} - V_{P1} \tag{3.40}$$

The potential gradient is pushing the minority carriers (electrons in *p*-type semiconductor) towards the lightly doped side and the holes in the opposite direction. For minority carriers the potential barrier is reflecting like an imperfect mirror. When electron–hole pairs are generated in the lightly doped epitaxial layer in figure 3.8 and start to diffuse, the electrons are reflected back, while the holes are attracted to the *p*-well and the substrate. The potential difference can be expressed from (3.38), (3.39) and (3.40) as

$$V_{bi} = \frac{kT}{q}\ln\left(\frac{N_{A1}}{N_{A2}}\right). \tag{3.41}$$

The potential difference depends only on the acceptor concentrations and not on the doping profile. When $N_{A1} = N_{A2}$ the potential becomes zero as in a uniformly doped, neutral semiconductor.

Example 3.6. Calculate the potential difference at 300 K between an epitaxial layer and a *p*-well, boron doped to 10^{14} and 10^{17} cm^{-3}, correspondingly.
Solution: From (3.41) we get

$$V_{bi} = \frac{kT}{q}\ln\left(\frac{N_{A1}}{N_{A2}}\right) = \frac{1.38 \times 10^{-23} \times 300}{1.6 \times 10^{-19}}\ln\left(\frac{10^{17}}{10^{14}}\right) = 0.18\text{V}$$

Compared with the built-in voltage of the *pn* junction the potential barrier is much smaller and is not 100% effective as a reflecting mirror of minority carriers. Due to their thermal energy some charge carriers can diffuse over it. The diffusion current over the barrier can be estimated using thermionic emission theory [1] similarly to charge transfer in PPDs in chapter 2 as

$$I = I_{max}\exp\left(-\frac{qV_{bi}}{kT}\right) \tag{3.42}$$

In (3.42) the saturation current $I_0 = A^*T^2S_A$ is replaced with the maximum current I_{max} because the supply of charge is limited by the charge present in the lightly doped side of the barrier. The maximum current is much smaller than I_0 and can be found from the condition of zero barrier. Substituting (3.41) in (3.42) gives

$$\frac{I}{I_{max}} = \frac{N_{A2}}{N_{A1}} \tag{3.43}$$

Equation (3.43) means that the minority carrier flow is suppressed by the ratio of the doping concentrations on both sides. For the values in example 3.6 the electron current is suppressed by a factor of 1000, which means that 0.1% of the electrons are not reflected and lost.

The reflective barrier in the form of a deep *p*-well (DPW) can be useful in more complex image sensors using NMOS and PMOS transistors in the pixel [14]. Figure 3.9 shows a deep *p*-well preventing charge generated underneath it from reaching the *n*-well. Without it, the *n*-well, which is normally biased to the highest

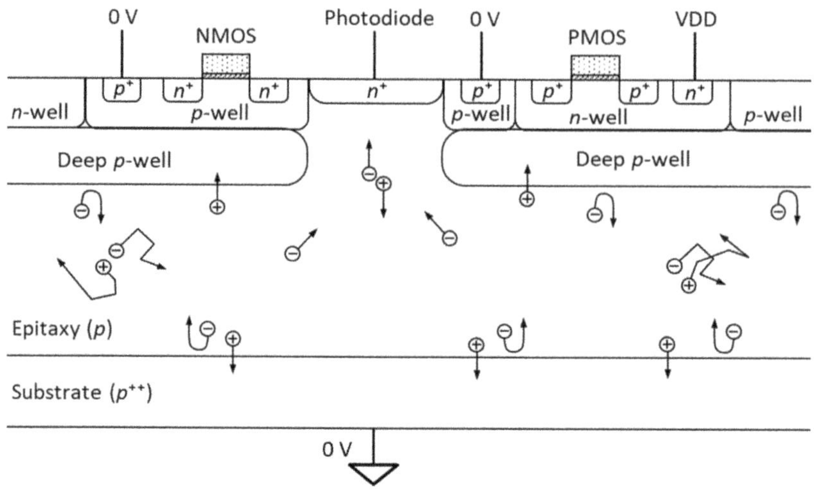

Figure 3.9. Deep p-well protecting from parasitic charge collection in a PMOS with an aperture for charge collection in a photodiode.

voltage on chip, will compete for charge with the photodiode and could collect most of it.

The DPW is biased to substrate potential via the transistor p-wells because the two overlap, but it is perfectly possible to leave it floating. Since the DPW is heavily doped and electrically neutral, the n-well on top has very little influence, and the DPW acquires its potential from the p-wells and the neutral epitaxial silicon below.

The aperture in the DPW allows charge to collect on a photodiode while parasitic charge collection elsewhere is suppressed. After diffusing, charge generated under the DPW reaches the depleted region under the photodiode and is promptly collected. Charge generated above the deep p-well will be lost because it is collected either in the n-well or in the drain and source of the NMOS transistors. To minimise this charge loss, the transistor wells and the p-well should occupy a small fraction of the thickness of the epitaxial layer.

3.4 Back-side illumination

3.4.1 Front and back-side illumination

Image sensors using front-side illumination (FSI) are built like a typical integrated circuit. The starting material is usually a p-type silicon wafer having an epitaxial layer with thickness between 5 and 40 μm. The high-quality epitaxial silicon is grown on top of a silicon substrate, which is heavily boron-doped, typically to 10^{19} cm^{-3}. The standard substrate thickness for 8″ (200 mm diameter) wafers is 725 μm. During the growth, the epitaxial layer is boron-doped, but with much lower concentration in the range 10^{14}–10^{15} cm^{-3}.

The example in figure 3.10 uses a 5 μm thick epitaxial silicon, 4 metallisation layers (M1 through M4), and PPD pixels on 5 μm pitch. The electrical connections to the sensor are made with wire bonds to the top-level metal M4. Each metallisation

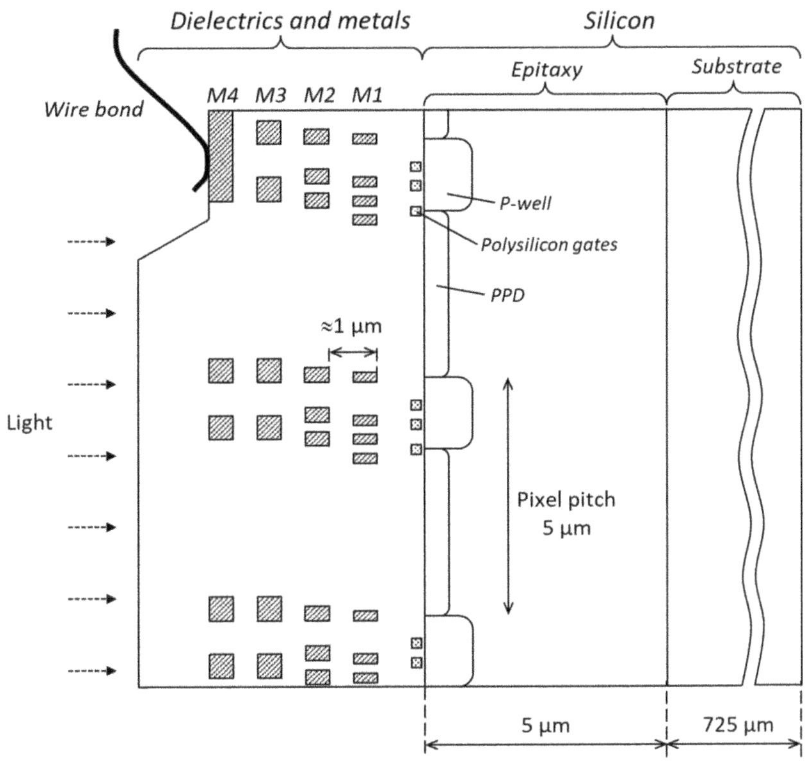

Figure 3.10. Cross-section of an FSI image sensor on 5 μm pixel pitch. Everything except the wire bond (a 25 μm diameter metal wire) is approximately drawn to scale. AR coatings, colour filters and micro-lenses on the front side are not shown.

layer adds approximately 1 μm of dielectrics, usually a combination of SiO_2 and Si_3N_4, and a passivation layer on top of M4 is also implemented.

Before interacting with the silicon, light must pass through all the front-side material and is subject to reflections and scattering from the metal layers. The metal lines are made as thin as possible to minimise reflections and cover mostly the *p*-wells, so that short wavelengths can reach the PPD unimpeded. Blue and green wavelengths are absorbed almost completely in the typical *p*-well with depth of 1 μm and this further reduces the QE. Various methods to reduce light losses have been developed for FSI imagers, including micro-lenses, light pipes, and an optimised process with only two metal layers over the pixels [15]. Despite this, FSI imagers struggle with achieving good QE, especially at shorter wavelengths, and the issues get worse as the pixel pitch decreases, due to the higher fraction of metal coverage.

Much better QE is possible if the light enters the silicon from the bottom of the epitaxial layer, completely avoiding the metal tracks. To do this, the substrate must be removed, and the wafer flipped over, as shown in figure 3.11, creating a back-side illuminated (BSI) sensor. The same sensor can exist in both FSI and BSI variants.

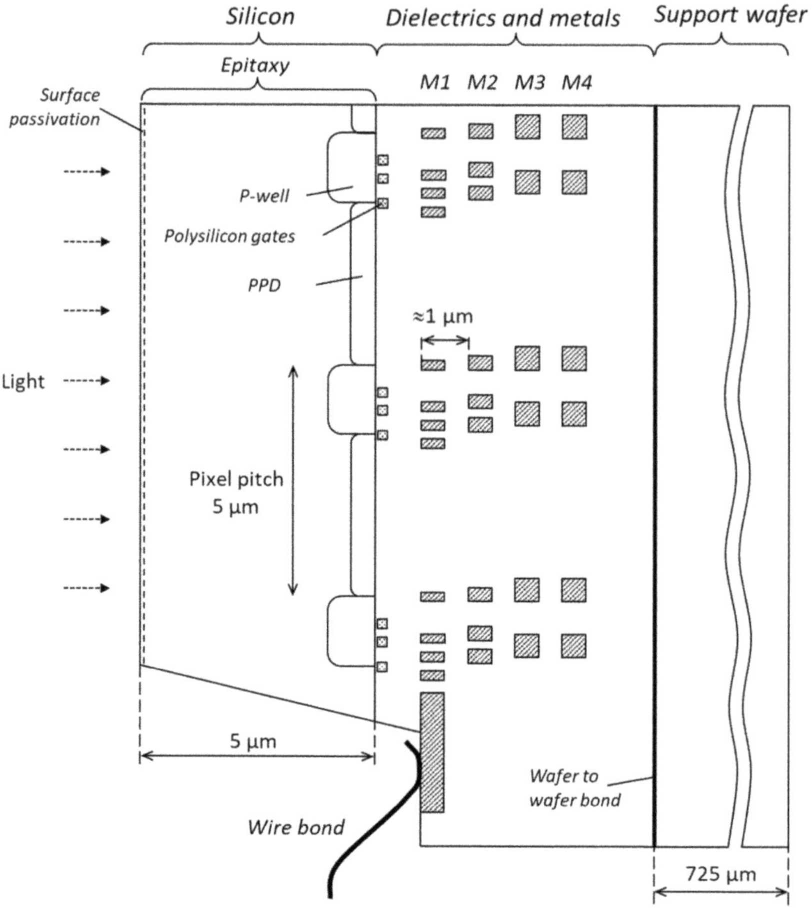

Figure 3.11. Cross-section of a BSI version of the sensor in figure 3.10. AR coatings, colour filters and micro-lenses are not shown.

The manufacture of BSI sensors involves many additional steps and is complex. The starting point is a finished FSI wafer, which is then fused to a support wafer using molecular bonding. Most of the substrate is removed by grinding, followed by a chemical etch. Using a mixture of hydrofluoric, nitric, and acetic acids, the etch rate slows down as the boron concentration decreases [15], providing a natural etch stop at the boundary between the substrate and the epitaxial layer. The boron out-diffusion from the substrate, created during the epitaxial growth and subsequent high temperature steps, and usually a fraction of a micron deep, is also removed. The newly created back surface acquires a very thin native oxide layer and is passivated using a dedicated process to ensure controlled conditions at the Si–SiO_2 interface. Furthermore, the epitaxial silicon and the first level dielectric are etched away at the chip's periphery so that the metal pads for connections can be exposed. Unlike in FSI imagers, the connections are made to the first level metal M1.

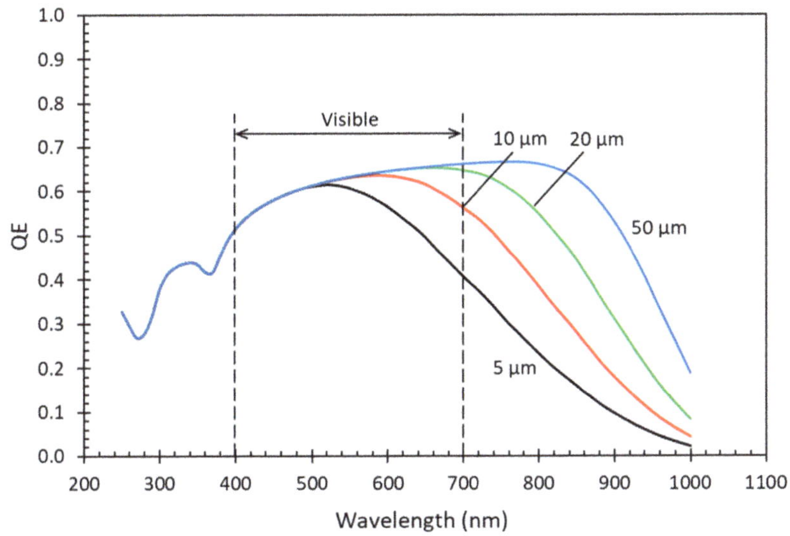

Figure 3.12. Theoretical QE of BSI silicon sensors with different thickness, calculated using reflectivity and absorption data from [17] and formula (3.10). An anti-reflective coating is not used.

The bottom of the epitaxial layer is now the entry surface for the light. In contrast to FSI sensors, most of the photogenerated charge is created away from the PPD and the *p*-wells, and this is particularly true for short wavelengths.

BSI sensors can achieve nearly the theoretical QE shown in figure 3.12 [11], limited by the thickness and the reflectivity of the epitaxial silicon. With an appropriate anti-reflective coating (ARC), the QE can exceed 90% over a selected wavelength band [16]. Thanks to the high QE, many high-performance CIS in mobile phones and science applications use BSI technology.

3.4.2 Back-side interface

The key to the success of the BSI technology is the quality of the surface passivation. The freshly ground and polished silicon surface acquires a very thin (about 2 nm) native oxide, which tends to be positively charged. This creates a partially depleted layer which attracts photogenerated electrons to the back surface, where they recombine at interface traps. This is particularly detrimental for short wavelengths since most of the charge is generated within this depletion layer and cannot escape. Furthermore, interface traps at the back surface generate extra dark current.

Figure 3.13 compares the potential profiles and the interface conditions for FSI and BSI sensors. In the FSI sensor, the PPD pinning implant at the front and the p/p^{++} reflective barrier at the back keep the photogenerated electrons away from any Si–SiO_2 interfaces and within the epitaxial layer. Without surface passivation, some of the electrons in the BSI sensor can recombine at the back side, leading to poor QE and higher dark current. However, the back surface is not an infinite sink of electrons and has a limited capacity for recombination, and not every electron reaching it is lost.

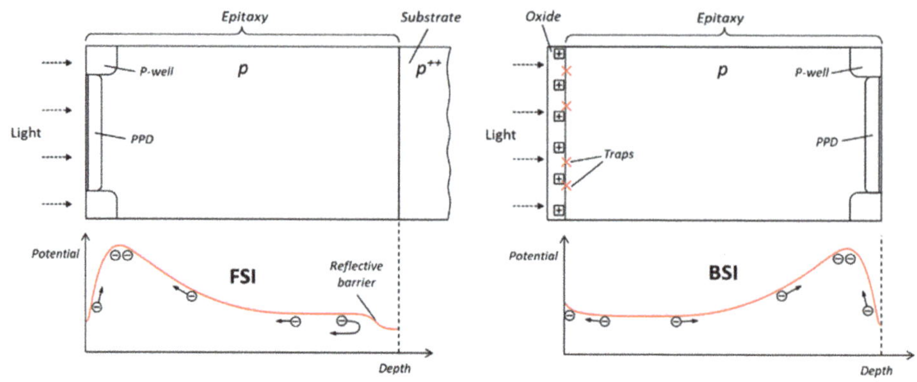

Figure 3.13. Potential profiles and electron motion in a FSI sensor and in a BSI sensor without back surface passivation.

The purpose of the passivation is to provide a potential barrier to electrons, so that they cannot reach the back surface. However, some of the charge generated *within* the barrier can still recombine, therefore the width of the barrier must be very small. Having looked at the shallow $p+$ pinning implant in PPDs in section 3.2.5, the issue with the back surface passivation may look very similar. There is, however, one important difference: the passivation is done *after* all the metallisation is complete. As the metals would not survive temperatures above approximately 450 °C, the usual high temperature (above 900 °C) implant anneal is out of question. The BSI implant is less effective in suppressing surface recombination than the pinning implant because the high temperature passivation techniques, such as forming gas anneal, cannot be used. As with the pinning implant, the passivating implant must be very shallow because the absorption length of silicon at 400 nm wavelength is only 100 nm.

Starting with the SRH surface recombination rate used in section 3.2.5, equation (3.35) can be simplified by using a constant surface density of the interface traps N_{st} (measured in cm^{-2} and not $cm^{-2}\,eV^{-1}$ as N_{ss}). Since integration over the bandgap is not required, the surface recombination rate for $E_t \cong E_i$ in the general case is [10]

$$U_{bs} = \frac{\sigma v_{th} N_{st}(pn - n_i^2)}{p + n + 2n_i} \tag{3.44}$$

In (3.44) the electron and hole concentrations are at the surface. In almost neutral p-type silicon $p \gg n$ and $p \gg n_i$, and using that $n = n_0 + \Delta n$, where Δn is the excess electron concentration, we can write[6]

$$U_{bs} = \frac{\sigma v_{th} N_{st}[p(n_0 + \Delta n) - n_i^2]}{p} \tag{3.45}$$

[6] Here the subscripts are omitted for simplicity since we only deal with p-type silicon, and $n \equiv n_p$ and $n_0 \equiv n_{p0}$.

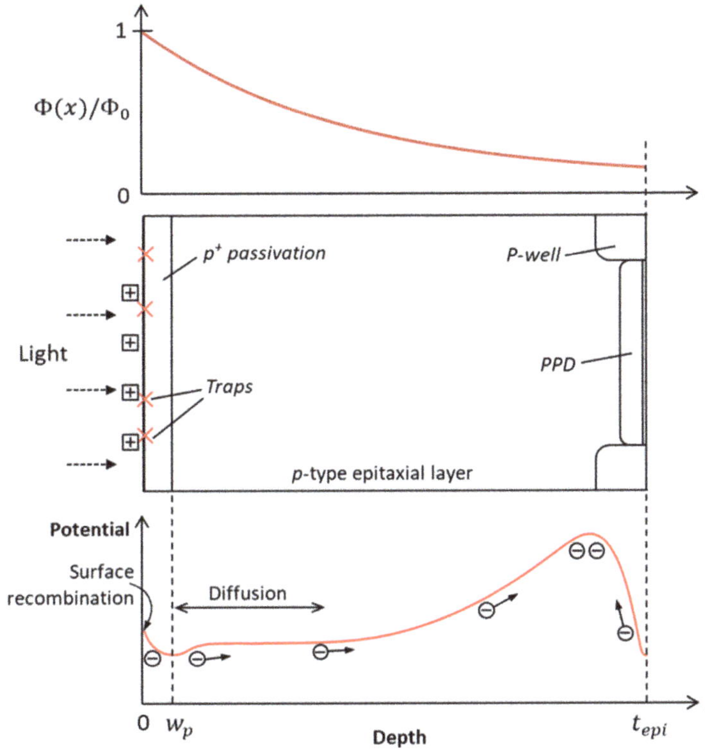

Figure 3.14. Potential diagram in a BSI PPD sensor with an imperfect back surface passivation. The thickness of the passivation has been exaggerated for clarity.

Using that in p-type silicon the excess hole concentration is much smaller than the doping concentration, we have $pn_0 \approx n_i^2$, and equation (3.45) becomes:

$$U_{\mathrm{bs}} = \sigma v_{\mathrm{th}} N_{\mathrm{st}} \Delta n \tag{3.46}$$

The term $S_n = \sigma v_{\mathrm{th}} N_{\mathrm{st}}$ is called surface recombination velocity and is measured in units of cm s^{-1}.

In figure 3.14 the passivation is a shallow $p+$ implant with thickness w_p, and the epitaxial layer with thickness t_{epi} is partially depleted. The fixed positive charge at the back side creates conditions which allow some of the photogenerated electrons to recombine at surface traps, but some will diffuse out towards the PPD and will get collected. We need to find the signal leaving the passivation layer and add it to the rest of the signal generated between w_p and t_{epi}, which is fully collected by drift and diffusion, as shown in section 3.1. This will give us a measure of the QE, including the recombination at the back surface.

First, the current from most of the epitaxial layer is found by integrating (3.2) with the reflection R taken into account:

$$J_{\mathrm{epi}} = q(1 - R)\int_{w_p}^{t_{\mathrm{epi}}} \alpha \Phi_0 e^{-\alpha x} dx = q\Phi_0(1 - R)(e^{-\alpha w_p} - e^{-\alpha t_{\mathrm{epi}}}) \tag{3.47}$$

Next, we need to calculate the diffusion current out of the passivation layer collected as signal, given as

$$J_{\mathrm{diff}} = qD_n\left(\frac{\partial n}{\partial x}\right)_{x=w_p} \tag{3.48}$$

To find the electron concentration we solve the stationary diffusion equation (3.5) with the following boundary conditions: at the surface ($x = 0$) the steady-state electron flux equals the surface recombination rate

$$D_n\frac{\partial n}{\partial x} = S_n(n - n_0) \tag{3.49}$$

and at $x = w_p$ the excess electron concentration Δn is zero. Using the solution in ([1] p 802) with the simplifications $L_n \gg w_p$ and $\alpha^2 L_n^2 \gg 1$, (3.48) becomes

$$J_{\mathrm{diff}} = q\Phi_0(1 - R)\left(\frac{\alpha D_n + S_n(1 - e^{-\alpha w_p})}{\alpha\left(D_n + S_n w_p\right)} - e^{-\alpha w_p}\right) \tag{3.50}$$

Using that $L_n \gg w_p$ is justified because the passivation thickness should be less than 0.1 μm, while the diffusion length is about a micron at high boron dopant concentrations. For wavelengths below 400 nm the absorption coefficient α increases to more than 10^5 cm^{-1} (<100 nm absorption length), therefore $\alpha^2 L_n^2$ is at least a hundred. Both conditions are valid only for the passivation layer, spanning from $x = 0$ to $x = w_p$.

The QE can be obtained from the sum of (3.47) and (3.50) as the factor multiplying $q\Phi_0$:

$$\eta = (1 - R)\left(\frac{\alpha D_n + S_n(1 - e^{-\alpha w_p})}{\alpha\left(D_n + S_n w_p\right)} - e^{-\alpha t_{\mathrm{epi}}}\right) \tag{3.51}$$

We can see from (3.51), that when there is no surface recombination ($S_n = 0$), or the passivation layer is infinitely thin ($w_p = 0$) the QE becomes $\eta \approx (1 - R)(1 - e^{-\alpha t_{\mathrm{epi}}})$ as expected.

It is useful to put some typical numbers in (3.51) to see what the effect is in practice. In a passivation implant doped to $\approx 10^{19}$ cm^{-3} the electron diffusion coefficient is around 2 cm^2 s^{-1} due to the low mobility [18]. In a Si–SiO$_2$ interface without a thermal anneal N_{st} could be as high as 10^{12} cm^{-2}, which would make the recombination velocity $S_n = 10^4$ cm s^{-1} for the typical $\sigma = 10^{-15}$ cm^2 and $v_{\mathrm{th}} = 10^7$ cm s^{-1}.

Figure 3.15 shows that the QE is barely affected at $S_n = 10^4$ cm s^{-1} for a 50 nm thick back-side passivation. The QE deteriorates at larger S_n, but only at wavelengths below 500 nm because the passivation is very thin. This may look OK, but we must be wary because the surface recombination rate (3.46) used in this derivation is for a nearly neutral semiconductor. Depending on the fixed positive

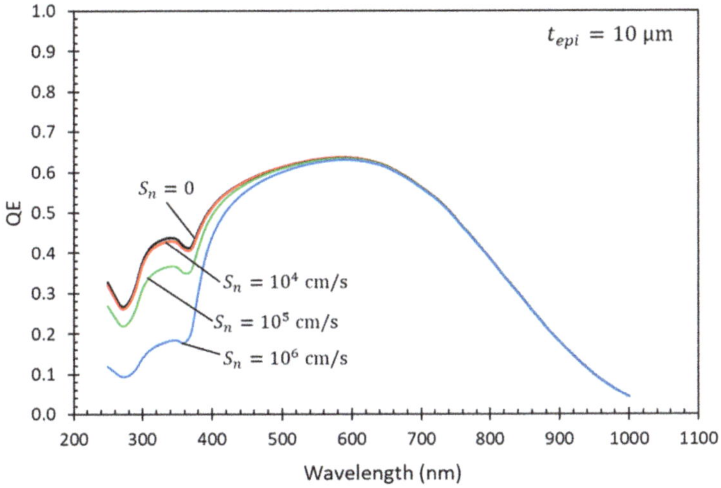

Figure 3.15. Calculated QE of a BSI silicon sensor with 10 μm epitaxial thickness for different surface recombination velocities using (3.51), for w_p = 50 nm and D_n = 2 cm^2 s^{-1}. Reflectivity and absorption data from [17] are used, and there is no anti-reflective coating.

charge at the back surface, the silicon could be partially depleted, and the recombination velocity could be much higher.

In partial depletion the surface recombination velocity S_{n_pd} is given by [10]

$$S_{n\_pd} = S_n \frac{N_A}{p + n + 2n_i} \tag{3.52}$$

where N_A is the acceptor concentration of the passivating implant and p and n are the carrier concentrations at the surface. A maximum $S_{n\_pd}^{\max}$ is achieved when $p \approx n \approx n_i$

$$S_{n\_pd}^{\max} = S_n \frac{N_A}{4n_i} \tag{3.53}$$

which means that in extreme cases the recombination velocity could be orders of magnitude higher than expected.

3.4.3 BSI technologies

Once the substrate has been removed, there are three main methods for back-side passivation [15]: (1) implantation as discussed in the previous section; (2) delta doping; and (3) dielectric with fixed negative charge. All methods aim to achieve high and stable QE (called 'QE pinning' in analogy to surface pinning in CCDs [11]), free from the shortcomings of earlier developments—QE hysteresis, non-uniformity, and high dark current.

Shallow doping with low energy implants has been used for decades. For p-type epitaxial wafers the implant is boron or BF_2 at energies around 10 keV. Since high temperatures are not allowed, the implant is activated by a laser anneal. The beam from a powerful UV laser, chosen for the short absorption length in silicon, is stepped in a randomised fashion across the back side of the wafer. This is necessary because the beam size is much smaller than the wafer, and the effects from the non-uniform anneal can sometimes be seen as a pattern in the measured QE and dark current. Nevertheless, passivation by implantation offers good performance for mainstream CIS, is relatively affordable and is widely used.

Delta doping [19] is a technique involving atomic layer deposition of extremely thin layers using molecular beam epitaxy at temperatures below 450 °C. A monolayer of boron is deposited, followed by the growth of nanometres-thick epitaxial silicon to encapsulate it, which is then oxidised for passivation. This creates a highly doped boron layer with surface density of 2×10^{14} cm^{-2} just nanometres from the newly formed silicon surface. This is ideal for both surface passivation and for the creation of electric field pushing the electrons away from the interface, and allows near-theoretical QE.

In the range between 100 and 300 nm the absorption length of silicon falls below 5 nm, which is less than 10 atomic distances. Delta doping is the only method capable of providing high QE in UV imagers due to the extremely sharp doping profile it can generate. The method is used in some scientific imagers requiring deep UV sensitivity [16]. Despite its excellent performance, delta doping is not widely adopted for commercial CIS because of the high cost of the process and the equipment, and the slow throughput.

The third method relies on the fixed negative charge in deposited Al_2O_3 layers on top of the thin native oxide. The charge can have surface density of 10^{12}–10^{13} cm^{-2} [20] and can create a hole accumulation layer in *p*-type silicon. The holes in this case are not provided by an implant, but work in the same way to create field-assisted passivation.

3.5 Depletion depth and potential gradients

3.5.1 Depletion depth as a 3D effect

In chapter 1 we derived the formula (1.45) for the depletion depth of a one-sided n^+p junction. The depletion depends on the applied reverse voltage V_r as $W \propto \sqrt{V_{bi} + V_r}$, where V_{bi} is the built-in voltage, but does not appear to depend on the *size* of the diode. This is because the formula is valid for an infinitely wide diode, or in practical terms, for a width of the *n*+ cathode that is substantially larger than the thickness of the diode. For the device in figure 3.16, an infinitely wide diode would correspond to the cathode occupying the whole of the available width of 20 μm.

Intuitively, a very small diode (for example, 1 μm wide) should not be capable of generating the same depletion depth as a much larger diode, so it is reasonable to expect some dependence. The technology CAD (TCAD) simulation in figure 3.17 confirms this—the depletion from small diodes is indeed smaller than the 'infinite' one and increases as the width is increased. This is also visible from the potential profiles in depth for different diode widths, taken at the centre of the diode.

If we had two diodes next to each other with different sizes, but biased at the same reverse voltage, the larger diode would exert higher potential in the depletion underneath. This means that a *potential gradient* from the smaller to the bigger diode

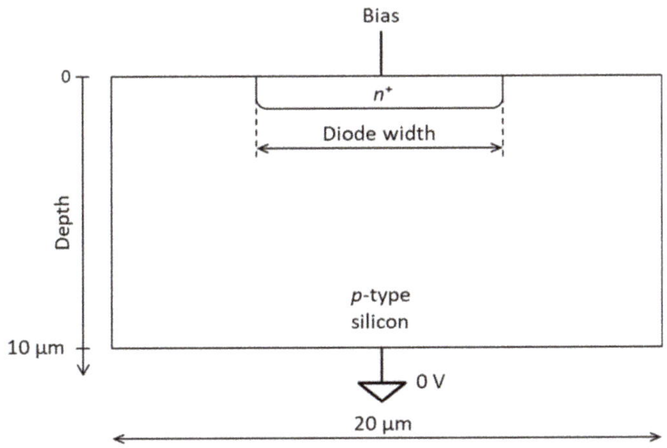

Figure 3.16. An n^+p junction with the width of the cathode as a parameter.

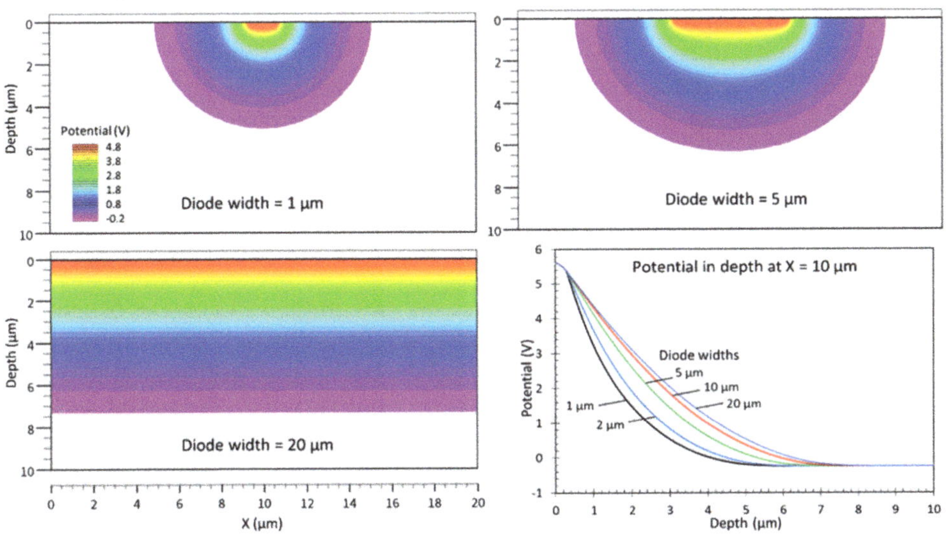

Figure 3.17. Potential distribution in 2D for the diode in figure 3.16 for different widths of the n^+ cathode, biased at 5 V.

is created. Figure 3.18(a) illustrates this by plotting the potential at a fixed depth in the depletion as a function of the width of the diode.

The highest electric field, calculated from the magnitude of the potential gradient, is at the smallest widths. Here it is about 450 V cm^{-1}, taken from the 0.45 V change between 1 μm and 2 μm diode size increment. Although small compared to the electric field in a *pn* junction (which can be higher than 10^4 V cm^{-1}), this can still be very useful: for mobility $\mu_n = 1400$ cm^2 V^{-1} s^{-1} an electron would traverse 6.3 μm in a nanosecond.

The idea of different diode sizes can be taken a step further by using a single diode with trapezoidal shape, as shown in figure 3.18(b). The potential gradient developed by this structure would depend on the two widths and the length, and also on the

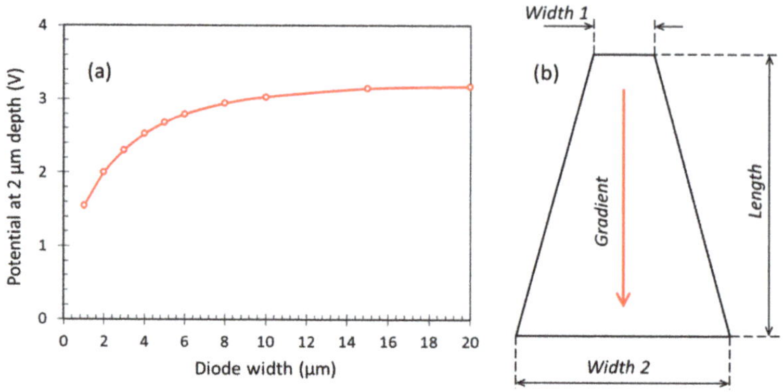

Figure 3.18. Potential at a depth of 2 µm in the centre of the diode in figure 3.17 (a); trapezoidal diode (b).

resistivity of the epitaxial silicon and the applied bias. Even though some information can be gained from a 2D model as in figure 3.17, this effect is inherently three-dimensional, and is best studied with 3D tools.

A similar effect can be achieved by using a diode with a fixed width, in which the depth of the n^+/p junction increases towards one end. This implies that several implants with different energies and masks are required, which is a significant complication. The advantage of the shaped structures is that they can be realised with a single implant, using a single mask, which is both simpler and cheaper.

3.5.2 Potential gradients in PPDs

Creating a potential gradient towards the transfer gate in the PPD can be beneficial for speeding up the charge collection and transfer [21]. This is particularly useful for large PPDs which are most susceptible to image lag. Pixels containing several trapezoidal shapes forming 'fingers' [22] have been developed and are in widespread use. Using this technique, a 100 µm long, narrow pixel [23] has been demonstrated, which uses the proximity of the PPD to the *p*-well to enhance the gradient created by the PPD shape. Without the 3D potential shaping, such large pixels would have very long charge transfer times.

A sophisticated example [24] is shown in figure 3.19, where implants with increasing width and depth are used together to create a sizeable potential gradient within the PPD. The maximum achieved electric field is around 500 V cm^{-1}, which is sufficient to bring down the charge collection time to below 10 ns. Another benefit is that the collected charge is stored just next to the transfer gate, which helps reduce the transfer time too.

3.6 Punch-through

Normally the depletion around the drain and source *pn* junctions in a MOSFET is completely contained in the transistor well. This is necessary to prevent charge from draining parasitically to the transistors instead of being collected in the PPD. Naturally, any charge generated within the *p*-well will get collected by the source and the drain and will never reach the PPD.

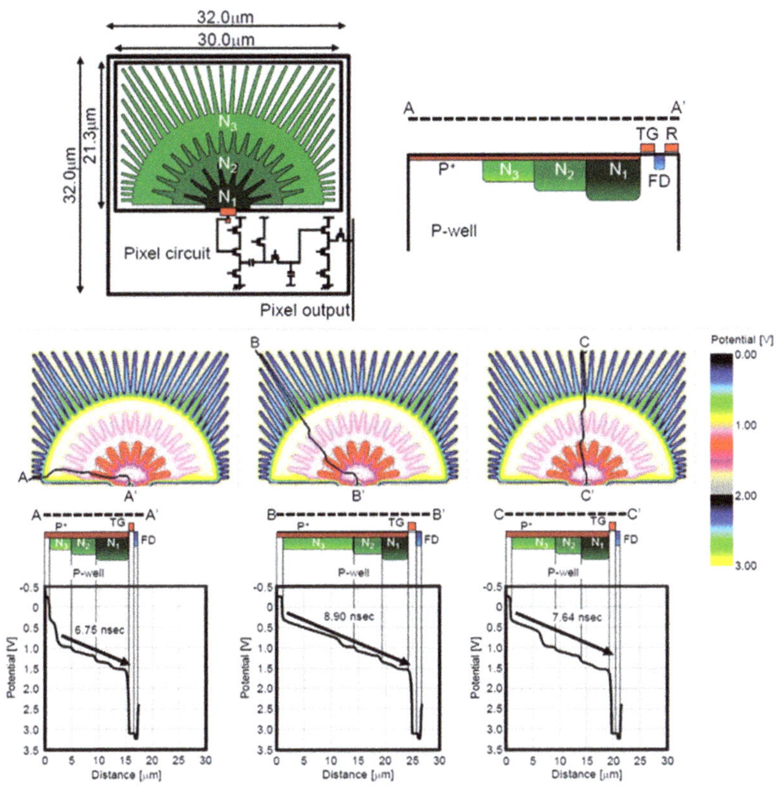

Figure 3.19. Potential gradients in a PPD for quick charge collection and transfer. Reprinted with permission from [24].

If the depletion can reach the bottom of the p-well it would rapidly expand under it because the epitaxial layer is lightly doped. In this case we say that the depletion punches through the p-well, and the effect is called punch-through (PT). This is usually undesirable, and measures are taken to prevent it by making the p-well heavily doped and deep enough, with good margin.

The PT effect is not always unwanted and can find some uses, with one example of a photodiode in a p-well shown in figure 3.20. By carefully designing the n-type doping profile of the diode (shown in figure 3.21(a)) the depletion region can be fully contained within the p-well for low bias voltages. Below 2 V the potential at the bottom of the p-well in figure 3.21(b), at around 1 μm in depth, is near the substrate potential—sure indication that the p-well is holding on.

Punch-through occurs when the diode depletion exceeds W_{PT} at bias above 5 V and rapidly expands to the full depth of the device. By using the PT effect, the depletion abruptly grows by an order of magnitude, from <1 μm (the depth of W_{PT}) to 10 μm. This is accompanied by an increase of the photosensitivity by the same factor. At diode bias below 2 V the photogenerated charge under the p-well is reflected into the epitaxial layer and not collected. At bias above 5 V, suddenly, all the charge is collected. A very important detail in achieving this is the high resistivity

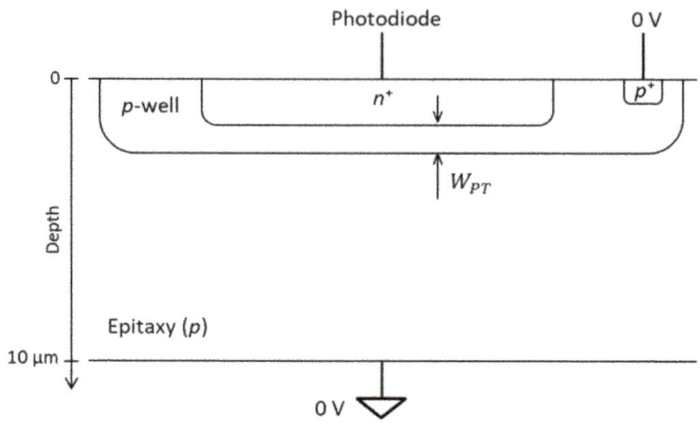

Figure 3.20. Photodiode in a *p*-well.

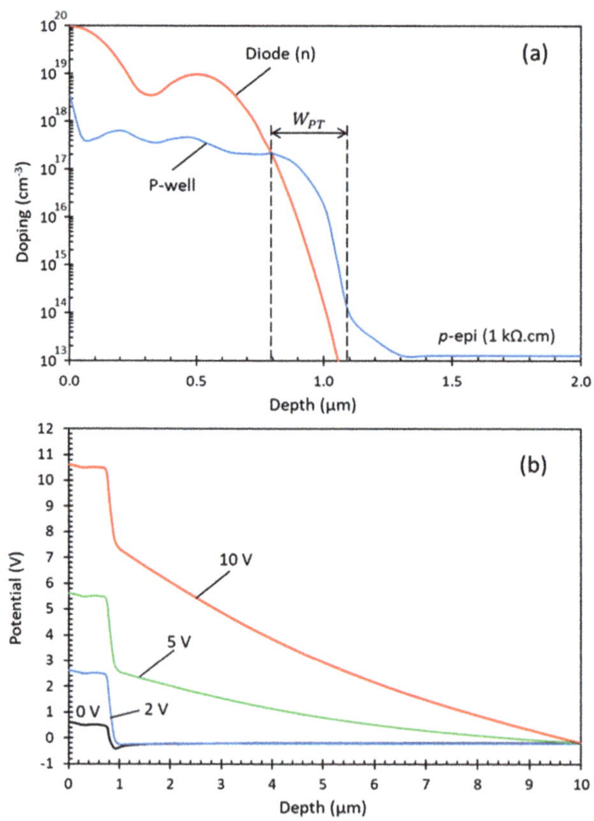

Figure 3.21. Doping concentration along the centre line of the photodiode (a); potential profiles in depth for increasing photodiode bias.

epitaxial layer, allowing large depletion depth at low bias voltages. This is the functionality of an *electronic shutter*, allowing the exposure to be controlled. PT-based electronic shutters have been developed for CCDs [25] where they are practical due to the higher available voltages.

In practice, a device using PT as an electronic shutter is not easy to design. The depth of the junction between the *n*-type diode and the *p*-well is determined by the difference between two large doping concentrations. Precise control of the implantation and annealing are needed to achieve the desired doping profile and correct operation. Further complications could be caused by the need for relatively high voltages and the high resistivity epitaxial layer for better shutter efficiency.

An alternative method to build shutter functionality is to drain away the signal without changing the photodiode voltage. In figure 3.22 the photodiode is surrounded by two drains placed in *p*-wells. Punch-through is achieved when the drain voltage V_D is pulsed sufficiently high, the drain depletions expand into the epitaxial layer below and become much larger than the photodiode's depletion. In this condition the charge that would normally be collected by the photodiode is 'stolen' by the two drains, and the shutter is closed. Once the drain voltage goes back to zero the depletions shrink to within the *p*-wells' volume and the shutter is open.

Charge generated in the *p*-wells does not reach the photodiode and is lost regardless of the drain bias, therefore they should be physically small. Also, the sizes and the depths of the n^+ drain and the *p*-well should be carefully designed, so that the punch-trough happens only in the vertical direction as shown in figure (3.23), and not sideways towards the photodiode.

A PPD with vertical overflow drain shutter using the PT effect has been developed too [13]. Here, the PPD is built on an *p*-type epitaxial layer grown on a n^{++} substrate instead of the usual p^{++} one. This creates a *pn* junction at the bottom of the epi layer. A deep *p*-well is implanted under the PPD to create a potential barrier to the substrate. When sufficiently high voltage is applied to the substrate relative to the epitaxial layer (held at 0 V), the deep *p*-well is punched through.

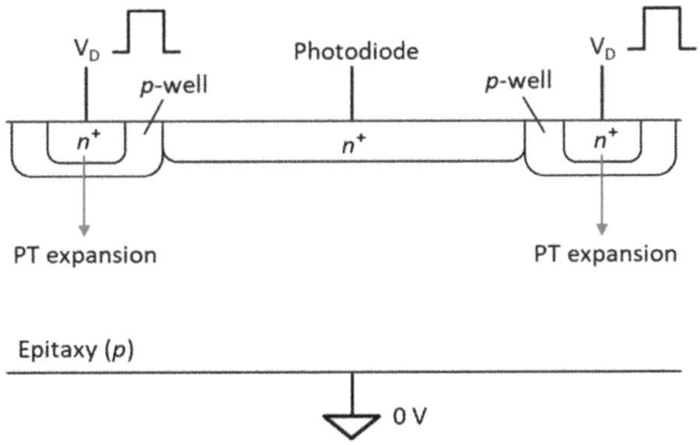

Figure 3.22. Electronic shutter using a drain in a *p*-well and punch-through.

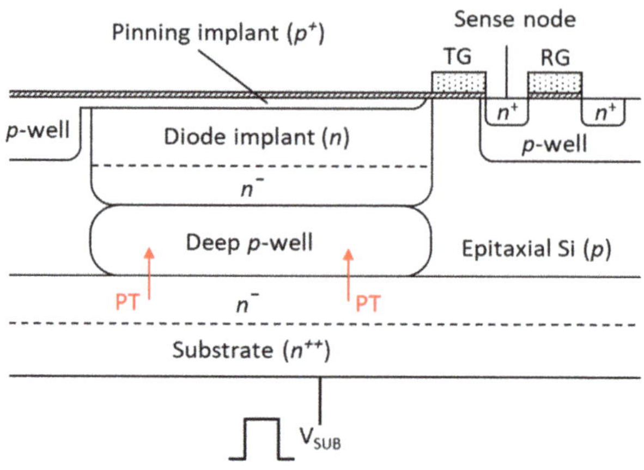

Figure 3.23. PPD with vertical overflow drain shutter [13].

In this condition the photogenerated charge and the charge stored in the PPD are drained to the substrate.

3.7 Field-induced junctions

High resistivity silicon, *p*-type doped to 10^{13}–10^{15} cm^{-3}, is very easy to bring to inversion. As we saw in chapter 1, the *n*-MOSFET threshold at such resistivities is negative. If the passivating dielectric has large fixed positive charge, the silicon below can be in inversion permanently, without any voltage applied. Such inversion layers can be useful—examples are the back-surface passivation in section 3.4.3 and the field-induced junction in photodiodes using nano-structured black silicon [26].

More often, surface inversion layers are undesirable and can lead to surprising effects. Figure 3.24(a) demonstrates such an effect in an array of photodiodes, where they become weakly connected by the thin inversion layer of electrons. Experimentally, the device behaves as one giant photodiode instead of an array of independent ones—the image is a blur, and the project is a failure.

The way to solve this, shown in figure 3.24(b), is borrowed from the field of radiation damage effects in semiconductors, which routinely deals with inversion layers caused by ionising radiation. Adding a narrow, highly doped *p*-well between the diodes cuts off the conduction path and eliminates the weak connections between them. The downside is that the diode's capacitance increases due to the proximity of the *p*-well, which is shaped as a grid and surrounds them on all sides. The *p*-well does not need to be connected to ground in every pixel; its low resistivity is sufficient to ensure constant potential when connected only at the chip's periphery.

It is usually a good idea to add measures to avoid such an effect even in lower resistivity silicon because the surface conditions are hard to predict. A variety of methods with increasing robustness are presented in [27].

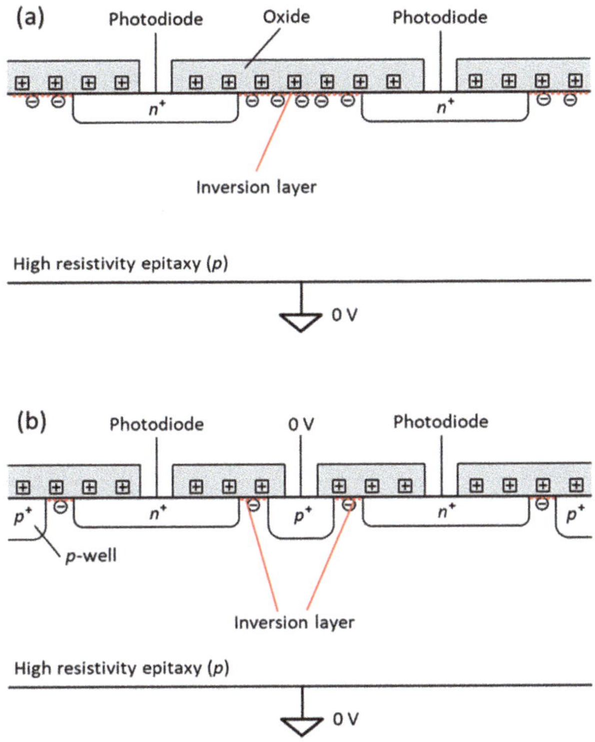

Figure 3.24. Field-induced leakage between photodiodes in high resistivity silicon.

Chapter summary

1. Photogenerated charge is fully collected from the epitaxial layer even in partial depletion, provided that the electron diffusion length is sufficiently long, which it usually is.
2. Very little charge is collected from the highly doped substrate due to the short electron lifetime and diffusion length.
3. The dominant sources of dark current are traps in the depletion region and at the surface, both having $\exp(-E_g/2kT)$ temperature dependence. The diffusion dark current is not trap-related, becomes dominant at high temperatures and has $\exp(-E_g/kT)$ dependence.
4. The surface dark current can be reduced by orders of magnitude by 'pinning' the surface to substrate potential with high hole concentration (in *p*-type silicon).
5. Reflective barriers prevent minority charge carriers from entering a highly doped region from a lower-doped region of the same type. They can be used to reduce parasitic charge collection.
6. Back-side illumination provides the highest QE because light reflections from metals and thick passivation layers are avoided. The back surface of the

sensor must have very thin passivation to minimise light absorption while preventing recombination and dark current generation.

7. Potential gradients can be created by the difference in the width of a photodiode. The small electric field generated in this way can be used to speed up charge collection and transfer in PPDs.
8. The punch-through effect can be used as an electronic shutter.
9. Undesirable field-induced junctions and surface leakage can be controlled by breaking the conduction path with highly doped regions.

References

[1] Sze S 1981 *Physics of Semiconductor Devices* 2nd edn (New York: Wiley)

[2] Richter A, Glunz S W, Werner F, Schmidt J and Cuevas A 2012 Improved quantitative description of Auger recombination in crystalline silicon *Phys. Rev.* B **86** 165202

[3] Janesick J, Andrews J, Tower J, Grygon M, Elliott T, Cheng J, Lesser M and Pinter J 2007 Fundamental performance differences between CMOS and CCD imagers: Part II *Proc. of SPIE, 669003 (San Diego)*

[4] MacKenty J 2012 Performance and calibration of the HST wide field camera 3 *Proc. of SPIE, 84421V (Amsterdam)*

[5] Andor Neo 5.5 sCMOS specifications Andor https://andor.oxinst.com/assets/uploads/products/andor/documents/andor-neo-scmos-specifications.pdf)

[6] Pichler P and Point I 2012 *Defects, Impurities, and Their Diffusion in Silicon* (Vienna: Springer)

[7] McGrath D, Tobin S, Goiffon V, Magnan P and Le Roch A 2018 Dark current limiting mechanisms in CMOS image sensors *IS&T Int. Symp. on Electronic Imaging 2018, Image Sensors and Imaging Systems (Burlingame, CA)*

[8] Brunetti A M, Musolino M, Strangio S and Choubey B 2020 Pixel design driven performance improvement in 4T CMOS image sensors: dark current reduction and full-well enhancement *IEEE Trans. Electron Devices* **67** 409–12

[9] Seo M, Kawahito S, Yasutomi K, Kagawa K and Teranishi N 2014 A low dark leakage current high-sensitivity CMOS image sensor with STI-less shared pixel design *IEEE Trans. Electron Devices* **61** 2093–7

[10] Grove A S 1967 *Physics and Technology of Semiconductor Devices* (New York: Wiley)

[11] Janesick J 2001 *Scientific Charge-Coupled Devices* (Bellingham, WA: SPIE Press)

[12] Gradshteyn I S and Ryzhik I M 2007 *Table of Integrals, Series, and Products* 7th edn (Amsterdam: Elsevier)

[13] Teranishi N 2016 Effect and limitation of pinned photodiode *IEEE Trans. Electron Devices* **63** 10–5

[14] Snoeys W 2013 Monolithic pixel detectors for high energy physics *Nucl. Instr. Meth. Phys. Res.* A **731** 125–30

[15] Lahav A, Fenigstein A, Strumb A and Rizzolo S 2020 Backside illuminated (BSI) complementary metal-oxide semiconductor (CMOS) image sensors *High Performance Silicon Imaging* ed D Durini 2nd edn (Cambridge: Woodhead Publishing) pp 95–117

[16] Nikzad S *et al* 2017 High-efficiency UV/optical/NIR detectors for large aperture telescopes and UV explorer missions: development of and field observations with delta-doped arrays *J. Astron. Telesc. Instrum. Syst.* **3** 036002

[17] Green M A 2008 Self-consistent optical parameters of intrinsic silicon at 300 K including temperature coefficients *Solar Energy Mater. Solar Cells* **92** 1305–10

[18] Law M, Solley E, Liang M and Burk D 1991 Self-consistent model of minority-carrier lifetime, diffusion length, and mobility *IEEE Electron. Dev. Lett.* **12** 401–3

[19] Hoenk M *et al* 2009 Delta-doped back-illuminated CMOS imaging arrays: progress and prospects *Proc. of SPIE 74190T (San Diego)*

[20] Hoex B *et al* 2008 On the c-Si surface passivation mechanism by the negative-charge-dielectric Al_2O_3 *J. Appl. Phys.*, **104** 113703

[21] Shin B, Park S and Shin H 2010 The effect of photodiode shape on charge transfer in CMOS image sensors *Solid-State Electron.* **54** 1416–20

[22] Cao X *et al* 2015 Design and optimisation of large 4T pixel *Int. Image Sensor Workshop (Vaals, The Netherlands)*

[23] Kalgi A K, Crouwels A, Dierickx B, Verbruggen W and Aken D V 2019 Fast charge transfer in 100µm long PPD pixels *Int. Image Sensor Workshop (Snowbird, UT)*

[24] Miyauchi K *et al* 2014 Pixel structure with 10 nsec fully charge transfer time for the 20M frame per second burst CMOS image sensor *Proc. of SPIE, 902203 (San Francisco, CA)*

[25] Reich R K *et al* 1993 Integrated electronic shutter for back-illuminated charge-coupled devices *IEEE Trans. Electron Devices* **40** 1231–7

[26] Juntunen M A, Heinonen J, Vähänissi V, Repo P, V D and Savin H 2016 Near-unity quantum efficiency of broadband black silicon photodiodes with an induced junction *Nat. Photon.* **10** 777–81

[27] Goiffon V, Cervantes P, Virmontois C, Corbière F, Magnan P and Estribeau M 2011 Generic radiation hardened photodiode layouts for deep submicron CMOS image sensor processes *IEEE Trans. Nucl. Sci.* **58** 3076–84

Konstantin D Stefanov

Chapter 4

Noise and readout techniques

4.1 Noise in image sensors

Noise is rarely an exciting subject. Usually, we all care about the signal and meticulously calculate its amplitude in the various parts of the system. This is, however, only half the story because noise is there too, unavoidable, and ever-present. The system performance depends on the signal-to-noise (SNR) ratio, where the signal and the noise are equal partners. Not paying enough attention to the noise could be a costly mistake, because, to paraphrase Lewis M Branscomb, 'Nature loves the noise as much as the signal'.

4.1.1 Thermal and reset noise

Thermal (or Johnson) noise is generated in conductors, such as resistors, due to the random thermal motion of the charge carriers within them. It takes the form of random voltage generated across the two ends of a resistor. The thermal noise voltage has zero average and therefore produces no DC voltage[1], but has non-zero root mean square (RMS).

The charge carriers are almost always electrons, but in a *p*-type semiconductor holes carry out the same role. Thermal noise is present at any temperature above absolute zero and does not depend on whether a current is flowing through the resistor, or on the material[2] the resistor is made of. Capacitors do not exhibit thermal noise because their impedance is only reactive, without any active (resistive) component.

The open-circuit, RMS thermal noise voltage of a resistor with resistance R is given by the formula

[1] Generating a DC voltage would mean that resistors can generate power out of nothing, which clearly does not happen.

[2] Here we talk only about thermal noise. There are some materials, for example those used in carbon resistors, that exhibit higher 'excess' noise through different mechanisms.

doi:10.1088/978-0-7503-3235-4ch4

$$\overline{v_n} = \sqrt{4kTRB_n} \tag{4.1}$$

In (3.38) k is the Boltzmann constant, T is the absolute temperature, and B_n is the *noise power* bandwidth in units of Hertz, which is different from the amplitude bandwidth. The term $e_n = \sqrt{4kTR}$ is called *voltage noise density* and is frequency independent for thermal noise; this is why it is 'white'—there is no spectrum and therefore no 'colour' in it, in analogy to visible light. The RMS noise voltage of any white noise (including shot noise which is also white) is given by multiplying the noise voltage density e_n by the square root of the noise bandwidth:

$$\overline{v_n} = e_n\sqrt{B_n} \tag{4.2}$$

This definition makes the noise voltage density have the strange units of volt per root Hertz (V/$\sqrt{\text{Hz}}$, or V/rtHz). This is a consequence of the way we calculate the noise—we work with *noise power*, which is the mean squared noise voltage instead of the instantaneous noise amplitude.

Example 4.1. Calculate the thermal voltage noise density of a 1 kΩ resistor at 20 °C (293 K) and the RMS noise voltage over 1 MHz noise bandwidth. The Boltzmann constant is 1.38×10^{-23} J K^{-1}.
Solution: Using that $e_n = \sqrt{4kTR}$

$$e_n = \sqrt{4kTR} = \sqrt{4 \times 1.38 \times 10^{-23} \times 293 \times 1000} = 4.02\ \text{nV}/\sqrt{\text{Hz}}$$

The RMS noise voltage is calculated from (4.2):

$$\bar{v}_n = e_n\sqrt{B_n} = 4.02 \times 10^{-9} \times \sqrt{10^6} = 4.02\ \mu\text{V RMS}$$

The noise power (the squared RMS noise voltage) is proportional to the noise power bandwidth, which we will investigate for the humble, but regularly encountered RC low pass filter shown in figure 4.1(a). This filter has amplitude (i.e. signal) bandwidth, defined as the frequency at which the output amplitude decreases by $\sqrt{2}$ (equal to −3 dB), compared to much lower frequencies:

$$B_{-3\text{dB}} = \frac{1}{2\pi RC} \tag{4.3}$$

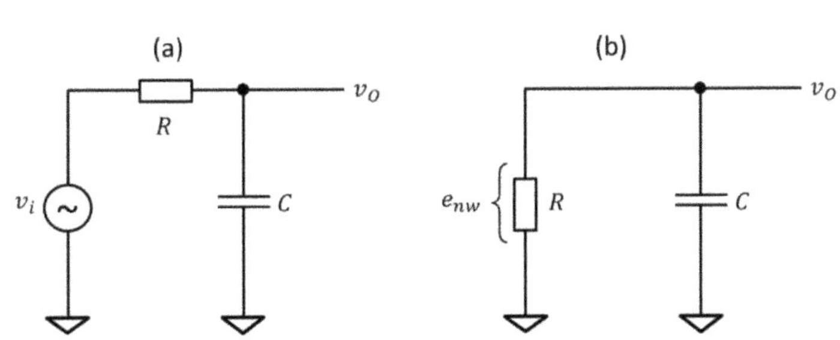

Figure 4.1. (a) RC low pass filter; (b) thermal noise from a resistor as an input to RC low pass filter.

The output amplitude as a function of frequency is given by [1]

$$v_o = \frac{v_i}{\sqrt{1 + (2\pi fRC)^2}} \tag{4.4}$$

If the input to the filter is a noise voltage with power $\overline{v_i^2}$, the noise power at the output is

$$\overline{v_o^2} = \frac{\overline{v_i^2}}{1 + (2\pi fRC)^2} \tag{4.5}$$

The term multiplying the input noise power $\overline{v_i^2}$ in (4.5) is the *power transfer function* $|H(f)|^2$ which determines how much of the input power will make it to the output. In the case when both the noise density and the power transfer function are frequency-dependent, the noise power is calculated as

$$\overline{v_n^2} = \int_0^\infty e_n^2(f)|H(f)|^2\, df \tag{4.6}$$

Equation (4.6) is used extensively later in this chapter to calculate the output noise of various readout circuits. For white noise $e_n^2(f)$ is frequency-independent and therefore constant $e_n^2(f) = e_{nw}^2$; any frequency dependence means that the noise has 'colour'.

To calculate the noise power bandwidth let us suppose that white noise voltage v_i with density e_{nw} is applied to the input of the RC low pass filter. Since e_{nw} is constant, we can write (4.6) as

$$\overline{v_o^2} = e_{nw}^2 \int_0^\infty |H(f)|^2\, df = e_{nw}^2 \int_0^\infty \frac{df}{1 + (2\pi fRC)^2} df = \frac{e_{nw}^2}{4RC} \tag{4.7}$$

The term multiplying the noise density is the noise power bandwidth $B_n = 1/4RC$. We can immediately see that B_n is higher than the signal bandwidth B_{-3dB} (4.3) by a factor of $\pi/2 = 1.57$. Now, if the input noise voltage is not coming externally but is generated by the resistor itself, as in figure 4.1(b), we get

$$\overline{v_o^2} = \frac{e_{nw}^2}{4RC} = \frac{4kTR}{4RC} = \frac{kT}{C} \tag{4.8}$$

Equation (4.8) has far-reaching consequences and is of course describing the notorious *kTC* noise. It is telling us that in a RC network the RMS thermal noise voltage is

$$\overline{v_n} = \sqrt{\frac{kT}{C}} \tag{4.9}$$

and intriguingly, that it does not depend on the resistance, despite the resistor actually generating the noise. This happens because the noise power is proportional to the resistance, but the noise bandwidth is inversely proportional to it, so the resistance cancels out.

The kTC noise can be found where resistors, capacitors and switches (which invariably have some resistance) are used, which is essentially most things electronic. In image sensors, kTC noise plays a fundamental part in determining their readout noise and in selecting the optimal readout technique.

Conventional readout circuits in 3T and 4T CIS, shown in figure 4.2(a), use a transistor to reset the sense node before the charge is collected or transferred. The reset causes kTC noise, also called *reset noise*, which dominates the noise performance of 3T CISs.

When switched on, the reset transistor (figure 4.2(b)) operates in the linear regime and behaves as a resistor, with resistance R_{on} in the kilo-Ohm range. For AC signals R_{on} and the sense node capacitance C_{sn} are connected in parallel to the gate of the source follower (SF) as in figure 4.2(c). This is identical to figure 4.1(b) which we used to derive the formula for the kTC noise, therefore we expect to see the same at the output, assuming unity SF gain.

Figure 4.3 shows an idealised signal at a sense node subjected to periodic reset. The high noise during reset is interrupted when the reset transistor switches off, and

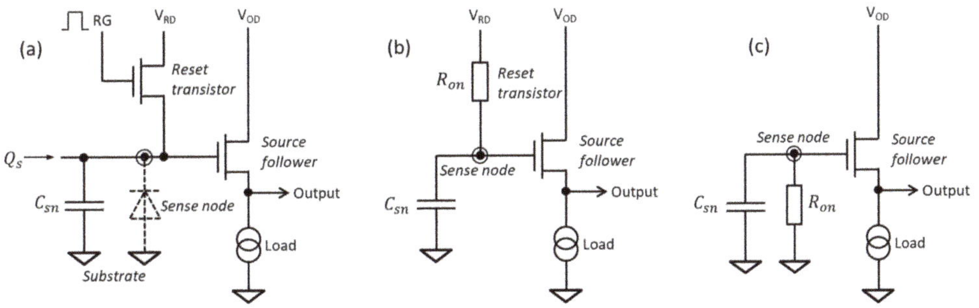

Figure 4.2. (a) Conventional CIS output circuit using a reset transistor; (b) equivalent circuit during reset; (c) small signal model of the sense node during reset. V_{RD} and V_{OD} are the DC supplies to the reset transistor and the source follower, respectively, and RG is the pulsed voltage to the gate of the reset transistor.

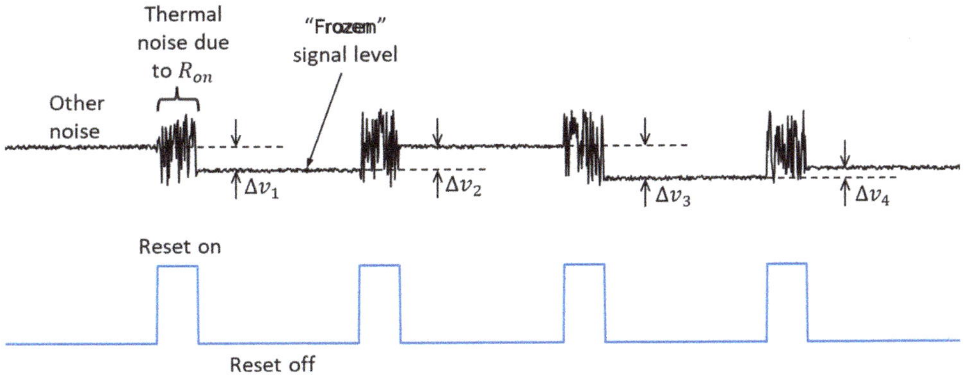

Figure 4.3. Sense node voltage with periodic reset.

the signal becomes 'frozen' at the instantaneous noise voltage at that moment. The photogenerated signal is expected to appear in between resets but is not shown here for clarity. If we just sample the signal in between the resets, the differences between successive samples Δv will be the difference between the instantaneous noise voltages at reset turn-off. This is equivalent to merely sampling during reset at random, therefore the output signal will contain the full reset noise as given by (4.9) with $C = C_{sn}$.

If the reset noise is not large and the application can tolerate it, we can live with it, but in most CISs the reset noise is prohibitively high and can dwarf all the other noise sources in the system. To quantify it in a convenient form, we can convert the RMS noise given by (4.9) to equivalent noise charge (ENC) using that $\overline{Q_n} = C_{sn}\overline{v_n}$, and then express Q_n in electrons RMS by dividing by the elementary charge:

$$\overline{Q_n} = \frac{C_{sn}}{q}\sqrt{\frac{kT}{C_{sn}}} = \frac{\sqrt{kTC_{sn}}}{q} \tag{4.10}$$

Table 4.1 gives the expected reset noise for a wide range of sense node capacitances. Perhaps counter-intuitively, higher reset noise in volts corresponds to lower reset noise in electrons! This is not a mistake, but is telling us that the noise, when expressed in volts RMS may not give us the best measure; using electrons is more fundamental because it is easier to relate to the number of photons received by the sensor.

As C_{sn} decreases, the reset noise in electrons RMS decreases too, but is still considerable even for very small capacitances. A brief look at the datasheets of commercial sensors shows that their noise can be at least an order of magnitude lower. The reason for that is the correlated double sampling (CDS) technique, described in section 4.2, which allows the reset noise to be fully suppressed in 4T and other image sensors using charge transfer. CDS does not work well in 3T pixels and their noise performance suffers in comparison.

Table 4.1. Reset noise at 20 °C expressed in volts RMS and equivalent noise charge in electrons RMS for sense node capacitances from 1 fF to 1 pF.

C_{sn} (fF)	CVF (μV/e$^-$)	Reset noise (μV RMS)	Reset noise Q_n (e$^-$ RMS)
1	160	2010.8	12.6
2	80	1421.9	17.8
5	32	899.3	28.1
10	16	635.9	39.7
20	8	449.6	56.2
50	3.2	284.4	88.9
100	1.6	201.1	125.7
200	0.8	142.2	177.7
500	0.32	89.9	281.0
1000	0.16	63.6	397.4

4.1.2 Shot noise

Shot noise comes in two varieties, photonic and electronic. Both are caused by the discrete nature of their carriers—photons and electrons, respectively.

Starting with the photonic shot noise first, let us consider a sensor element which under steady illumination receives on average N photons per fixed time interval. The number of photons received in each time interval statistically fluctuates and is described by the Poisson distribution. If the average number of photons in a time interval is N, the probability P that k photons are registered in a chosen interval of the same length is

$$P(k) = \frac{N^k e^{-N}}{k!} \tag{4.11}$$

The standard deviation σ_N is given by:

$$\sigma_N = \sqrt{N} \tag{4.12}$$

It is the mechanism of random, independent arrival of photons that is responsible for the Poisson statistics. Any correlation between the emission or the arrival times of the photons would show itself by deviations from equations (4.11) and (4.12), as observed in some quantum systems.

If every photon generates a single photoelectron, as for visible light in silicon, the number of collected electrons 'inherits' the Poisson distribution from the photons. The resulting signal will exhibit shot noise with standard deviation given by (4.12).

Even if the incoming photons are perfectly described by Poisson statistics, if the mechanism by which they are converted to electrons is not truly random, the resulting standard deviation will be different from (4.12). This is observed for energetic photons (x-rays, gammas) capable of generating multiple electron–hole pairs because the events of carrier generation are not independent and are correlated by the law of energy conservation. Therefore, the standard deviation of the number of generated electrons from a photon with energy E_{ph} and electron–hole creation energy E_{w} is lower than the one expected from the Poisson distribution $\left(\sqrt{E_{\text{ph}}/E_{\text{w}}}\right)$ and is given by:

$$\sigma_N = \sqrt{F\frac{E_{\text{ph}}}{E_{\text{w}}}} \tag{4.13}$$

The constant F is called Fano factor [2], is less than one, and depends on the semiconductor material. For silicon and $E_{\text{ph}} > \approx 50$ eV, $F = 0.115$ and this results in a substantially supressed standard deviation. This much reduced noise is called 'Fano noise' and is the limit of the energy resolution for soft x-rays commonly used in detector calibration [3].

Electronic shot noise is a consequence of the fact that electric current consists of discrete electric charges, being electrons or holes. The statistical fluctuations of the number of charge carriers passing through a conductor per unit time is seen as noise in the measured current.

The RMS current noise in a DC current I, measured in a noise bandwidth B_n is given by

$$\overline{i_{n(rms)}} = \sqrt{2qIB_n} \tag{4.14}$$

The term $i_n = \sqrt{2qI}$ is the *current noise density*, measured in units of amperes per root Hertz (A/$\sqrt{\text{Hz}}$). Similarly to the thermal noise, the shot noise is white and follows Gaussian statistics, which transitions to Poisson distribution at very low currents. Shot noise is independent on temperature and exists only when there is a current flow, unlike thermal noise, which is proportional to $\sqrt{T}$ but does not depend on whether a current is flowing or not.

Shot noise is characteristic in forward and reverse currents in *pn* junctions and dark current in image sensors. To generate shot noise, charge carriers must arrive independently of each other, implying that they must be either independently generated, as in dark current from individual traps, or they are crossing a barrier and cannot freely travel in both directions, as in forward current in *pn* junctions.

Any correlation in charge carrier generation or arrival reduces the statistical fluctuations of the current, and therefore results in less noise. Shot noise is strongly suppressed in conductors, such as metallic wires and resistors, due to long-range correlation between electrons. There is virtually no shot noise in simple resistive circuits. This is why there is no shot noise in the circuit in figure 4.4(a), but in figure 4.4(b) there is.

Noise from independent mechanisms and sources adds up in quadrature, meaning that the noise powers are added, not the RMS voltages. For example, a shot noise current flowing through a resistor generates RMS noise voltage drop v_{ns} across it. This adds to the thermal noise voltage v_{nw} and the total noise voltage across the resistor becomes

$$\overline{v_n} = \sqrt{\overline{v_{nw}^2} + \overline{v_{ns}^2}} \tag{4.15}$$

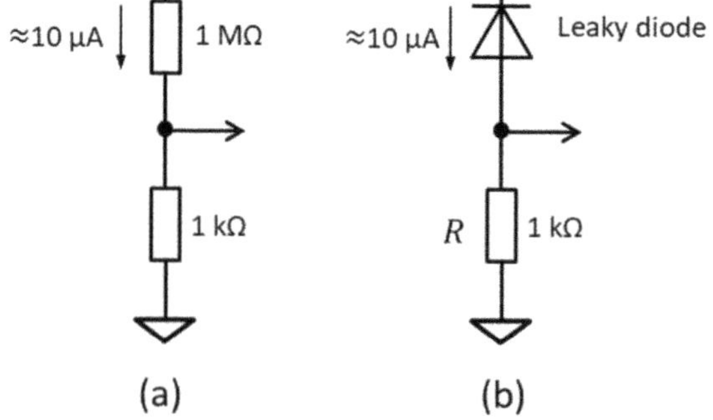

Figure 4.4. (a) No shot noise; (b) current flowing through the diode and the resistor has shot noise determined by (4.14).

Summation in quadrature can be expanded to multiple noise sources and takes the form

$$\overline{v_{\mathrm{n}}} = \sqrt{\overline{v_{\mathrm{n}1}^2} + \overline{v_{\mathrm{n}2}^2} + \dots + \overline{v_{\mathrm{n}N}^2}} \tag{4.16}$$

in the case of N noise sources.

Example 4.2. Calculate the total noise voltage across the 1 kΩ resistor at 20 °C (293 K) over 1 MHz noise bandwidth for figure 4.4(b).
Solution: There are both shot and thermal noise voltages which should be added in quadrature. The shot noise voltage is found by multiplying the resistance by the RMS noise current (4.14): $\overline{v_{ns}} = R\sqrt{2qIB_{\mathrm{n}}}$

$$\bar{v}_{\mathrm{ns}} = 1000 \times \sqrt{2 \times 1.6 \times 10^{-19} \times 10 \times 10^{-6} \times 10^{6}} = 1.79\,\mu\mathrm{V\ RMS}$$

Repeating example 4.1, the thermal noise is $\overline{v_{\mathrm{nw}}} = 4.02\ \mu\mathrm{V}$ RMS. The total noise is therefore

$$\bar{v}_{\mathrm{n}} = \sqrt{1.79^2 + 4.02^2} = 4.4\,\mu\mathrm{V\ RMS}$$

4.1.3 1/*f* and random telegraph noise

Another important noise in electronic components is $1/f$ noise, named after the frequency dependence of its noise power. It is also called flicker noise, or pink noise if you are into audio. Because the noise power falls as $1/f$, the voltage noise density e_{n} is proportional to $1/\sqrt{f}$ and the use of (4.6) for calculating the RMS noise voltage over the bandwidth is mandatory.

Flicker noise is not limited to electronics but is present also in a huge variety of physical systems [4]. The experimentally measured noise power often deviates from the pure $1/f$ dependence in parts or the whole of the spectrum and can be expressed as $\propto 1/f^{\gamma}$, where γ is between 0.8 to 1.2.

In the presence of both white and $1/f$ noise the total noise power density e_{n}^2 can be written as

$$e_{\mathrm{n}}^2 = e_{\mathrm{nw}}^2 + e_{\mathrm{nf}}^2/f \tag{4.17}$$

where the parameter e_{nf} describing the $1/f$ noise has dimensions of volts. Because thermal (which is white) noise is ever-present, the parameter e_{nf} can be expressed from the white noise density and the frequency f_{nc} at which they have equal powers, such that $e_{\mathrm{nw}}^2 = e_{\mathrm{nf}}^2/f_{\mathrm{nc}}$. This leads to

$$\therefore e_{\mathrm{nf}} = e_{\mathrm{nw}}\sqrt{f_{\mathrm{nc}}} \tag{4.18}$$

and allows us to rewrite (4.17) in the more convenient form

$$e_{\mathrm{n}}^2 = e_{\mathrm{nw}}^2\left(1 + f_{\mathrm{nc}}/f\right) \tag{4.19}$$

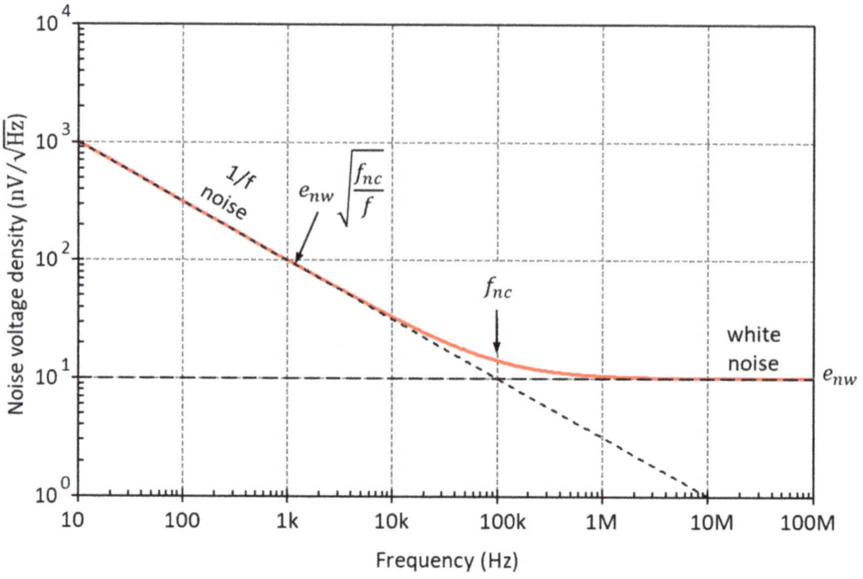

Figure 4.5. Noise density spectrum with white noise ($e_{\text{nw}} = 10\ \text{nV}/\sqrt{\text{Hz}}$) and $1/f$ noise ($f_{\text{nc}} = 100$ kHz).

The frequency f_{nc} is called noise corner frequency and is shown in figure 4.5 where the noise density is plotted against the frequency on a log-log scale. Notice how the $1/f$ noise density increases by an order of magnitude for every two orders of magnitude of frequency decrease.

The noise density is often plotted against the frequency as in figure 4.5 when measured experimentally. A good sign of a low noise device is its noise corner frequency—the lower, the better.

Both $1/f$ and random telegraph signal (RTS) noise are present in MOS transistors. They are subject to intensive research and thousands of papers on their origins and effects have been published. In brief, $1/f$ noise is caused by the fluctuations of the number of carriers in the transistor channel due to capture and release by interface states, fluctuations of the carrier mobility, or both [5]. Experimentally it has been established that the carrier fluctuation model is a better fit for *n*-MOSFETs, which are our main interest, and that their $1/f$ noise has only a weak temperature dependence [6].

In very small transistors it has been observed that the $1/f$ spectra look 'bumpy' [7], due to the capture and release of carriers by individual traps. The effect is also visible in the time domain as signal 'jumps' between two or more levels and is well known as RTS, giving rise to random telegraph noise (RTN). The noise power spectrum of a single trap has Lorentzian spectrum and $1/f^2$ dependence. When there are multiple RTS sources in the transistor channel, their combined effect can be approximated with $1/f$ noise [5].

Even though the effects of RTN in image sensors have been considered [7–9], in the following sections we only deal with white (thermal) and $1/f$ noise.

4.1.4 MOSFET noise

MOS transistors are an integral part of the pixel, whether used for buffering, amplification or switching. Noise originating in pixel MOSFETs usually dominates the noise performance of the sensor except at high signal levels, when photon shot noise becomes prevalent.

The dominant noise sources in MOSFETs are thermal noise, due to their channel conductance, and 1/*f* noise, which depends strongly on the processing technology. Determining the contribution from these noise sources in conjunction with the transfer function of the chosen readout is something we need to do very often.

The technique used in noise analysis is to think of the electronic devices to be 'noiseless', but to add their observed noise to the circuit as external voltages and currents. The MOSFET is a voltage-controlled device generating current, therefore we can consider that the thermal and 1/*f* noise show in the drain current as *current noise*. Those current noise sources are in parallel with the drain current of the noiseless MOSFET and generate voltage noise on the load resistance. Figure 4.6(a) shows how this technique is applied to a MOSFET amplifier. The gain at low frequencies is $g_m R_L$ and the input voltage v_i is amplified to $v_o = v_i g_m R_L$ at the output. Transistor noise is represented by a current noise source with density i_n generating RMS noise voltage across the load resistor, which is the output noise voltage:

$$\overline{v_n} = R_L i_n \sqrt{B_n} \tag{4.20}$$

For simplicity, equation (4.20) assumes that the current noise is white.

The output noise is what we measure experimentally, and knowing it allows us to do one more step. Commonly in noise analysis the noise is represented at the input of the amplifier, or is 'referred to the input'. This is done because the comparison between signal and noise becomes independent of the amplifier gain—both are at the input and therefore are amplified the same. The noise is referred to the input by 'transferring' it mathematically from an output noise voltage or current to an input-referred noise voltage, by dividing it by the noise gain[3] of the amplifier. This is

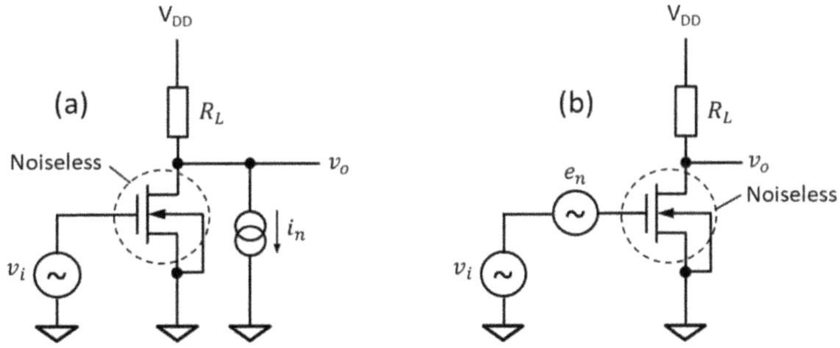

Figure 4.6. Equivalent noise schematic of a MOSFET amplifier (a); referring the output noise current to an input noise voltage.

[3] The noise gain can be different from the signal gain; an example is the inverting opamp amplifier [1].

shown in figure 4.6(b) where the current noise source at the output is replaced by an input noise voltage with density e_n. The MOSFET amplifies this additional voltage together with the signal, and the output noise voltage is the same for both circuits.

To consider the effects of the different noise sources we need to have models for their noise density spectra. MOSFETs operate by modulating the conduction of the channel, therefore thermal noise is generated there. If the channel is considered as a chain of infinitely small resistor segments connected in series, with each generating thermal noise, the drain current noise density in saturation is given by [10]

$$i_{\mathrm{nw}}^2 = S_{\mathrm{I_d}} = \frac{8kT}{3} g_{\mathrm{m}} \tag{4.21}$$

Unless the MOSFET is operated at very low currents in subthreshold mode, the shot noise in the drain current is negligible. The voltage noise density referred to the input is obtained by using the basic MOSFET formula linking the drain current to the gate–source voltage $i_D = g_m v_{GS}$. We can use it to describe any AC drain current, noise or otherwise, as if it is generated by a change in the gate–source voltage. For the amplifiers in figure 4.6, working in a common source configuration we obtain the input-referred voltage noise density as

$$e_{\mathrm{nw}}^2 = S_{\mathrm{V_g}} = \frac{i_{\mathrm{n}}^2}{g_{\mathrm{m}}^2} = \frac{8kT}{3g_{\mathrm{m}}} \tag{4.22}$$

Compared to (4.1), we can infer that the thermal noise of the MOSFET is equivalent to the resistance

$$R_{\mathrm{n}} = \frac{2}{3g_{\mathrm{m}}} \tag{4.23}$$

The factor 2/3 is a consequence from the distributed channel conductance and usually does not deviate much from 2/3, although it has some dependence on the operating conditions.

The drain current noise density of the $1/f$ noise in saturation is commonly written as [11]

$$i_{\mathrm{nf}}^2 = \frac{K_{\mathrm{F}}}{C_{\mathrm{ox}}^2 WL} \frac{g_{\mathrm{m}}^2}{f} \tag{4.24}$$

where K_F is a coefficient depending mostly on the process technology, and also on the temperature and the transistor bias.

The quadrature sum of the two noise currents is:

$$i_{\mathrm{n}}^2 = \frac{8kT}{3} g_{\mathrm{m}} + \frac{K_{\mathrm{F}}}{C_{\mathrm{ox}}^2 WL} \frac{g_{\mathrm{m}}^2}{f} \tag{4.25}$$

The total input-referred noise voltage density can be found from (4.25) by dividing by the transconductance squared

$$e_n^2 = \frac{i_n^2}{g_m^2} = \frac{8kT}{3g_m} + \frac{K_F}{C_{ox}^2 WLf} \tag{4.26}$$

Equation (4.26) is central for the following noise calculations and is widely used for image sensor design.

4.1.5 Source follower noise

The first transistor in the signal chain dominates the SNR, and therefore the noise characteristics of the in-pixel source follower are covered here in greater detail. In contrast to the voltage-driven circuits described in section 4.1.4, the in-pixel source follower has charge as an input signal. Charge can be represented as a current source, characterised with very high (essentially infinite) impedance. This high impedance gate drive causes the noise characteristics of the source follower to deteriorate, as we will see below.

First, we start with figure 4.7(a), showing a voltage-driven noiseless SF and its inherent transistor noise as a noise current i_n in parallel with the channel. Our goal is to find out how this noise current can be referred to the input as an input noise voltage, since the input signal is voltage. We are not making any assumptions of the nature of the noise current—it can be of any type.

The input signal v_i in figure 4.7(a) comes from a low impedance source. The input capacitance C_d is the sum of all the capacitances between the gate and the substrate,

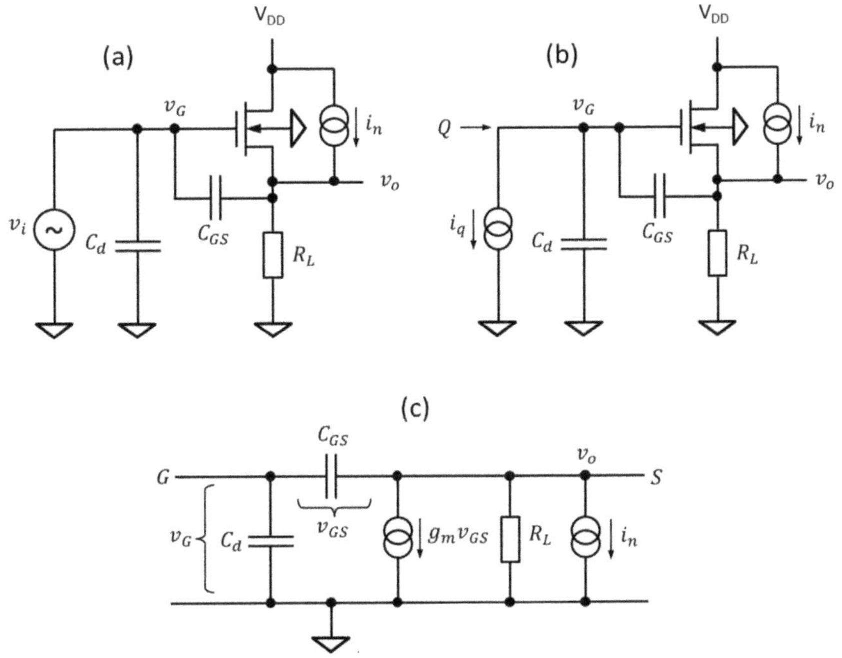

Figure 4.7. Source follower driven by a low impedance voltage signal (a); high impedance current signal (b); and its equivalent schematic for noise analysis (c).

such as the sense node, gate–drain (since the drain is at AC ground), and the parasitic wiring capacitances. All the capacitances between the gate and the channel are summed in C_{GS}, considered separately from the substrate capacitance, because the MOSFET source is not an AC ground.

Let us consider a small change Δi_n in the instantaneous drain noise current, which will cause the output voltage to change by Δv_{no}. The change of the gate–source voltage is equal to Δv_{no} because the gate is at AC ground, as v_i is a low-impedance source. Due to the inherent negative feedback of the SF circuit, an increase of the output voltage will result in a decrease of the gate–source voltage, and therefore of the drain current Δi_D. The AC output voltage is the sum of the noise-induced and the modulated drain current-induced voltage drops on the load, in anti-phase:

$$\Delta v_{no} = \frac{\Delta i_n}{g_L} - \frac{\Delta i_D}{g_L} = \frac{\Delta i_n}{g_L} - \frac{g_m \Delta v_{GS}}{g_L} = \frac{\Delta i_n}{g_L} - \frac{g_m \Delta v_{no}}{g_L} \tag{4.27}$$

Here we have used that the load conductance is the inverse of its resistance, $g_L = 1/R_L$. Solving (4.27) for Δv_{no} gives

$$\Delta v_{no} = \frac{\Delta i_n}{g_m + g_L} \tag{4.28}$$

which implies that the effective output conductance is $g_m + g_L$, the parallel combination of the transistor transconductance and the load conductance. If we divide Δv_{no} by the source follower gain G_{SF} we get the input-referred noise voltage

$$v_{ni} = \frac{\Delta v_{no}}{G_{SF}} = \frac{i_n}{\frac{g_m}{g_L + g_m}(g_L + g_m)} = \frac{i_n}{g_m} \tag{4.29}$$

This result is in line with the method used for the common source circuit in figure 4.6.

Now, considering the SF in figure 4.7(b) we can think of the gate as floating because it is driven by a current source with an infinite impedance. Therefore, the change of the gate–source voltage caused by a change of the output voltage gets attenuated by the capacitive divider formed by C_d and C_{GS}, as can be seen in the equivalent schematic in figure 4.7(c)

$$\Delta v_{GS} = \Delta v_{no} \frac{C_d}{C_d + C_{GS}} \tag{4.30}$$

Using the same method as in equation (4.27), the change of the output noise voltage caused by the transistor noise current is

$$\Delta v_{no} = \frac{\Delta i_n}{g_m \left(\frac{C_d}{C_d + C_{GS}} \right) + g_L} \tag{4.31}$$

Comparing with (4.28) we see that the effective transistor transconductance is reduced by a factor of $C_d/(C_d + C_{GS})$.

Since the input to the SF is charge, the transistor noise current should be referred to the input as an equivalent noise charge to allow like-for-like comparison. The ENC is the noise voltage referred to the input, multiplied by the input capacitance

$$Q_{\mathrm{ni}} = \frac{v_{\mathrm{no}} C_{\mathrm{in}}}{G_{\mathrm{SF}}} \tag{4.32}$$

The input capacitance C_{in} of the source follower was derived previously in chapter 1 and is

$$C_{\mathrm{in}} = C_{\mathrm{d}} + C_{\mathrm{GS}}(1 - G_{\mathrm{SF}}) \tag{4.33}$$

Substituting (4.31) and (4.33) into (4.32) we get

$$Q_{\mathrm{ni}} = \frac{i_{\mathrm{n}}}{G_{\mathrm{SF}}} \left[\frac{C_{\mathrm{d}} + C_{\mathrm{GS}}(1 - G_{\mathrm{SF}})}{g_{\mathrm{m}}\left(\frac{C_{\mathrm{d}}}{C_{\mathrm{d}} + C_{\mathrm{GS}}}\right) + g_{\mathrm{L}}} \right] = \frac{i_{\mathrm{n}}}{\frac{g_{\mathrm{m}}}{g_{\mathrm{m}} + g_{\mathrm{L}}}} \left[\frac{(C_{\mathrm{d}} + C_{\mathrm{GS}})(C_{\mathrm{d}} + C_{\mathrm{GS}}(1 - G_{\mathrm{SF}}))}{(g_{\mathrm{m}} + g_{\mathrm{L}})\left(C_{\mathrm{d}} + C_{\mathrm{GS}}\frac{g_{\mathrm{L}}}{g_{\mathrm{m}} + g_{\mathrm{L}}}\right)} \right] \tag{4.34}$$

Using that $1 - G_{\mathrm{SF}} = g_{\mathrm{L}}/(g_{\mathrm{L}} + g_{\mathrm{m}})$, we finally get

$$Q_{\mathrm{ni}} = \frac{i_{\mathrm{n}}}{g_{\mathrm{m}}}(C_{\mathrm{d}} + C_{\mathrm{GS}}) \tag{4.35}$$

Formula (4.35) means that the effective input capacitance for the ENC is $C_{\mathrm{d}} + C_{\mathrm{GS}}$, higher than the actual input capacitance given by (4.33). Therefore, the input-referred noise of a SF driven by a high-impedance signal is higher compared to the low-impedance driven SF by the factor

$$\alpha_{\mathrm{nSF}} = \frac{C_{\mathrm{d}} + C_{\mathrm{GS}}}{C_{\mathrm{in}}} = \frac{C_{\mathrm{d}} + C_{\mathrm{GS}}}{C_{\mathrm{d}} + C_{\mathrm{GS}}(1 - G_{\mathrm{SF}})} \tag{4.36}$$

This factor is never less than unity and can be considered as an 'excess noise factor', or a 'noise multiplier' [12] in pixel source followers. Despite its lower input capacitance, the SF does not have correspondingly lower ENC, due to the reduced transistor transconductance. This is in accordance with the general principle that feedback cannot improve the SNR [13], also known as 'no free lunch'.

4.2 Correlated double sampling

4.2.1 Reset noise suppression

Correlated double sampling (CDS) is a fundamental method of image sensor readout with the primary purpose of supressing the reset noise. Simply put, CDS works by subtracting the signal level from the corresponding reset level (or the other way around), but there are several ways to do implement it.

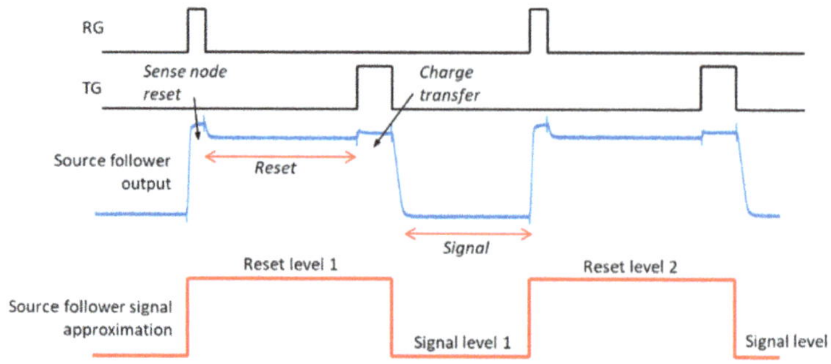

Figure 4.8. Output signal from a 4T pixel and its approximation.

The example in figure 4.8 is the output from two consecutive readouts of a 4T pixel, showing two reset and signal levels. Not all of the SF output is occupied by the two levels; a certain time is allocated for the reset of the sense node and the charge transfer. This time is not usable for determining the signals and is wasted, therefore, it should be made as short as possible. It is usually acceptable to ignore it, and in this case the SF signal can be approximated with a square wave.

A discussed in section 4.1.1 and in figure 4.3, *Signal level 1* 'inherits' the instantaneous noise voltage from the previous reset operation, 'frozen' in the *Reset level 1*. Therefore, subtracting them as a pair would remove the reset noise. The term 'correlated' in CDS means that the order of samples is important, and we cannot subtract just any reset and signal levels. Subtracting the wrong signal-reset pair, such as *Signal level 1* from *Reset level 2*, does not remove the reset noise.

Because of the subtraction, the DC component of the signal is removed too. In this way, the output offset from the source follower, which is different pixel-to-pixel, is removed. Also, since the subtraction is done with a certain time period, the CDS works a high-pass filter because low frequencies get attenuated. To fully suppress the reset noise, the electronic gains for the reset and signal levels must be the same. If they are not, the CDS efficiency becomes less than one, and some of the reset noise will sneak through.

A downside of the signal-reset subtraction is that the other noise contributions, such as the thermal noise superimposed on the output signal, increase by $\sqrt{2}$. If, for argument's sake, the reset noise didn't exist, the reset level would be the same for each and every signal, and only the signal level would have to be sampled. Our goal is to find the optimal CDS method to remove the reset noise, but in a way that produces the highest signal-to-noise ratio.

4.2.2 Double sampling

Double sampling (DS) is the simplest form of CDS and is extensively used. A simplified schematic of the DS circuit working on the signal coming from an SF is shown in figure 4.9(a). The reset and the signal levels are sampled instantaneously, for a very short time as shown in the timing diagram in figure 4.9(b), and then subtracted [14]. When the reset sample is taken with a positive sign and the signal

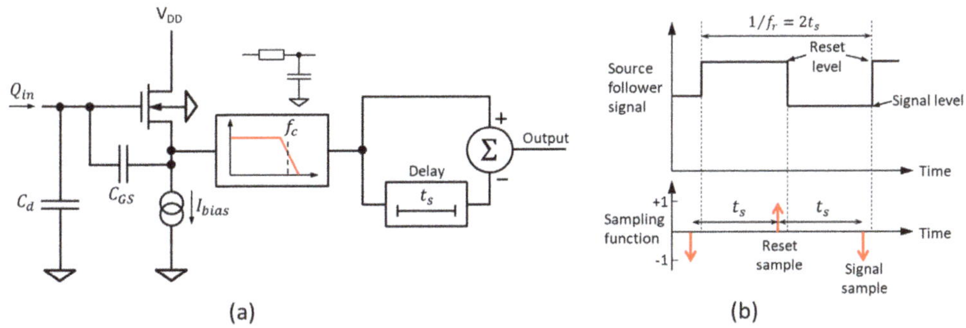

Figure 4.9. Double sampling: simplified schematic (a); and timing diagram (b).

sample with a negative sign, the resulting output voltage is positive, which is what is normally desired for subsequent amplification and digitisation.

The sample points must be fixed with respect to the signal and are normally near the end of the reset and the signal intervals to allow the longest possible time for them to settle to their final values. Ensuring maximum settling time means that the signal period is double the sampling period and equals $2t_s$.

The power transfer function of the double sampler can be found from the Fourier transform of the sampling function $h(t)$. For the ideal DS $h(t)$ is the difference between two delta functions, separated by the sampling period t_s:

$$h(t) = \delta(t + t_s/2) - \delta(t - t_s/2) \tag{4.37}$$

By using that $\int_{-\infty}^{\infty} f(t)\delta(t - t_0)dt = f(t_0)$, the Fourier transform of (4.37) is

$$H_{\mathrm{DS}}(f) = \int_{-\infty}^{\infty} [\delta(t + t_s/2) - \delta(t - t_s/2)]e^{-j2\pi ft}dt = 2j\sin(\pi f t_s) \tag{4.38}$$

The power transfer function is the squared modulus of $H_{\mathrm{DS}}(f)$:

$$|H_{\mathrm{DS}}(f)|^2 = 4\sin^2(\pi f t_s) = 2(1 - \cos(2\pi f t_s)) \tag{4.39}$$

The power transfer function extends into infinity as shown in figure 4.10, however, real-world systems have finite bandwidths, and this limits the overall frequency response. In the simplest case the signal bandwidth is limited by a single-pole low-pass filter with cut-off frequency f_c, therefore (4.39) is multiplied by the filter's power transfer function

$$|H_{\mathrm{LPF}}(f)|^2 = \frac{1}{1 + (f/f_c)^2} \tag{4.40}$$

The product of (4.39) and (4.40) results in a response approaching zero at both frequencies extremes—low frequencies are supressed by the double sampling and high frequencies by the low pass filter. The filter may not be a separate circuit, as the output impedance of the source follower and its load capacitance can produce the same effect as a RC circuit.

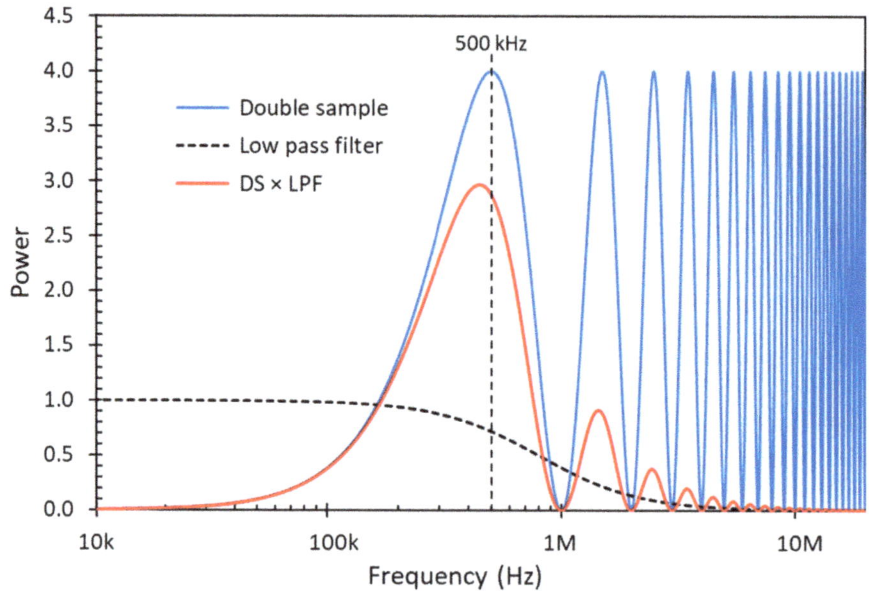

Figure 4.10. Power transfer function of the double sampling circuit for $f_s = 500$ kHz with a low pass filter with $f_c = 800$ kHz, giving $2\pi f_c t_s = 5$ for 1% settling error.

Using (4.19), (4.39) and (4.40) in (4.6) the noise power at the output of the CDS circuit is

$$\overline{v_n^2} = \int_0^\infty e_{nw}^2\left(1 + \frac{f_{nc}}{f}\right)\frac{4\sin^2(\pi f t_s)}{1 + (f/f_c)^2}df \tag{4.41}$$

It is useful to break (4.41) into its constituent white and $1/f$ noise terms because there is an exact solution for the white noise but only an approximate one for $1/f$ noise. For white noise the RMS output noise voltage is [15]

$$\overline{v_{nw}} = e_{nw}\sqrt{\pi f_c\left(1 - \exp\left(-2\pi f_c t_s\right)\right)} \tag{4.42}$$

Normally the cut-off frequency f_c should be higher that the signal frequency, so that the signal settles sufficiently before it is sampled. If $2\pi f_c t_s > 5$ (so that the settling error is less than $e^{-5} \approx 1\%$) the exponential term in (4.42) becomes small enough and can be ignored, therefore the RMS noise voltage simplifies to

$$\overline{v_{nw}} = e_{nw}\sqrt{\pi f_c} \tag{4.43}$$

Equation (4.43) is surprisingly simple and tells us that the noise power bandwidth (the term under the square root) of the double sampler is $B_n = \pi f_c$.

We can compare this to the output noise voltage after a RC low-pass filter (4.7), for which we have $f_c = 1/(2\pi RC)$ and noise bandwidth equal to $1/4RC = (\pi/2)f_c$. This is a useful comparison because the low-pass filter in (4.40) is normally created by the resistances and capacitances of MOSFETs and their loads. The noise power bandwidth of the double sampler is double that of the RC filter, and therefore the RMS noise voltage at the output is $\sqrt{2}$ times larger. This, of course, makes perfect sense because the signal is sampled twice and the samples subtracted, so that the standard deviation increases by $\sqrt{2}$.

If the cut-off frequency is significantly reduced, the noise voltage will drop due to the exponential term in (4.42). This may look like a way of reducing noise, however, it is deceptive because the signal amplitude decreases too, and faster, by a factor of $\left(1 - \exp\left(-2\pi f_c t_s\right)\right)$ due to the insufficient signal settling time [16]. As the cut-off frequency f_c is reduced, the SNR would therefore decrease as $\sqrt{1 - \exp\left(-2\pi f_c t_s\right)}$.

The $1/f$ part of (4.41) can be solved approximately for $2\pi f_c t_s \gg 1$, which is fine because we want $2\pi f_c t_s > 5$ for a negligible signal settling error. The solution is [15]

$$\overline{v_{\mathrm{nf}}} = e_{\mathrm{nw}}\sqrt{2f_{\mathrm{nc}}\left(\gamma + \ln\left(2\pi f_c t_s\right)\right)} \tag{4.44}$$

where $\gamma = 0.577$ is the Euler's constant. Adding (4.43) and (4.44) in quadrature gives the total output RMS noise voltage

$$\overline{v_{\mathrm{n}}} = e_{\mathrm{nw}}\sqrt{\pi f_c + 2f_{\mathrm{nc}}\left(\gamma + \ln\left(2\pi f_c t_s\right)\right)} \tag{4.45}$$

If we use that t_s equals half the period of the output signal, as in figure 4.9, so that $t_s = 1/(2f_r)$ with f_r the signal's frequency, we can write

$$\overline{v_{\mathrm{n}}} = e_{\mathrm{nw}}\sqrt{\pi f_c + 2f_{\mathrm{nc}}\left(\gamma + \ln\left(\pi f_c / f_r\right)\right)} \tag{4.46}$$

Equation (4.46) is widely used to calculate the noise performance in CIS. It includes transistor parameters such as the white and $1/f$ noise densities determined by the geometry, the operation point and the CMOS process, and the readout and the −3 dB cut-off frequencies, selectable for a specific readout scheme. From formula (4.46) we can obtain the input-referred noise charge at the in-pixel source follower by using its voltage gain G_{SF}, conversion gain G_c (CVF) and the noise multiplier α_{nSF}, as described in section 4.1.5.

Example 4.3. Calculate the input-referred noise in electrons RMS for a DS circuit with signal frequency $f_r = 500$ kHz, $f_c = 800$ kHz, SF voltage noise density $e_{\mathrm{nw}} = 30$ nV/$\sqrt{\mathrm{Hz}}$ and $f_{\mathrm{nc}} = 50$ kHz. The SF has gain $G_{\mathrm{SF}} = 0.85$, $C_d = 2.5$ fF and $C_{\mathrm{GS}} = 2$ fF.
Solution: First, we calculate the output RMS noise voltage from (4.46):

$$\bar{v}_{\mathrm{n}} = 30 \times \sqrt{3.14 \times 8 \times 10^5 + 2 \times 5 \times 10^4 \times (0.577 + \ln(3.14 \times 8 \times 10^5 / 5 \times 10^5))} = 49.6\,\mu\mathrm{V}$$

If we were to calculate the noise components separately, we would find that the white noise contributes 47.6 μV and the $1/f$ noise 14 μV. The input-referred noise charge in electrons RMS is calculated similarly to (4.32) by dividing the output noise voltage by the SF gain, multiplying by the SF excess noise factor and the input capacitance, and dividing by the elementary charge:

$$\overline{Q_{\text{in}}} = \alpha_{\text{nSF}}\frac{\overline{v_{\text{n}}}}{G_{\text{SF}}}\frac{C_{\text{in}}}{q}$$

The input capacitance of the SF can be calculated from (4.33) and α_{nSF} from (4.36)

$$C_{\text{in}} = C_{\text{d}} + C_{\text{GS}}(1 - G_{\text{SF}}) = 2.5 + 2 \times (1 - 0.85) = 2.8\ \text{fF}$$

$$\alpha_{\text{nSF}} = \frac{C_{\text{d}} + C_{\text{GS}}}{C_{\text{d}} + C_{\text{GS}}(1 - G_{\text{SF}})} = \frac{2.5 + 2}{2.5 + 2 \times (1-0.85)} = 1.61$$

Finally, the input-referred ENC is

$$\bar{Q}_{\text{in}} = \frac{1.61 \times 49.6 \times 10^{-6} \times 2.8 \times 10^{-15}}{0.85 \times 1.6 \times 10^{-19} \times 0.85} = 1.64\ \text{e}^{-}$$

For reference, the CVF at the output of the SF is

$$G_{\text{c}} = \frac{q}{C_{\text{in}}}G_{\text{SF}} = \frac{1.6 \times 10^{-19}}{2.8 \times 10^{-15}} \times 0.85 = 48.6\ \mu\text{V/e}^{-}$$

The calculated noise represents the best-case scenario when all other noise sources in the system are negligible.

4.2.3 Dual slope integrator

The dual slope integrator (DSI), sometimes called differential averager (DA), is another widely used CDS technique, originating in CCD signal processing [17]. Schematically it can be represented as an analogue integrator (figure 4.11(a)) which receives signal from the straight and the inverted version of the source follower's output in turns. Instead of two brief signal samples, the DSI integrates (in other words, averages) the reset and the signal levels throughout their entire duration t_{int}, and subtracts the averages due to the sign inversion accomplished with an electronic switch, as depicted in figure 4.11(b).

The integrating function can be realised with a classical analogue inverting integrator [1] with output voltage V_{o} described by (4.47). The integrator acts as a low-pass filter on the input signal V_{i} with time constant $\tau = R_{\text{int}}C_{\text{int}}$.

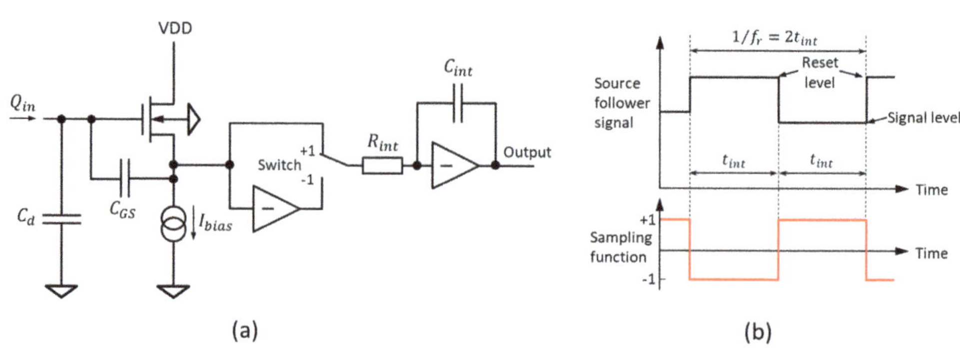

Figure 4.11. Dual slope integrator: simplified schematic (a); and timing diagram (b).

$$V_o = -\frac{1}{R_{int}C_{int}}\int V_i dt \tag{4.47}$$

In between each reset-signal integration cycle the output of the circuit must be zeroed by discharging the integration capacitor C_{int}. This is assumed to take a very short time and is not shown in the timing diagram for simplicity. Since we have to treat the signal and the reset in the same manner, the two integration periods are equal.

A characteristic of the DSI is that it suppresses both low frequencies, due to the subtraction of the two levels by the sign inversion, and high frequencies, due to the integration. Therefore, the DSI does not need external bandwidth limiting as the double sampler because it is a low-pass filter by itself. Another important feature of the DSI, as will be shown in section 4.2.4, is that it is the *optimum* signal processing technique when only white noise is present and gives the highest theoretically possible SNR.

The power transfer function of the DSI is given by [17]

$$|H_{DSI}(f)|^2 = \frac{4\sin^4(\pi f t_{int})}{(\pi f t_{int})^2} \tag{4.48}$$

Similarly to the DS in (4.39), the response at very low frequencies is proportional to f^2, resulting in the elimination of the DC component and the attenuation of low frequencies. The denominator in (4.48) is responsible for the low-pass function due to the $1/f^2$ overall dependence at higher frequencies.

Figure 4.12 compares the power transfer functions of the double sampler and the dual slope integrator for the same sampling frequency. It is obvious that, with an

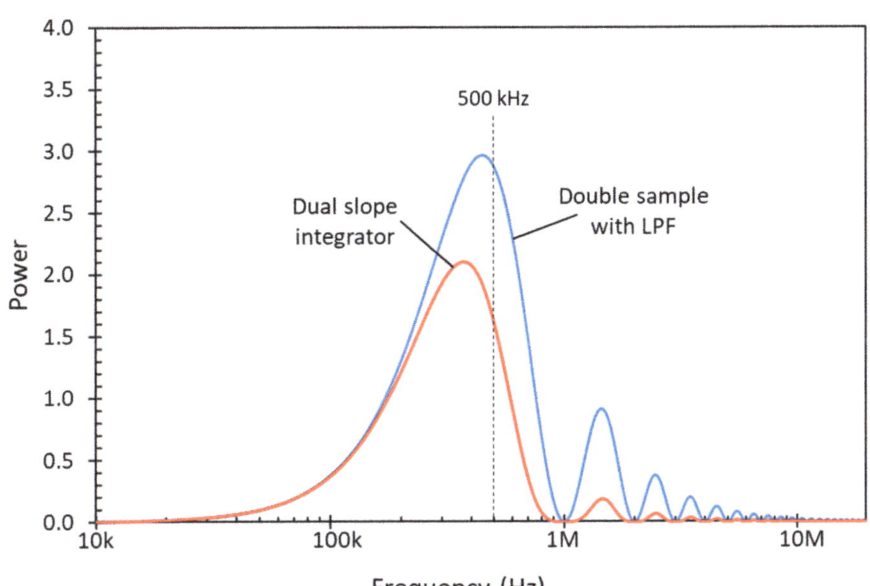

Figure 4.12. Power transfer function of the dual slope integrator and the double sampler for $f_s = 500$ kHz and a low pass filter for the double sampler with $f_c = 800$ kHz, corresponding to figure 4.10.

identical input, the DSI will have lower noise due to the smaller area under the curve of the power transfer function.

The output noise power from a DSI circuit can be found by integrating the noise power spectrum multiplied by the power transfer function, as we did for the double sampler:

$$\overline{v_n^2} = \int_0^{\infty} e_{nw}^2 \left(1 + \frac{f_{nc}}{f}\right) \frac{4\sin^4(\pi f t_{int})}{(\pi f t_{int})^2} df \tag{4.49}$$

Similarly, we can separate (4.49) into white and $1/f$ noise parts, and using that $\int_0^{\infty} \frac{\sin^4 x}{x^2} dx = \pi/4$, for the white noise only the output RMS noise voltage is

$$\overline{v_{nw}} = \frac{e_{nw}}{\sqrt{t_{int}}} \tag{4.50}$$

Equation (4.50) shows that the lowest noise is achieved at the longest possible t_{int}, which is half the signal period. Therefore, $t_{int} = 1/(2f_r)$, and the noise voltage can be written as

$$\overline{v_{nw}} = e_{nw}\sqrt{2f_r} \tag{4.51}$$

The white noise bandwidth of the DSI is $2f_r$ and is considerably smaller than the noise bandwidth of the double sampler (πf_c from (4.43)), therefore its SNR is correspondingly better. This is because $f_c > f_r$ must be observed to avoid attenuating the signal.

Using that $\int_0^{\infty} \frac{\sin^4 x}{x^3} dx = \ln 2$, the $1/f$ part of (4.49) gives

$$\overline{v_{nf}} = 2e_{nw}\sqrt{f_{nc} \ln 2} \tag{4.52}$$

It is interesting to point out that the $1/f$ noise part does not depend on the readout frequency. Summing (4.51) and (4.52) in quadrature gives the total output noise of the DSI with white and $1/f$ noise input

$$\overline{v_n} = e_{nw}\sqrt{2f_r + 4f_{nc} \ln 2} \tag{4.53}$$

We have already established that the DSI should have lower noise than the double sampler, now let's see what the difference is.

Example 4.4. Calculate the input-referred noise in electrons RMS for a DSI circuit with signal frequency $f_r = 500$ kHz, source follower voltage noise density $e_{nw} = 30$ nV/$\sqrt{\text{Hz}}$ and $f_{nc} = 50$ kHz. The SF has gain $G_{SF} = 0.85$, $C_d = 2.5$ fF and $C_{GS} = 2$ fF, the same as in example 4.3.

Solution: First, we calculate the output RMS noise voltage from (4.53):

$$\bar{v}_n = 30 \times \sqrt{2 \times 5 \times 10^5 + 4 \times 5 \times 10^4 \times \ln 2} = 32\,\mu\text{V}$$

Comparing with example 4.3 we see that the noise of the DSI is 32/49.6 = 65% of the noise of the double sampler, a substantial improvement. Since α_{nSF}, G_{SF} and C_{in} are the same as in example 4.3 we can simply scale the result to get the input-referred noise of the DSI = 1.06 e^- RMS.

The DSI is clearly better than the double sampler, but is rarely used in practice. The reason is that it is more complicated than the sampler (which can be made with a couple of switches and capacitors) and more difficult to implement in practice as an analogue circuit. Nevertheless, the DSI lives on in its digital implementation which achieves nearly the same performance, as we will see in section 4.2.5.

4.2.4 Optimal signal processing

A common task in signal processing is to recover a signal in the presence of noise. Written in a general form, the output signal from a signal processing algorithm $y(t)$ is generated by the convolution of the input signal $x(t)$ with a sampling function $h(t)$

$$y(t) = x(t)*(t) = \int_{-\infty}^{\infty} x(\tau)h(t - \tau)d\tau \qquad (4.54)$$

Equation (4.54) describes a huge variety of signal processing methods, including the double sampler and the dual slope integrator. Theory tells us that when both the signal and the noise are of known type, there is an optimal processing method, dubbed 'matched filter', which has the highest attainable SNR [18]. If the noise has power density $|N(f)|^2$ the maximum SNR is achieved with a sampling function having a Fourier transform[4]

$$H(f) = \frac{X^*(f)}{k_{MF}\,|N(f)|^2} \qquad (4.55)$$

Here $X^*(f)$ is the complex conjugate of the Fourier transform of the input signal and k_{MF} is a real constant with physical dimension. If $N(f)$ is frequency-independent, as in white noise, the Fourier transform of the sampling function is the complex conjugate of the signal's Fourier transform, multiplied by a constant.

Therefore, the sampling function $h(t)$ of the matched filter is the scaled version of the mirrored signal waveform. This tells us that the sampling function should look the same as the signal, therefore if the signal is shaped as a square wave, the sampling function should be a square wave too. Intuitively this makes sense if you imagine superimposing two identical waveforms by sliding them on top of each other.

Since the pixel signal looks like a square wave, and the DSI sampling function is a square wave too, signal theory claims that no other method can achieve higher SNR than the DSI, provided that the noise is white [17].

---

[4] The Fourier transform is a complex function. The power density is defined as $|N(f)|^2 = N(f)N^*(f)$ and is real.

With a mixture of white and $1/f$ noise the sampling function of the matched filter changes under the frequency dependence of the noise in $N(f)$. By using the inverse Fourier transform of (4.55) we will find that $h(t)$ is no longer of the same shape as the input signal. Therefore, the DSI is not the optimal processing method and its noise performance is worse than the ideal matched filter.

What can replace the DSI when we have both white and $1/f$ noise? It can be shown that the sampling function must apply more weight to the signal on both sides of the charge transfer [19, 20] to improve the SNR. Unlike white noise, which is statistically stationary, and its variance is time-independent, for $1/f$ noise the signal variance increases as the logarithm of the observation time [21]. Samples taken closer in time show higher correlation, therefore applying more weight to them makes sense. As the proportion of $1/f$ noise increases, the sampling function develops steeper edges following a $1/t$ dependence, as shown in figure 4.13.

This type of function is very difficult to realise with analogue circuits. Supposing that a signal processor using the optimal sampling function can be built, we can establish how much better it is than the DSI for the mixture of white and $1/f$ noise present in CIS. Following the mathematics in [9] and [17], figure 4.14 shows how the noise performance of the ideal matched filter overtakes the DSI as the ratio f_{nc}/f_r, and therefore the relative part of the $1/f$ noise increases. There is a nearly logarithmic dependence on the f_{nc}/f_r indicating that the matched filter can significantly outperform the DSI only when there is a very large amount of $1/f$ noise, or the readout frequency is very low, and in both cases $f_{nc} \gg f_r$.

The DSI is quite effective anyway in suppressing $1/f$ noise because of the high-pass filter characteristic of its power transfer function (4.48), which approaches zero in a f^2 trend at low frequencies. From figure 4.14 we can see that even for $f_{nc}/f_r = 1$ the SNR of the DSI is only 7% worse than the matched filter.

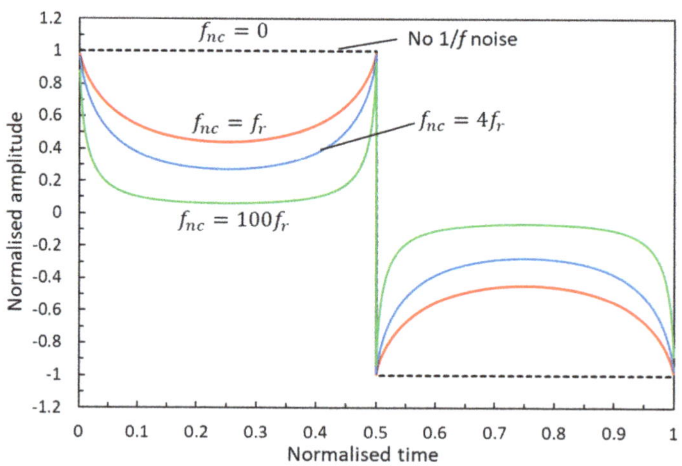

Figure 4.13. Calculated shape of the sampling function for increasing $1/f$ noise, indicated by the increase of the noise corner frequency f_{nc} from zero (no $1/f$ noise) to 100 times the readout frequency. Adapted from [19].

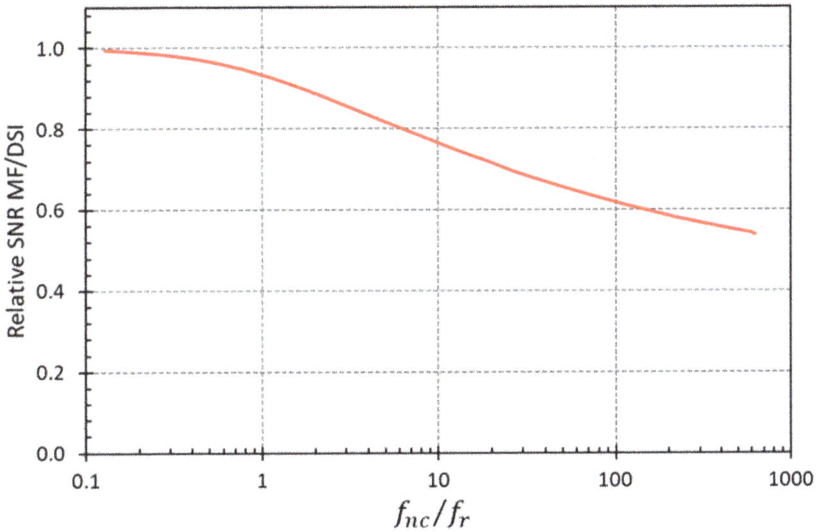

Figure 4.14. Relative SNR between the matched filter (MF) and the DSI for increasing fraction of 1/*f* noise. Adapted from [19].

4.2.5 Digital CDS and multiple sampling

The double sampler and the DSI implement CDS by subtracting or integrating voltages using analogue circuitry. There is nothing to stop us from doing this digitally, except the increase in the complexity of the CIS, and get digital CDS (DCDS). Moreover, if we want to implement some non-trivial sampling functions like in figure 4.13, DCDS may be the only practical option.

The simplest way to have DCDS is to substitute the analogue samples in figure 4.9(b) with ADC conversions. Provided that no additional noise is introduced by the ADC, the noise performance should be identical to (4.46). There are certain advantages to be had by doing CDS digitally, especially if the ADC is highly linear and has wide input voltage range. Increased linearity, wider dynamic range and better matching due to fewer analogue amplifiers and components are all possible.

The real strength of the digital techniques, however, lies in signal oversampling, called digital correlated multiple sampling (DCMS). DCMS offers the optimal signal processing that can be tailored for the specific noise spectrum. As with any other CDS method, the purpose of DCMS is to remove the reset noise in a way that results in the highest SNR [22, 23]. The output signal is oversampled, each sample is converted to digital code and the subsequent processing is done digitally.

Here we consider that the ADC samples continuously at regular intervals t_s without pauses or breaks. The sampling is synchronous with the signal, so that it always occurs at the same places in the signal waveform. We are interested in the parts of the signal that have settled sufficiently and ignore the samples outside the settled periods, and also ignore samples during sense node reset and charge transfer. Figure 4.15 shows N valid sample pairs during the reset and the signal time intervals and M samples during the transitions, which are ignored. Both the reset and the

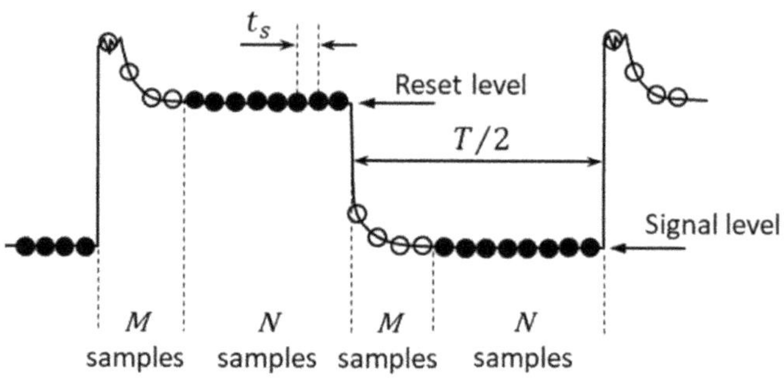

Figure 4.15. Timing diagram of a DCMS processor.

signal periods must be treated the same and have the same number of valid samples for the best noise performance. The total number of samples in the signal period can be determined from the following relationship, where $N + M$ is the digital half-period of the signal:

$$(N + M)t_s = \frac{T}{2} \tag{4.56}$$

The most straightforward DCMS method is to average the signal and the reset level samples and to subtract them. This is called differential averaging (DA) and is described by the function:

$$y = \frac{1}{N}\sum_{i=0}^{N-1}(x[i] - x[i - N - M]) \tag{4.57}$$

In (4.57) y is the output signal from the DA and $x[i]$ is the i-th sample of the input. As with the analogue CDS, we assume that the input signal has a limited bandwidth with cut-off frequency f_c, single-pole frequency response and a power transfer function given by (4.40). The total power transfer function of the differential averager with such bandwidth-limited input signal is [24]

$$|H_{DA}(f)|^2 = \frac{1}{N^2}\frac{4\sin^2(N\pi f t_s)\sin^2[(N + M)\pi f t_s]}{\sin^2(\pi f t_s)\left[1 + (f/f_c)^2\right]} \tag{4.58}$$

Since we are averaging N samples from the reset and the signal portions of the waveform, it is reasonable to expect that the noise is reduced by $\sqrt{N}$. Intuitively, it could appear that the noise can be reduced to nearly zero if the number of samples is large. Unfortunately, this is a mirage: since the signal period is fixed, so is the time during which samples can be taken. Therefore, increasing the number of samples can only come from increasing the sampling rate, thus increasing the bandwidth and the noise.

We can take this to the extreme and investigate what happens if N approaches infinity. To do this, M must be zero so that N can be maximized by removing the low-pass frequency response and making the signal as fast as possible, so we have

$M \to 0$, $f_c \to \infty$ and $N \to \infty$. By using that $Nt_s = t_{\text{int}}$ and $t_s \to 0$ (which follows when $N \to \infty$ and the signal period is fixed), and also that $\sin x \approx x$ for $x \to 0$, we can write

$$|H_{\text{DA}}(f)|^2 \approx \frac{4\sin^4(N\pi f t_s)}{N^2 \sin^2(\pi f t_s)} \approx \frac{4\sin^4(N\pi f t_s)}{N^2(\pi f t_s)^2} = \frac{4\sin^4(\pi f t_{\text{int}})}{(\pi f t_{\text{int}})^2} \tag{4.59}$$

Despite the number of samples being infinite the power transfer function (4.59) is not zero, therefore the noise is not zero too. Instead, we find that (4.58) approaches the power transfer function of the DSI (4.48) for the same conditions. This should not come as a surprise, because digitally averaging N samples taken at interval t_s is functionally the same as analogue integration over the time Nt_s.

In real applications the number of samples is finite, the signal bandwidth is limited, and $M \geqslant 0$. Using the formula (4.58) we can investigate the performance of the differential averager in several practical cases. In the most trivial case only one sample pair is taken, therefore $N = 1$ and $M = 0$ and we get

$$|H_{\text{DA}}(f)|^2 = \frac{4\sin^2(\pi f t_s)}{1 + (f/f_c)^2} \tag{4.60}$$

We can recognise that formula (4.60) is the power transfer function of the double sample CDS with a low-pass bandwidth-limited input signal, which is expected since the signal is sampled in the same way.

For arbitrary N, M and f_c we can use the constraint that N and M must fit within one half-period (4.56), and that the cut-off frequency f_c determines the settling time during which no useful samples are taken. When the analogue bandwidth is reduced, the noise is naturally reduced too. At the same time, the settling time increases and therefore the number of useful samples N goes down, so there must be a trade-off between the two.

The settling time, equal to $t_{\text{set}} = Mt_s$, can be expressed from the equation

$$A = A_0[1 - \exp(-t/\tau)] \tag{4.61}$$

describing the step response of a signal with a single-pole response and time constant $\tau = 1/(2\pi f_c)$. The settling time for the signal to reach amplitude A is

$$t_{\text{set}} = -\tau \ln\left(1 - \frac{A}{A_0}\right) = \tau\,|\ln\varepsilon|, \tag{4.62}$$

at which point it deviates from the final value A_0 by the tolerance $\varepsilon = 1 - A/A_0$. Therefore, M is

$$M = \frac{|\ln\varepsilon|}{2\pi f_c t_s} \tag{4.63}$$

We would like to know how many ADC samples are sufficient to get close to the ideal DSI, and what is better for noise reduction: lower f_c (and therefore smaller N), or the opposite? The last question is essentially asking whether an analogue RC filter is better than digital averaging.

The calculation in figure 4.16 shows the noise ratio between the differential averager and the DSI for white noise, and the number of samples N for increasing cut-off frequency f_c. This is done numerically using formula (4.58) for 1% settling tolerance, finding N and M from (4.56) and (4.63). While the noise ratio is always >1, we see that 15 samples are enough to bring the noise from the DA to within 10% of the DSI. Kawahito's work [15] on both thermal and $1/f$ noise input to the DA concludes that $N \cong 10$ is required for the noise performance to be very close to the DSI.

The second important observation in figure 4.16 is that the DA has better noise performance as the cut-off frequency increases, accompanied by the increase of the number of samples as the settling time gets shorter. This is a clear indication that the single-pole analogue filtering is inferior to digital averaging—bandwidth reduction does not reduce the noise as much as taking more samples. Therefore, the strategy to achieve good noise performance is to average as many samples as possible (with ⩾10 a good goal) while keeping the signal bandwidth as high as possible.

An additional complication to digital averaging is the well-known aliasing effect in analogue-to-digital conversion. The spectrum of the power transfer function can extend beyond the Nyquist frequency of the ADC and the higher frequency components can appear in the bandwidth of interest. The Nyquist frequency is the hard limit on how fast the signal can be. The calculation in figure 4.16 uses an

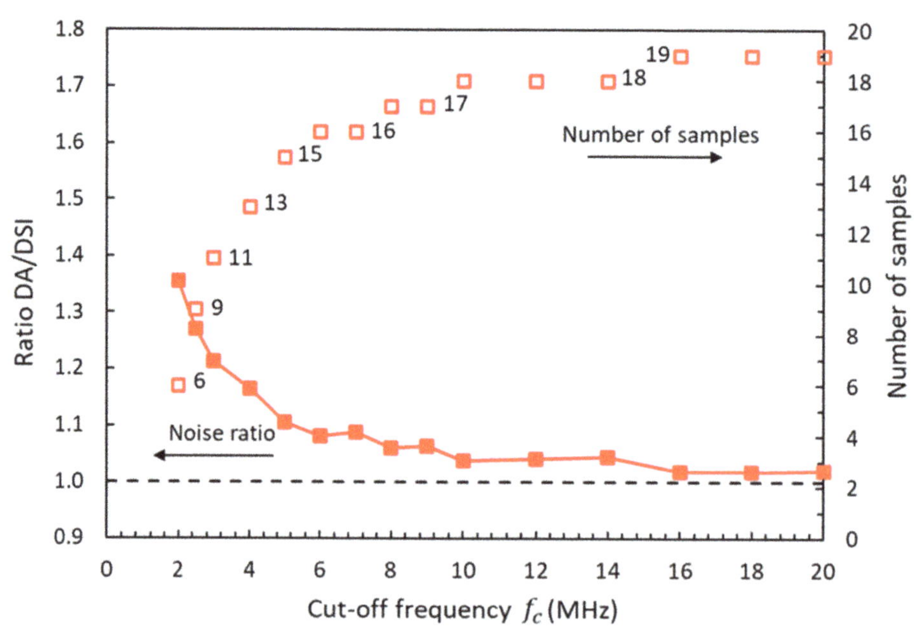

Figure 4.16. Numerically calculated SNR ratio (relative SNR) between the digital differential averager and the ideal DSI, for 40 MHz ADC sample rate and 1% settling tolerance. The calculation uses a brick-wall anti-alias filter at the Nyquist frequency of 20 MHz.

ideal (brick-wall) anti-aliasing filter set at half the ADC sampling rate. Without it, the noise performance begins to suffer due to aliasing as the cut-off frequency increases [24], because the single-pole low-pass response is not an effective anti-aliasing filter.

DCMS can offer improved noise performance for $1/f$ noise using weighted sampling [19]. Originally applied to CCDs [25], this method can be used in CIS, provided that the complexity of the digital circuitry can be accommodated.

4.2.6 Column-level noise

CIS readout circuits operate in a column-parallel fashion with amplification, switches and sample and hold capacitors implementing CDS for the pixels connected in a column. Figure 4.17 is a simplified schematic of a typical column circuit, using the analogue switch SW controlled by the S&H signal to store the signal sample on the capacitor C_S. The full circuit uses two capacitors for sampling of the reset and the signal levels (but only half is shown here for simplicity) and implements the double sampling CDS technique.

The total source-to-substrate capacitance of all source followers and the parasitic capacitance of the metal track connecting all the outputs in the column can be substantial. This column capacitance C_{col} can be high enough to limit the signal

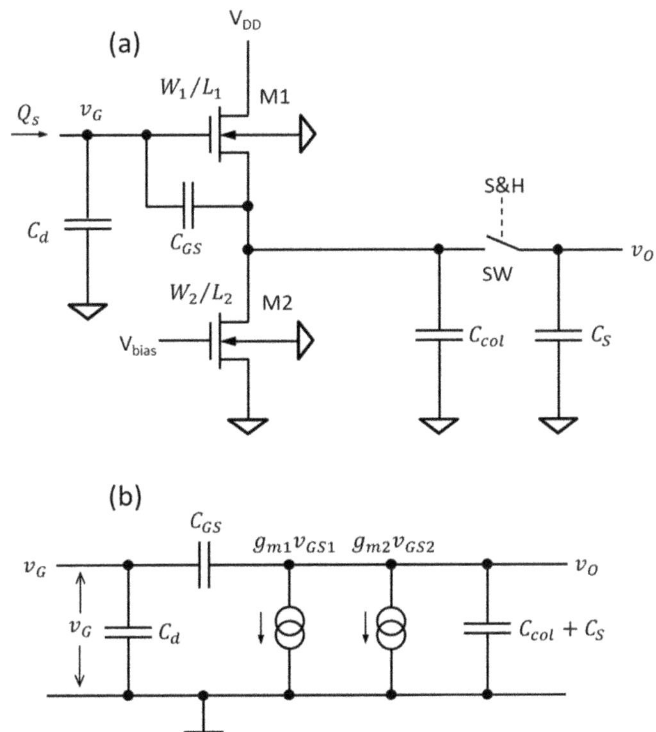

Figure 4.17. Simplified column CDS schematic (a) and its equivalent schematic for noise analysis with the switch SW closed (b).

bandwidth to the desired value without an additional low-pass filter. This is very common in CIS and is therefore worth describing in more detail, as it gives a very useful estimate for the readout noise [15].

As usual, we consider the thermal and the $1/f$ noise separately to make the calculations using (4.46) more manageable. Starting with the white noise, the square of the output noise current density is the quadrature sum of the noise currents of the SF and column load transistors. Using (4.21) we can write

$$i_{\mathrm{nw}}^2 = \frac{8kT}{3}(g_{\mathrm{m1}} + g_{\mathrm{m2}}) \tag{4.64}$$

The noise current can be translated to input-referred voltage noise density by using (4.22) and the SF excess noise factor (4.36).

$$e_{\mathrm{nw}}^2 = \alpha_{\mathrm{nSF}}^2 \frac{i_{\mathrm{nw}}^2}{g_{\mathrm{m1}}^2} = \alpha_{\mathrm{nSF}}^2 \frac{8kT}{3g_{\mathrm{m1}}^2}(g_{\mathrm{m1}} + g_{\mathrm{m2}}) \tag{4.65}$$

Equation (4.43) uses the output voltage noise density to calculate the noise from the double sampler. We can use it with the input-referred noise instead without problems, with the bonus that the SF voltage gain is already included. We also need to know the cut-off frequency f_{c}, which for the circuit in figure 4.17 is determined by the transconductance of the SF and the load capacitance as

$$f_{\mathrm{c}} = \frac{g_{\mathrm{m1}}}{2\pi(C_{\mathrm{col}} + C_{\mathrm{S}})} \tag{4.66}$$

The 'on' resistance of the analogue switch SW is usually much smaller than the SF output resistance and can be ignored. Therefore, substituting (4.65) and (4.66) into (4.43) we get the following for the input-referred noise voltage of the double sampling CDS:

$$\begin{aligned} \overline{v_{\mathrm{nw}}} &= \sqrt{\frac{8kT\alpha_{\mathrm{nSF}}^2}{3g_{\mathrm{m1}}^2}(g_{\mathrm{m1}} + g_{\mathrm{m2}})}\sqrt{\frac{\pi g_{\mathrm{m1}}}{2\pi(C_{\mathrm{col}} + C_{\mathrm{S}})}} \\ &= \alpha_{\mathrm{nSF}}\sqrt{\frac{4kT}{3(C_{\mathrm{col}} + C_{\mathrm{S}})}\left(1 + \frac{g_{\mathrm{m2}}}{g_{\mathrm{m1}}}\right)} \end{aligned} \tag{4.67}$$

Equation (4.67) shows the familiar kT/C dependence with the striking difference that the capacitance is of the combined column load, not of the sense node. This is a consequence from the transistor noise power being proportional to kT/g_{m1}, while the bandwidth is proportional to $g_{\mathrm{m1}}/(C_{\mathrm{col}} + C_{\mathrm{S}})$, so g_{m1} cancels. Equation (4.67) is very convenient because we do not need to know the SF white noise density and the cut-off frequency. Only a knowledge of the transconductance ratio is required and since $g_{\mathrm{m}} \propto \sqrt{I_{\mathrm{D}}(W/L)}$ we have $g_{\mathrm{m2}}/g_{\mathrm{m1}} = \sqrt{(W_2/L_2)/(W_1/L_1)}$. This ratio is around one in most cases, but could be made smaller to reduce the noise; that means the MOSFET M2 should be a transistor with $W_2 \ll L_2$ because for the SF we normally have $W_1 \cong L_1$.

Example 4.5. Calculate the input-referred column-level white noise in electrons RMS for a double sampler circuit at 20 °C (293 K) with $C_{col} + C_S = 2$ pF, $g_{m2}/g_{m1} = 1$, source follower gain $G_{SF} = 0.85$, capacitances $C_d = 2.5$ fF and $C_{GS} = 2$ fF, the same as in example 4.3.

Solution: The input capacitance of the SF $C_{in} = 2.8$ fF and the excess noise factor $\alpha_{nSF} = 1.61$ were calculated in example 4.3. Using (4.67), the input-referred noise voltage is

$$\bar{v}_{nw} = 1.61 \times \sqrt{\frac{4 \times 1.38 \times 10^{-23} \times 293}{3 \times 2 \times 10^{-12}}(1 + 1)} = 118\ \mu\text{V RMS}$$

Since the noise voltage is input-referred and includes the SF gain, to convert to ENC we only have to divide by the input CVF, therefore

$$\bar{Q}_{nw} = \bar{v}_{nw}\frac{C_{in}}{q} = \frac{118 \times 10^{-6} \times 2.8 \times 10^{-15}}{1.6 \times 10^{-19}} = 2.07\ e^-$$

The noise in this example is higher than example 4.3 due to the inclusion of the column load transistor noise, which adds a factor of $\sqrt{2}$, and due to the small column load capacitance. Doubling $C_{col} + C_S$ to 4 pF would reduce the noise to 1.46 e^- RMS, provided that the decrease of bandwidth can be accommodated.

The expression for the input-referred ENC in electrons RMS can be written from (4.67) as

$$\overline{Q_{nw}} = \overline{v_{nw}}\frac{C_{in}}{q} = \frac{C_d + C_{GS}}{q}\sqrt{\frac{4kT}{3(C_{col} + C_S)}\left(1 + \frac{g_{m2}}{g_{m1}}\right)} \tag{4.68}$$

considering that the product of C_{in} (1.107) and α_{nSF} (4.36) is $\alpha_{nSF}C_{in} = C_d + C_{GS}$. Formula (4.68) is somewhat simpler because C_{in} and G_{SF} do not feature, and shows that the ENC depends on a few intrinsic and parasitic capacitances, as well as on both transconductances.

Next, using (4.24) we can write the output noise current density coming only from $1/f$ noise as:

$$i_{nf}^2 = \frac{K_F}{C_{ox}^2 W_1 L_1}\frac{g_{m1}^2}{f} + \frac{K_F}{C_{ox}^2 W_2 L_2}\frac{g_{m2}^2}{f} \tag{4.69}$$

Normally the column load transistor M2 has much larger area than the source follower, so we have $W_2L_2 \gg W_1L_1$. Since g_{m2} is similar of less than g_{m1} for reduction of the white noise, the second term in (4.69) can usually be ignored. Transferring the noise current to input-referred voltage density, keeping the same notations as in (4.17) gives

$$e_{nf}^2 = \frac{\alpha_{nSF}^2 i_{nf}^2}{g_{m1}^2} = \frac{\alpha_{nSF}^2 K_F}{C_{ox}^2 W_1 L_1} \tag{4.70}$$

Using $e_{nw}^2 = e_{nf}^2/f_{nc}$, the $1/f$ noise part of (4.46) becomes:

$$\overline{v_{nf}} = e_{nw}\sqrt{2f_{nc}\left(\gamma + \ln\left(\pi f_c/f_r\right)\right)} = e_{nf}\sqrt{2\left(\gamma + \ln\left(\pi f_c/f_r\right)\right)} = \\ = \frac{\alpha_{nSF}}{C_{ox}}\sqrt{\frac{2K_F\left(\gamma + \ln\left(\pi f_c/f_r\right)\right)}{W_1L_1}} \quad (4.71)$$

Formula (4.71) is not as clean as the expression for the thermal noise because it contains the cut-off frequency, which depends on the column capacitance and the SF transconductance as in (4.66). This dependence, however, is very weak because it is under the logarithm. Similarly to (4.68) we can convert to input-referred ENC and get

$$\overline{Q_{nf}} = \overline{v_{nf}}\frac{C_{in}}{q} = \frac{C_d + C_{GS}}{qC_{ox}}\sqrt{\frac{2K_F\left(\gamma + \ln\left(\pi f_c/f_r\right)\right)}{W_1L_1}} \quad (4.72)$$

Example 4.6. Calculate the ENC due to $1/f$ noise for the parameters in example 4.5. Use $C_{ox} = 4.93$ fF µm^{-2} (7 nm thick gate oxide), $f_c = 800$ kHz, $f_r = 500$ kHz, $W_1 = 0.6$ µm, $L_1 = 0.8$ µm, $K_F = 10^{-31}$ V^2 F^2 cm^{-2}.
Solution: Substituting into (4.71)

$$\overline{v_{nf}} = \frac{1.61}{4.93\times10^{-7}}\times\sqrt{\frac{2\times10^{-31}\times(0.577 + \ln(3.14\times8\times10^5/5\times10^5))}{0.6\times10^{-4}\times0.8\times10^{-4}}} = 31\ \mu\text{V}\quad\text{RMS}$$

$$\bar{Q}_{nf} = \bar{v}_{nf}\frac{C_{in}}{q} = \frac{31\times10^{-6}\times2.8\times10^{-15}}{1.6\times10^{-19}} = 0.54\ \text{e}^-$$

This example is less straightforward because K_F may not be readily available, but it can be calculated from (4.26) if e_{nw} (or the transconductance) and the corner frequency are known. Here we have taken $e_{nw} = 30$ nV/$\sqrt{\text{Hz}}$ and $f_{nc} = 100$ kHz. The $1/f$ noise contribution is small because the corner frequency is much lower than the readout frequency.

Similar considerations for the column noise can be made when DSI or DCMS is used instead of the double sampler [23]. Formulas (4.51) and (4.52) can be used to derive the column-level noise when the DCMS sampling function approaches the DSI. Using (4.65) and (4.51), the ENC due to white noise in the ideal DSI is

$$\overline{Q_{nw}} = \frac{4(C_d + C_{GS})}{q}\sqrt{\frac{kTf_r}{3g_{m1}^2}(g_{m1} + g_{m2})} \quad (4.73)$$

Equation (4.73) can be simplified further by assuming that $g_{m1} \gg g_{m2}$ and becomes:

$$\overline{Q_{nw}} \approx \frac{4(C_d + C_{GS})}{q}\sqrt{\frac{kTf_r}{3g_{m1}}} \quad (4.74)$$

Similarly, the $1/f$ noise ENC from (4.70) and (4.52) can be expressed as [15]

$$\overline{Q_{\rm nf}} = \frac{2(C_{\rm d} + C_{\rm GS})}{qC_{\rm ox}}\sqrt{\frac{K_{\rm F}\ln 2}{W_1 L_1}} \tag{4.75}$$

4.2.7 MOSFET optimisation

The total input-referred ENC is given by the quadrature sum of the thermal and $1/f$ components

$$\overline{Q_{\rm n}^2} = \overline{Q_{\rm nw}^2} + \overline{Q_{\rm nf}^2} \tag{4.76}$$

The ENC is proportional to the MOSFET capacitances $C_{\rm d}$ and $C_{\rm GS}$ and therefore depends on the width W_1 and length L_1. The transconductance is also a function of both and the drain current as$g_{\rm m1} \propto \sqrt{(W_1/L_1)I_{\rm D}}$. Therefore, with all other parameters fixed, it may be possible to optimise the size of the in-pixel SF in order to achieve the lowest possible noise. Such multi-parameter optimisations have been performed in [26].

The minimum obtainable noise is ultimately limited by the $1/f$ noise because the thermal noise can be reduced sufficiently by limiting the bandwidth and the readout rate. Then, considering only $1/f$ noise, the optimisation problem becomes easier.

Returning to the breakdown of MOSFET gate capacitances in chapter 1, the gate–source capacitance of the source follower M1 in figure 4.17 is

$$C_{\rm GS} = \frac{2}{3}C_{\rm ox}W_1L_1 + C_{\rm e}W_1 \tag{4.77}$$

where $C_{\rm e}$ is the capacitance per unit length of gate-channel edge [27]. The gate capacitance to substrate $C_{\rm d}$ can be written as the sum of the sense node $C_{\rm sn}$, gate–drain edge $C_{\rm GDe} = C_{\rm e}W_1$ and the parasitic wiring $C_{\rm par}$ capacitances:

$$C_{\rm d} = C_{\rm sn} + C_{\rm e}W_1 + C_{\rm par} \tag{4.78}$$

Therefore, the $1/f$ noise ENC from (4.75) can expressed as a function of the channel width and length as in [27]

$$\overline{Q_{\rm nf}} = \frac{2\left(C_{\rm sn} + C_{\rm par} + 2C_{\rm e}W_1 + \frac{2}{3}C_{\rm ox}W_1L_1\right)}{qC_{\rm ox}}\sqrt{\frac{K_{\rm F}\ln 2}{W_1L_1}} \tag{4.79}$$

Finding the minimum of (4.79) can give the optimal W_1 and L_1 for the lowest $1/f$ noise. Such optimisations tend to favour small transistors with relatively long channel length [27, 28], as figure 4.18 demonstrates.

Equation (4.79) can be simplified by ignoring the edge capacitances, so that the transistor area $A_1 = W_1L_1$ is optimised instead. Following [15] and using $C_{\rm e} = 0$, (4.79) can be rewritten as:

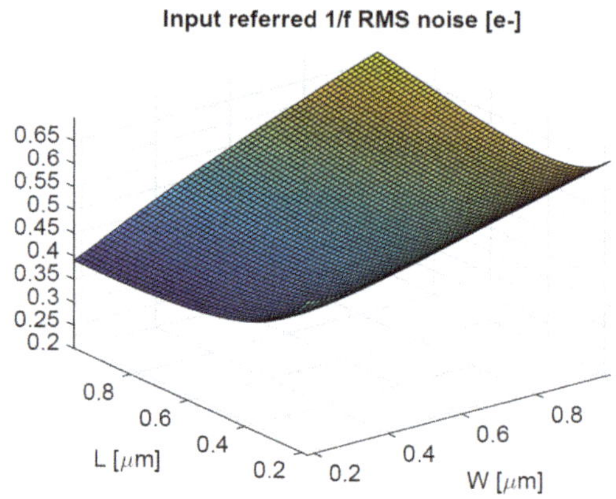

Figure 4.18. Calculated input-referred $1/f$ noise of the in-pixel source follower as a function of the channel width and length. Reprinted with permission from [27].

$$\overline{Q_{\mathrm{nf}}} = \frac{2\sqrt{K_{\mathrm{F}}\ln 2}\left(C_{\mathrm{sn}} + C_{\mathrm{par}}\right)}{qC_{\mathrm{ox}}\sqrt{A_{\mathrm{l}}}} + \frac{4\sqrt{K_{\mathrm{F}}\ln 2}}{3q}\sqrt{A_{\mathrm{l}}} \tag{4.80}$$

This minimum of $\overline{Q_{\mathrm{nf}}}$ can be found by solving the equation $d\overline{Q_{\mathrm{nf}}}/dA_{\mathrm{l}} = 0$, which gives

$$C_{\mathrm{sn}} + C_{\mathrm{par}} = \frac{2}{3}C_{\mathrm{ox}}W_1L_1 \tag{4.81}$$

Equation (4.81) means that the minimum $1/f$ noise in the case $C_{\mathrm{e}} = 0$ is achieved when $C_{\mathrm{d}} = C_{\mathrm{GS}}$ [15]. This result follows the capacitance matching principle which has been established in the past for CCDs [13]. The minimum $1/f$ noise is then

$$\overline{Q_{\mathrm{nf}}^{\mathrm{min}}} = \frac{8\sqrt{W_1L_1K_{\mathrm{F}}\ln 2}}{3q} \tag{4.82}$$

which for the parameters in example 4.6 gives $\overline{Q_{\mathrm{nf}}^{\mathrm{min}}} = 0.3\ \mathrm{e}^{-}$ RMS.

Chapter summary

1. Thermal and $1/f$ noise from the in-pixel source follower are the dominant noise source in CIS in the absence of dark current and photon shot noise.
2. Thermal noise in a MOSFET is caused by the channel resistance.
3. $1/f$ noise is caused by mobility or carrier number fluctuations in the MOSFET channel. It is the dominant noise source at low frequencies.
4. A source follower buffering a sense node has an excess noise factor due to high impedance gate drive, reducing its effective transconductance.
5. The gate–source capacitance of the source follower is reduced by a factor of $(1 - G_{\mathrm{SF}})$ due to the negative feedback causing the source to closely follow the gate input.

6. The purpose of the CDS is to remove the reset noise while producing the highest SNR.
7. The dual slope integrator is the optimal signal processing technique for CIS output signals and delivers the highest SNR when only white noise is present.
8. DCMS approximates the dual slope integrator when the number of samples is large, typically more than 10.
9. DCMS can outperform the DSI for large $1/f$ noise by using appropriate sample weighting.
10. The column capacitance to the source follower $C_{col} + C_S$ can serve as a bandwidth limiter, making the ENC of the white noise proportional to $\sqrt{kT/(C_{col} + C_S)}$.
11. The readout noise is ultimately limited by the $1/f$ noise which is dominant at the low readout rates necessary to sufficiently reduce the white noise.
12. The channel width and length of the in-pixel source follower can be optimised for achieving the lowest readout noise.

References

[1] Horowitz P and Hill W 2015 *The Art of Electronics* 3rd edn (New York: Cambridge University Press)

[2] Fano U 1947 Ionization yield of radiations. II. The fluctuations of the number of ions *Phys. Rev.* **71** 26–9

[3] Janesick J 2007 *Photon Transfer DN λ* (Bellingham, WA: SPIE Press)

[4] Caloyannides M 1974 Microcycle spectral estimates of $1/f$ noise in semiconductors *J. Appl. Phys.* **45** 307

[5] Vandamme L K J and Hooge F N 2008 What do we certainly know about $1/f$ noise in MOSTs? *IEEE Trans. Electron Devices* **55** 3070–85

[6] Chang J, Abidi A A and Viswanathan C R 1994 Flicker noise in CMOS transistors from subthreshold to strong inversion at various temperatures *IEEE Trans. Electron Devices* **41** 1965–71

[7] Janesick J, Andrews J and Elliott T 2006 Fundamental performance differences between CMOS and CCD imagers: part I *Proc. SPIE 6276 (Orlando, FL)*

[8] Leyris C, Martinez F, Valenza M, Hoffmann A, Vildeuil J C and Roy F 2006 Impact of random telegraph signal in CMOS image sensors for low-light levels *2006 Proc. of the 32nd European Solid-State Circuits Conf. (Montreux, Switzerland)*

[9] Hopkinson G and Lumb D 1982 Noise reduction techniques for CCD image sensors *J. Phys. E: Sci. Instrum.* **15** 1214–22

[10] Jordan A and Jordan N 1965 Theory of noise in metal oxide semiconductor devices *IEEE Trans. Electron Devices* **12** 148–56

[11] Nemirovsky Y, Corcos D, Brouk I, Nemirovsky A and Chaudhry S 2011 $1/f$ Noise in advanced CMOS transistors *IEEE Instrum. Meas. Mag.* **14** 14–22

[12] Yadid-Pecht O, Mansoorian K, Fossum E and Pain B 1997 Optimization of noise and responsivity in CMOS active pixel sensors for detection of ultra low light levels *Proc. SPIE* **3019** 125–36

[13] Burt D 1991 CCD performance limitations: theory and practice *Nucl. Instr. Meth.* A **305** 564–73

[14] Kansy R 1980 Response of a correlated double sampling circuit to 1/*f* noise *IEEE J. Solid-State Circuits* **15** 373–5

[15] Kawahito S 2011 Architectures for low-noise CMOS electronic imaging *Single-Photon Imaging* ed P Seitz and A Theuwissen (Berlin: Springer) pp 197–217

[16] Janesick J 2001 *Scientific Charge-Coupled Devices* (Bellingham, WA: SPIE Press)

[17] Hegyi D and Burrows A 1980 Optimal sampling of charge coupled devices *Astron. J.* **85** 1421–4

[18] Levanon N and Mozeson E 2004 Matched filter *Radar Signals* (New York: Wiley) pp 20–33

[19] Stefanov K 2015 Digital CDS for image sensors with dominant white and 1/*f* noise *J. Instrum.* **10** 04003

[20] Wey H M and Guggenbühl W 1990 An improved correlated double sampling circuit for low noise charge-coupled devices *IEEE Trans. Circuits Syst.* **37** 1559–65

[21] Kleinpenning T and de Kuijper A 1988 Relation between variance and sample duration of 1/*f* noise signals *J. Appl. Phys.* **63** 43–5

[22] Kawai N and Kawahito S 2005 Effectiveness of a correlated multiple sampling differential averager for reducing 1/*f* noise *IEICE Electron. Express* **2** 379–83

[23] Kawahito S and Seo M-W 2016 Noise reduction effect of multiple-sampling-based signal-readout circuits for ultra-low noise CMOS image sensors *Sensors* **16** 1867

[24] Stefanov K and Murray N 2014 Optimal digital correlated double sampling for CCD signals *IET Electron. Letters* **50** 1022–4

[25] Gach J-L, Darson D, Guillaume C, Goillandeau M, Cavadore C, Balard P, Boissin O and Boulesteix J 2003 A new digital CCD readout technique for ultra-low-noise CCDs *Publ. Astron. Soc. Pacific* **115** 1068–71

[26] Fowler B 2011 Single photon CMOS imaging through noise minimization *Single-Photon Imaging* ed P Seitz and A Theuwissen (Berlin: Springer) pp 159–95

[27] Boukhayma A, Peizerat A and Enz C 2016 Noise reduction techniques and scaling effects towards photon counting CMOS image sensors *Sensors* **16** 514

[28] Boukhayma A, Peizerat A and Enz C 2016 Temporal readout noise analysis and reduction techniques for low-light CMOS image sensors *IEEE Trans. Electron Devices* **63** 72–8

IOP Publishing

CMOS Image Sensors

Konstantin D Stefanov

Chapter 5

Characterisation

5.1 Introduction

Electro-optical (EO) characterisation is fundamental to image sensor operation and development. It aims to experimentally measure the parameters of the sensor under different operating conditions. The results are used to select the optimal settings for the best sensor performance, and for a comparison with the estimations calculated during the design phase. Despite the highly powerful simulations and electronic design automation (EDA) tools, there is still no substitute to sensor characterisation. High quality, detailed characterisation can offer a wealth of information that can help optimise the sensor's operation and identify unexpected behaviour or design flaws. It also provides feedback to the simulation tools used in the design.

Characterisation requires a great deal of investigative and problem-solving skills, logical thinking, solid technical knowledge, and ingenuity. It is highly skilled work taking years to master, and the people able to do this well are highly respected. No matter how experienced somebody is, there are always surprises, seemingly inexplicable behaviour and new things to learn; this is what makes it challenging but rewarding.

The number of experimental parameters to be measured can be very large, and the number of possible device settings is nearly infinite. Therefore, the characterisation has to be balanced so that it provides the desired information but is achievable within a limited time and budget. Certain parameters, such as the saturation signal, readout noise and dark current are of high importance and are invariably measured, while others may be of less or no interest and can be omitted. Publications such as the EMVA 1288 standard [1] and several ISO standards on electronic still-picture imaging [2] provide important information on measuring techniques.

The following is a non-exhaustive list of the main parameters to be determined, in order of their usual importance:

doi:10.1088/978-0-7503-3235-4ch5

- Electrical characteristics for operation—readout rate, signal range, settling time, electrical transfer function (ETF), power dissipation;
- Responsivity, its linearity and uniformity, and saturation signal;
- System and conversion gain;
- Readout noise and dynamic range;
- Dark current and its uniformity;
- Image lag;
- Quantum efficiency (QE);
- Modulation transfer function (MTF).

The electrical characteristics are dealt with during the commissioning of the readout electronics. The starting point is the simulated performance of the on-chip readout, which is usually very reliable. Some adjustments may be needed to create correctly functioning imaging system, capable of powering and driving the sensor and supplying the appropriate optical signals. The system should be able to cope with the output readout rate, correctly capture the outputs of the sensor and have negligible noise. After all, we should be studying the sensor, not the quirks in the electronics and the bugs in the software.

A light source with controlled intensity, wavelength and timing is required for the determination of most characterisation parameters, except for the dark current. Very often the illumination must be synchronous to the sensor's drive signals, and in such cases, light emitting diodes (LED) can be used to provide light pulses with sub-microsecond duration.

A great deal of image sensor characterisation can be automated by software, and many parameters can be determined from the same data. Good software skills can be essential in analysing the data, especially when it involves processing a large number of recorded images.

5.2 Readout modes

There are two main modes of CIS readout—rolling shutter (RS) and global shutter (GS). The possible readout modes are tightly coupled to the pixel and sensor architecture.

The term 'sensor readout' is used to indicate the time when some part of the sensor is being read out. The term 'integration time' means the period when the sensor is *not* being read out but is integrating a light-generated signal. A sensor can be sensitive to light during both periods and the difference between them in terms of signal collection can be negligible. This definition of the readout and the integration time will be used for the image sensor characterisation.

Rolling shutter is the simplest readout mode and is the usual way to read 4T pixels because it uses CDS and delivers the best noise performance. RS can be continuous (shown in figure 5.1), which delivers the highest readout rate, or image integration over the whole sensor can be added between two consecutive readouts (figure 5.2). In both cases the pixels are continuously sensitive to light, except during the charge transfer from the PPD to the sense node. As we can see, in RS mode all rows are

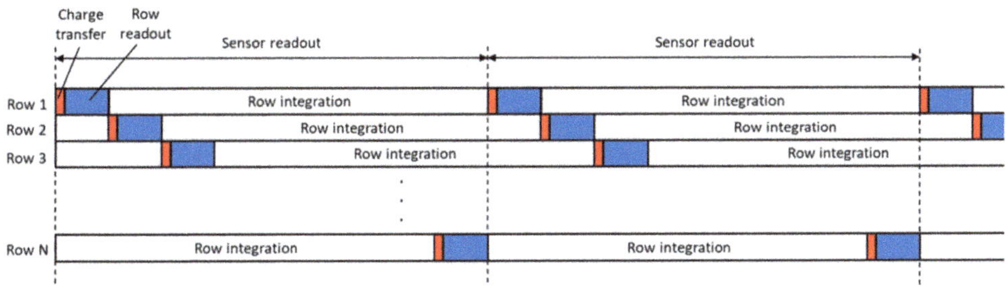

Figure 5.1. Rolling shutter, continuous 4T sensor readout.

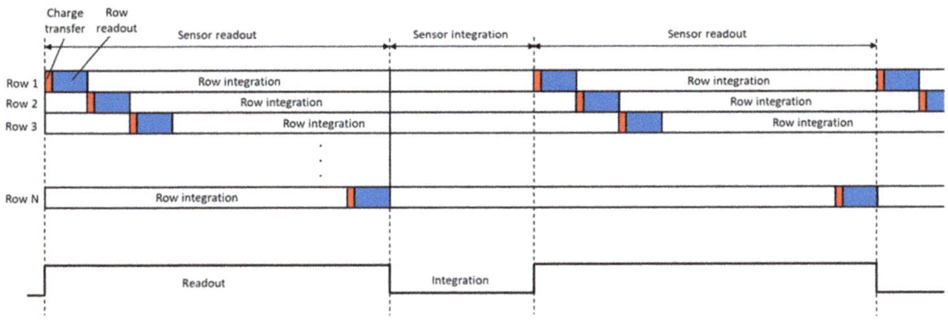

Figure 5.2. Rolling shutter readout with an additional sensor integration time.

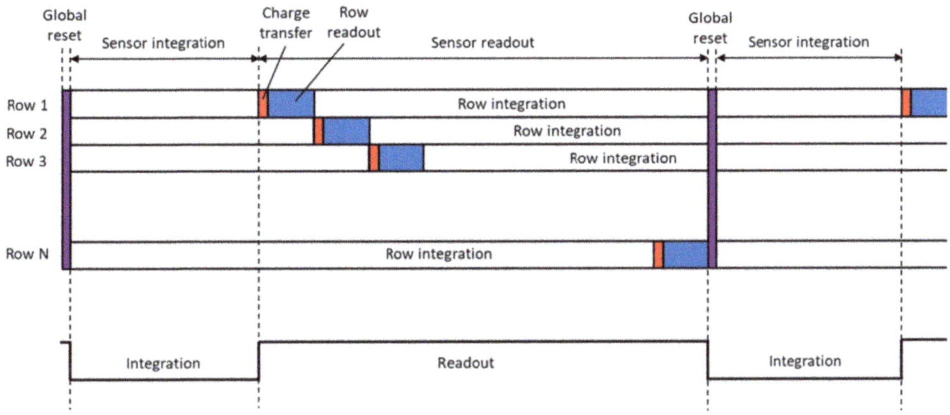

Figure 5.3. Rolling shutter with global reset using 5T pixels.

exposed to light for the same length of time, but *not at the same time*. This introduces distortions when imaging a moving scene and is a well-known effect.

To reduce motion distortions, a global reset is introduced with 5T pixels, but the readout is still in rolling shutter mode (figure 5.3). With global reset all pixels are emptied of charge at the beginning of the readout cycle, which helps reduce motion distortion but does not eliminate it. Global reset is particularly effective when most

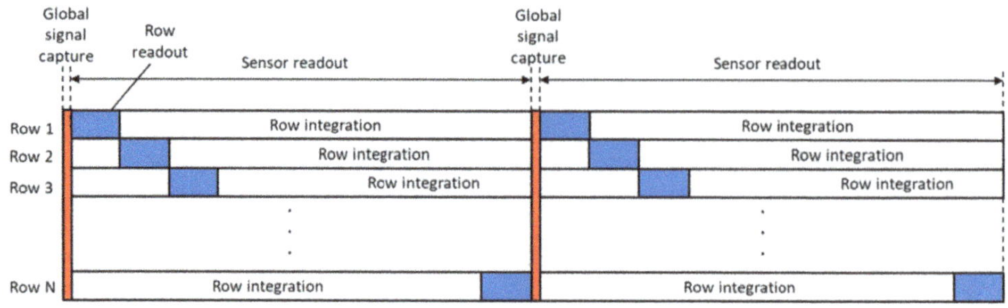

Figure 5.4. Global shutter readout mode.

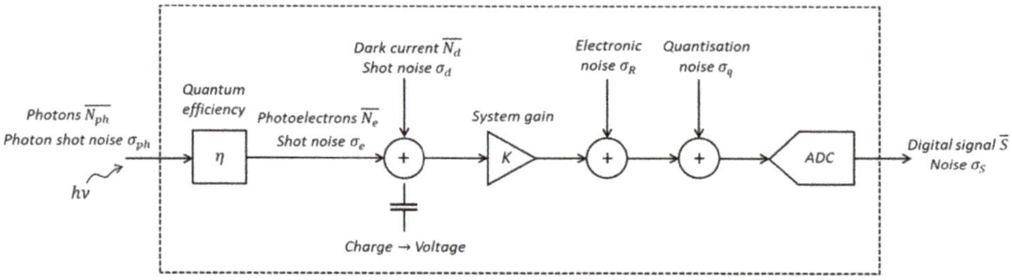

Figure 5.5. Imaging system model using a linear sensor, based on [1].

of the signal is generated during integration, for example by using a flash illumination. Similar effect can be achieved with the RS readout in figure 5.2. As the integration time increases, the proportion of time during which all rows are simultaneously receiving signal increases too. If the integration time is much longer than the readout time, the overall effect becomes close to having a global reset.

Global shutter is the preferred readout mode when imaging moving scenes because all pixels are sensitive to light simultaneously. This is accomplished by a sensor-wide, global signal capture from the photodiode to an in-pixel storage node (figure 5.4). The signal can be either charge or voltage and requires specialised pixel architecture. Following capture, the stored signals do not change under illumination and are read out row by row, simultaneously with the integration of the next image. Each pixel is photosensitive for the time duration between two applications of the global signal capture. Most GS sensors offer different readout schemes and can be read out in RS mode as well.

The majority of scientific CIS are designed for RS operation, while GS mode is frequently needed for machine vision applications. RS readout with pulsed illumination during an added integration time as in figure 5.2 is the main mode used for the characterisation methods described further on.

5.3 Principles of EO characterisation

It is very useful to consider the general block diagram of a linear image system in figure 5.5. Regardless of the sensor's architecture and the manufacturing technology, the commonalities in the processes of converting photons into digital signal make such a diagram representative of a wide range of systems.

The signal path starts with incoming photons with mean number $\overline{N_{ph}}$ received during the integration time, converted with quantum efficiency η to $\overline{N_e}$ photoelectrons:

$$\overline{N_e} = \eta \overline{N_{ph}} \tag{5.1}$$

Electrons created by dark current during the same time interval, with mean $\overline{N_d}$, are added to the photogenerated signal before the conversion to voltage. The mean output signal $\overline{S}$ is the digital code proportional to the total number of converted electrons:

$$\overline{S} = \frac{1}{K}(\overline{N_e} + \overline{N_d}) \tag{5.2}$$

The signal is in dimensionless analogue-to-digital units (ADU) corresponding to the generated ADC code, for example a 16-bit ADC can output code from 0 to 65535. Very often the equivalent notation 'digital number' (DN) is used instead of ADU.

The proportionality constant K is called *system gain* and is measured in the dimensionless units of e^-/ADU. The choice of units[1] is like this because we want to determine a physical value—the total number of electrons $\overline{N}$, from the output signal $\overline{S}$, which is just a digital code. From (5.2) this relationship is simply

$$\overline{N} = K\overline{S} \tag{5.3}$$

The system gain combines two very different, and virtually independent parts, into one:

- The *conversion gain* of the sensor, describing the voltage created by the charge-to-voltage conversion;
- Electronic gain in the system amplifying the voltage from the converted charge, and the digitisation by the ADC.

The conversion gain is a property of the conversion element, for example the sense node in the PPD, and is practically independent on the gain of the following electronic circuits. The conversion gain is a key design parameter and can be measured by knowing both the system and the electronic gain.

The digital output signal S is also called 'grey value' [1] which could be slightly confusing when applied to colour cameras, or with Gray code. To avoid this, the term 'signal value' meaning 'digital signal value' is used throughout this chapter.

Real-life systems exhibit some nonlinearity and therefore K is not a constant, but the model in figure 5.5 can still be used with due care. The system is not a 'black box' because we know its structure, although the exact physical implementation of the

[1] The inverse is used in [1], but it is much more common to use the present notation.

building blocks may be unknown. For the purposes of characterisation, it operates as a box where photons come in one end and ADC code comes out from the other.

The incoming photons have shot noise with standard deviation σ_{ph}, given by the Poisson statistics

$$\sigma_{ph} = \sqrt{N_{ph}} \tag{5.4}$$

If one interacting photon generates one photoelectron, which is the case for visible light in silicon, the standard deviation of the number of photoelectrons from (5.1) is

$$\sigma_e = \sqrt{\eta N_{ph}} \tag{5.5}$$

The dark current and signal electrons are treated in the same way by the charge-to-voltage conversion, and so is their shot noise. The readout noise with standard deviation σ_R represents all electronic noise in the system, generated in any part of the signal chain. The ADC quantisation noise has standard deviation $\sigma_q = 1/\sqrt{12}$ ADU and should be negligible in a well-designed system. All noise sources are considered to be statistically independent, therefore the output noise variance is the quadrature sum of all of them.

$$\sigma_S^2 = \frac{1}{K^2}(\sigma_e^2 + \sigma_d^2) + \sigma_R^2 + \sigma_q^2 \tag{5.6}$$

In equation (5.6) all noise sources are dimensionless, however, σ_e and σ_d are expressed in electrons, while σ_S, σ_R and σ_q are in ADU.

Characterising a sensor requires an experimental setup capable of providing light with controlled wavelength, intensity, uniformity and timing. Some measurements, for example the QE and the MTF are much more challenging than the 'easy ones' such as the linearity, and require sophisticated, and often expensive equipment. Many systems are not built to deal with a complete sensor characterisation. In its simplest form, a characterisation setup can comprise nothing more than a sensor and an LED in a light-tight box.

A full characterisation system (figure 5.6) would include a light source, monochromator, targets and optics, calibrated photodiodes, shutter, x-ray source and temperature control of the sensor.

The light source is usually a halogen lamp, but xenon arc or deuterium lamps can be used for UV wavelengths. The monochromator is used to select a narrow spectral line from the incoming broadband light for wavelength-dependent measurements such as the QE and the MTF. When narrowband light is not needed, LEDs with different dominant wavelengths can be used instead.

The integrating sphere provides a uniform output light beam from the source through multiple diffuse reflections off its inner wall. Various filters, targets (for MTF measurements) and focusing optics can then be used to project patterned or uniform light on the device under test (DUT). Sophisticated experimental apparatus covering wavelengths from vacuum UV to NIR, which is challenging to achieve with integrating spheres, has been developed [3] for demanding space-based imaging.

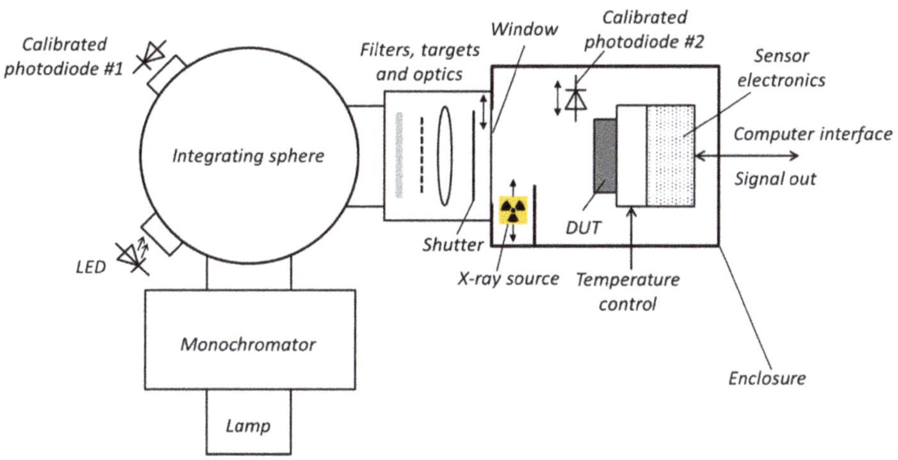

Figure 5.6. General diagram of a full optical characterisation setup.

The first photodiode monitors the stability of the light source and provides a reference measurement. Diodes with well-known photoresponse are available with traceable calibration curve, for example referenced to the US National Institute of Standards and Technology (NIST) calibration sources. The second photodiode is used to determine the photon flux reaching the sensor, which can be calculated using the photodiode size and the geometry of the setup.

A low energy x-ray source, typically ^{55}Fe, is often included in the setup to provide another method for system gain calibration. The x-ray source and the second calibrated photodiode can be moved in and out of the path of the optical beam. In its off-beam position the x-ray source is shielded and has no line of sight to the sensor. If a ^{55}Fe source is used, the glass cover of the sensor must be removed because the low energy x-rays cannot penetrate it.

The shutter's function is to block any light and ensure a measurement in complete darkness. It can be within the light-tight enclose too, and can be used to stop both light and x-rays. A normal photography shutter, made with thin steel blades, can easily stop the soft x-rays from a ^{55}Fe source. A transparent plastic shutter can be used to stop x-rays but not visible light.

Temperature control is often required, or as a bare minimum at least a temperature measurement of the sensor's package. Knowing the sensor's temperature is paramount for dark current measurements. When operating below approximately 15 °C, water vapour could start to condense on the sensor. Therefore, at lower temperatures vacuum enclosures with a window for the incoming light beam must be used.

5.4 Photoresponse, non-uniformity and nonlinearity

Determining the photoresponse of a sensor is normally the first thing to do and is also one of the most straightforward. In simple terms, it is just recording the output

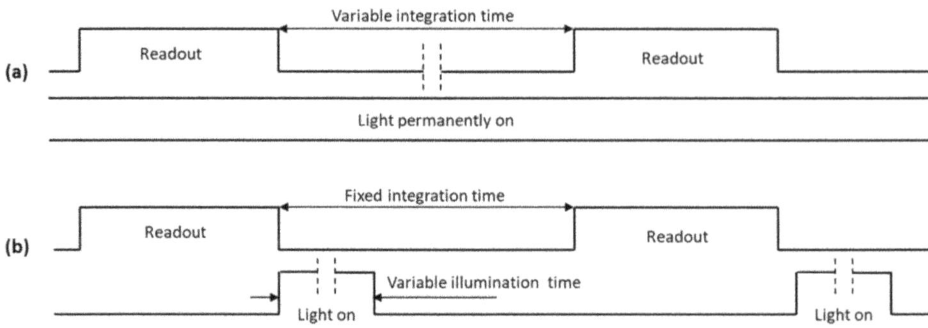

Figure 5.7. Timing diagrams for photoresponse measurement using constant illumination (a), and variable (pulsed) illumination (b). The readout time is fixed in both cases.

signal versus the illumination level. There are two ways to do this, as illustrated in figure 5.7:

(a) Constant illumination with variable integration time. The light source is permanently on, and the desired signal level is achieved by allowing the sensor to be light sensitive during a variable integration time.

(b) Variable illumination time with constant integration time. Light pulse with a controlled duration is applied to the sensor during a fixed integration time.

Both methods have their pros and cons. With constant illumination the light output can be very stable because the source is permanently on. On the other side, increasing the integration time in order to increase the signal adds dark current. It can be characterised and subtracted, but care should be taken because the device temperature may be lower at longer integration times, since the power dissipation from the readout is lower. Also, the sensor may be light sensitive during readout as well (normally in rolling shutter mode), therefore achieving very low signal levels is limited by the readout time. Global shutter mode largely solves this, but there could be some parasitic light sensitivity during readout. To achieve zero signal, the light source may have to be blocked by a mechanical shutter.

The main advantage of the variable illumination time is that the sensor is read out with a fixed time interval. The dark signal does not change and can be easily subtracted, and the device temperature is not modulated by the readout period. However, LEDs or lamps with a fast shutter must be used for pulsed illumination, and achieving a stable light output may be difficult, especially with short pulses. The light output may be subject to systematic effects, for example a LED warming up at longer illumination can produce a nonlinear signal.

The photoresponse is plotted as the averaged signal from the uniformly illuminated whole sensor (or many uniformly illuminated pixels) versus the:

(a) integration time (or the sum of the integration time and the readout time if the sensor is sensitive during readout), for constant illumination.

(b) Illumination time, when the integration time is fixed.

In both cases the signal mean at zero illumination is subtracted from the signal mean of all images taken with light. In addition to averaging over many pixels, averaging over many images taken at the same conditions is also used, and is recommended to reduce the statistical uncertainty from the shot noise when the number of pixels is small.

When measuring the photoresponse, the main things to watch out for are the signal offsets, dark current and light uniformity. Pixel and column offsets, showing as non-uniformity and fixed pattern spatial noise, could significantly impact the photoresponse at zero and low illumination levels. Dark current and especially hot pixels can also show up at low signals. When constant integration time is used, an image taken in darkness contains both the electrical non-uniformities and the dark current signal. Subtracting the mean of this dark image from the mean of all images taken under illumination removes both the offsets and the dark current. Usually the average of several (>10) dark images is used to reduce the noise fluctuations. In this way, zero illumination time corresponds to an image with zero mean signal.

Achieving uniform illumination is also important for the correct measurement of the photoresponse. Since it is averaged over many pixels, non-uniform illumination has the effect of scaling the response, relative to the actual, at all signals. If it is not possible to illuminate uniformly the whole sensor, a smaller but much more uniform part of the image can be selected as a region of interest (ROI).

The photoresponse in figure 5.8 has been obtained with constant integration time, using a pulsed LED illumination. The averaged dark image has been subtracted, therefore the signal at zero illumination is zero. If we know the system gain, the number of electrons can be calculated from the signal. At illumination times above

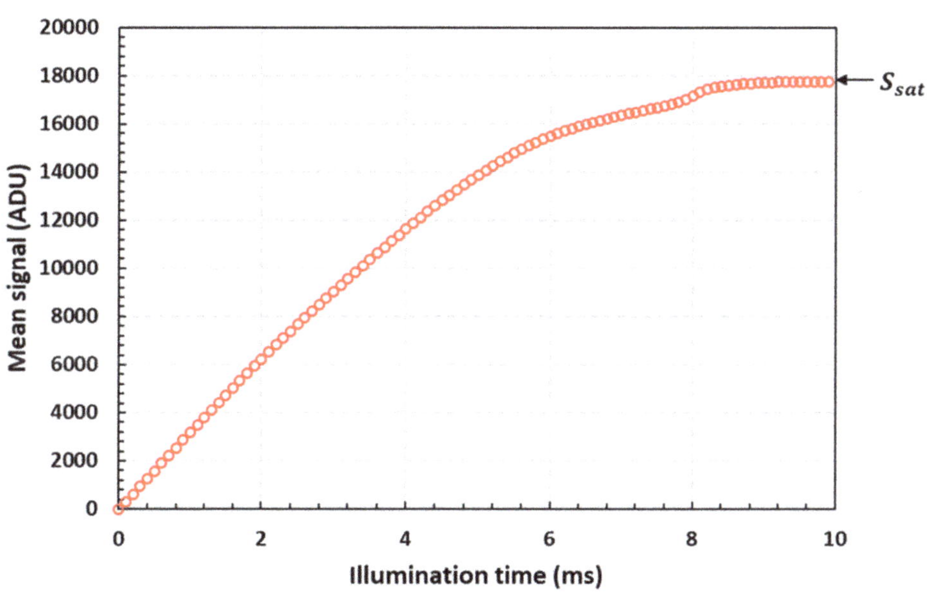

Figure 5.8. Photoresponse of a 4T CIS taken with constant integration time. The averaged signal from 48 000 pixels has been used.

6 ms the output signal begins to saturate and eventually reaches $S_{sat} \approx 18\,000$ ADU. Signal saturation could be due either to the PPD, the sense node, or the following electronics. Very often the sense node saturates first because its capacitance is too low to accommodate the full charge collected by the PPD. Here we see more complex behaviour in saturation, probably due to a combination of factors.

Under uniform illumination an ideal sensor is expected to produce the same signal from every pixel, after the shot noise has been sufficiently reduced by averaging. However, the pixels are not the same due to manufacturing variations, and their photoresponse is different. A measure for this is the photoresponse non-uniformity (PRNU), which gives rise to fixed pattern spatial noise.

The PRNU is calculated as the standard deviation of the signal σ_S over the whole sensor (or a large number of pixels) divided by the mean signal $\overline{S}$ from the same area, and expressed in percent:

$$\mathrm{PRNU} = 100 \times \frac{\sigma_S}{\overline{S}} \tag{5.7}$$

Needless to say, the pixel area must be uniformly illuminated, the DC offsets and the dark current must be removed, and the photon shot noise should be made negligible through averaging of sufficient number of images. The mean signal is normally chosen to be at 50% of saturation. Knowing the system gain is not needed to calculate the PRNU and the linearity of the sensor, which would be the same regardless of whether the signal is in electrons or ADU.

In practical terms, one needs to take many images in darkness and at the chosen illumination and average them on a pixel-by-pixel basis to suppress the photon shot noise. After that, the standard deviation of the difference between the two averaged images is calculated to give σ_S.

The image in figure 5.9 is from the same sensor having the photoresponse in figure 5.8, but has a piece of dust[2] in rows 48–52. The signal around the dust drops by about 1000 ADU, or 15%. The PRNU calculated from the data including the dust is 0.80%, while excluding it (which also excludes 8000 pixels out of 40 000) reduces the PRNU to 0.69%. This example is to show how careful one must be in order to obtain good quality measurements. Achieving uniform illumination and keeping the sensor clean only gets harder as the sensors get bigger.

We can easily see from figure 5.8 that the output signal does not increase linearly with the illumination time even below saturation. Linearity is required from most sensors and it is considered to be an essential property, but some are intentionally nonlinear, like the logarithmic image sensors [4] designed to achieve wide dynamic range.

The possible sources of nonlinearity could be either in the sensor or in the external electronics, such as amplifiers and ADCs. Normally, the external electronics is far more linear than the sensor. Complex imaging system models have been developed to include both sources of nonlinearity [5].

[2] Intentionally placed.

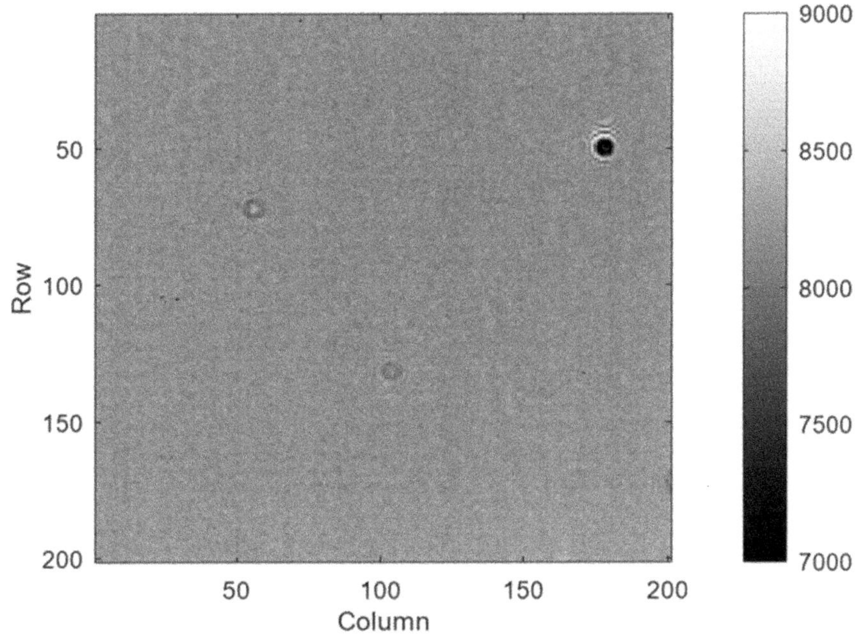

Figure 5.9. Image at 50% S_{sat} for PRNU calculation, showing several imperfections.

If sensor nonlinearity is involved, then two types can be distinguished [6]:

1. Electrical gain nonlinearity (also called V/V nonlinearity) of the in-pixel SF and the signal chain.
2. Conversion gain nonlinearity (also called V/e$^-$ nonlinearity), caused by the dependence of the sense node capacitance on its voltage.

Normally the electrical gain nonlinearity is below 1% [7], while the conversion gain nonlinearity can be above 10%.

The sense node capacitance can be written as a sum of two parts: a constant capacitance C_{sn0} caused by parasitic elements such as interconnects or intentionally added MOS capacitors, and a *pn* junction capacitance depending on the sense node voltage V_{sn} (derived in chapter 1):

$$C_{sn} = C_{sn0} + A\sqrt{\frac{\varepsilon_0 \varepsilon_{Si} q N_A}{2(V_{bi} + V_{sn})}} \tag{5.8}$$

It is obvious that if C_{sn0} is much larger than the *pn* junction capacitance, the sense node capacitance becomes less voltage-dependent, and therefore more linear. Also, the linearity is better if V_{sn} does not change much—this is true for very small signals, or when the signal is much smaller than the reset voltage.

In a linear sensor collecting N_e photoelectrons the output voltage V_{out} would simply be

$$V_{out} = G_{tot} V_{sn} = G_{tot} \frac{q N_e}{C_{sn}} \tag{5.9}$$

where G_{tot} is the total electronic gain. If C_{sn} is not constant, (5.9) can be generalised as

$$V_{out} = qG_{tot} \int \frac{dN_e}{C_{sn}(N_e)} \tag{5.10}$$

Using that $V_{out} = V_{LSB}S$, where S is the output signal in ADU, and that $V_{sn} = V_{out}/G_{tot}$, (5.8) can be substituted in (5.10):

$$V_{LSB}S = qG_{tot} \int \frac{dN_e}{C_{sn0} + A\sqrt{\frac{\varepsilon_0\varepsilon_{Si}qN_A}{2(V_{bi} + V_{LSB}S/G_{tot})}}} \tag{5.11}$$

Differentiating both sides gives

$$\frac{dS}{dN_e} = \frac{qG_{tot}}{V_{LSB}\left(C_{sn0} + A\sqrt{\frac{\varepsilon_0\varepsilon_{Si}qN_A}{2(V_{bi} + V_{LSB}S/G_{tot})}}\right)} \tag{5.12}$$

Equation (5.12) can be solved for N, but certainly there is no convenient linear relationship as (5.3). If all the constants on the right side of (5.12) are grouped together, the dependence can be written in a slightly easier to understand form:

$$\frac{dS}{dN_e} = \frac{1}{c_1 + \sqrt{\frac{c_2}{c_3 + S}}} \tag{5.13}$$

It is very useful to put some numbers into the equations above and see what the effect of the changing *pn* junction capacitance can be in a practical case.

Example 5.1. Calculate the nonlinearity due the sense node capacitance of a sensor with charge-to-voltage conversion factor (CVF) = 100 μV/e⁻ measured at reset for $V_{sn} = 2.5$ V, $V_{bi} = 0.85$ V and $C_{sn0} = 1.1$ fF. The maximum output signal is 1.5 V.
Solution: The sense node capacitance at $V_{sn} = 2.5$ V is

$$C_{sn} = \frac{q}{G_c} = \frac{1.6 \times 10^{-19}}{100 \times 10^{-6}} = 1.6\,\text{fF}$$

Therefore, the *pn* junction capacitance at $V_{sn} = 2.5$ V is 1.6—1.1 = 0.5 fF. We do not know the junction area A and the doping concentration N_A, but using (5.8) we can write

$$\sqrt{\frac{a}{V_{bi} + V_{sn}}} = 0.5\,\text{fF}$$

where the coefficient a includes all the unknows and can be found for $V_{sn} = 2.5$ V:

$$a = 0.5^2(0.85 + 2.5) = 0.8375\,\text{fF}^2\text{V}$$

At the maximum output signal the sense node voltage is $V_{sn} = 2.5 - 1.5 = 1.0$ V and the junction capacitance increases to

$$\sqrt{\frac{0.8375}{0.85 + 1.0}} = 0.67 \text{ fF}$$

The total sense node capacitance at $V_{sn} = 1.0$ V becomes $C_{sn} = 1.1 + 0.67 = 1.77$ fF. Therefore, the CVF nonlinearity is

$$\frac{[C_{sn}(1.0\text{ V}) - C_{sn}(2.5\text{ V})]}{C_{sn}(2.5\text{ V})} = \frac{1.77 - 1.6}{1.6} = 10.6\%$$

It is easy to verify that if $C_{sn0} = 0$ fF the nonlinearity would rise to 34%.

Calculating the nonlinearity from the photoresponse is not always straightforward, especially when the sensor exhibits significant nonlinearity. The nonlinearity is generally calculated as follows: first, the data are fitted with a straight line in the form $y = ax + b$. For each data point with index i the residual R_i between the data S_i and the linear fit y_i is calculated in percent:

$$R_i = 100 \times \left(\frac{S_i - y_i}{y_i} \right) \tag{5.14}$$

The absolute residuals are then averaged, and this gives the nonlinearity NL.

$$\text{NL} = \frac{1}{n} \sum_{i=1}^{n} | R_i | \tag{5.15}$$

Alternatively, some manufacturers define the residual as

$$R_i = 100 \times \left(\frac{S_i - y_i}{S_{\text{lin}}} \right) \tag{5.16}$$

where S_{lin} is the upper signal level used for the linear fit. Formula (5.16) gives lower nonlinearity at low signal levels than (5.14) because S_{lin} is larger than or equal to y_i.

The different ways to calculate the nonlinearity come from the different ways the data range for the linear fit can be chosen. Normally the data at very small signals (below 5%–10% S_{sat}) are excluded, because due to poor signal-to-noise ratio this range is rarely used for quantitative measurements. However, we expect a sensor to be most linear at small signals, therefore this range could be used in principle. Large signals near saturation, above 90% or 95% S_{sat}, are also excluded—the signal there is visibly nonlinear.

The EMVA 1288 standard uses the signal between 5% and 95% of the output saturation signal S_{sat} using a weighted linear fit [1], giving higher weight to smaller signals. This makes a lot of sense because we expect the sensor to be more linear at small signals. Figure 5.10(a) shows the photoresponse of a 4T pixel and two sets of residuals are plotted in figure 5.10(b) using the same limits of 5% and 95% S_{sat}. The weighted data fit is more linear at small signals and its nonlinearity over the fit range, calculated from (5.15), is 2.7%, while for the non-weighed data this is 4.1%.

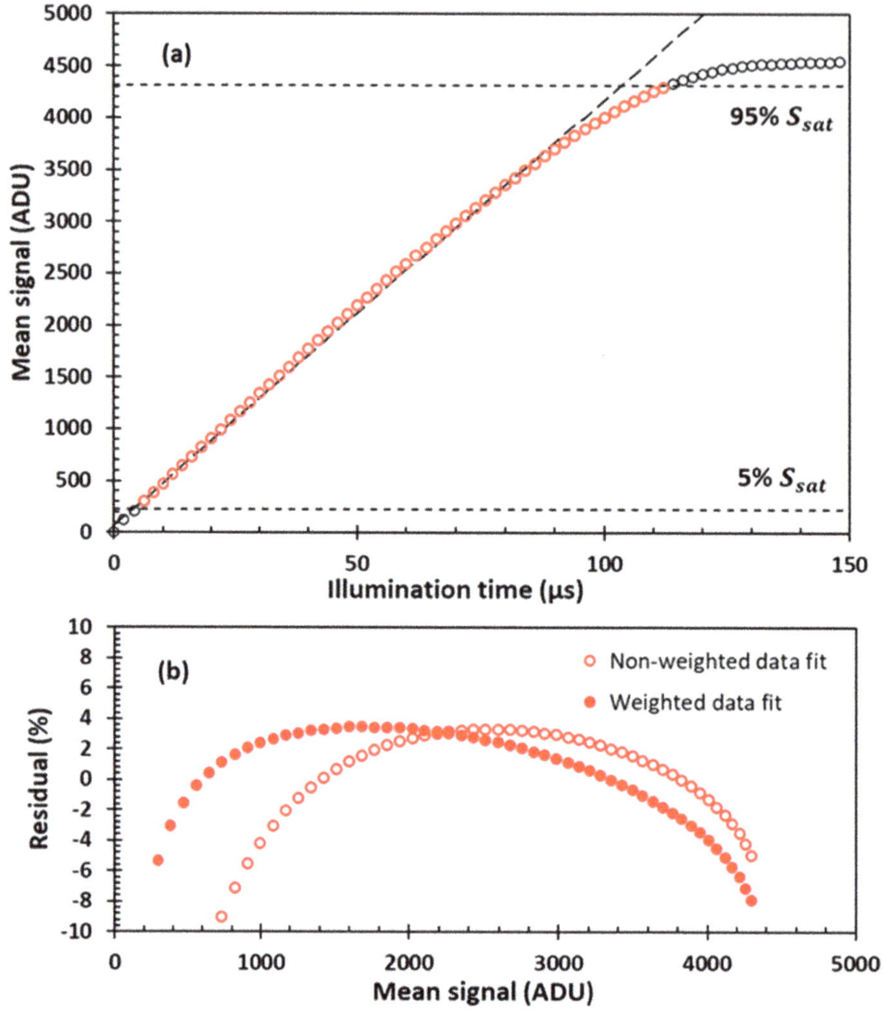

Figure 5.10. Obtaining a linear fit to the data between 5% and 95% S_{sat} using the data points in red (a); non-weighted and weighted residuals calculated according to [1] (b).

Due to the overall curvature of the photoresponse and the choice of the range for the fit the sensor appears more nonlinear at both low and high signals. We could choose a smaller signal range starting at zero signal for the linear fit, which should improve the calculated linearity for small signals, and explore the effect for large signals.

Figure 5.11 shows the residuals for the data in figure 5.10(a) using three different non-weighted linear fits, all starting at zero signal. The plot shows that the linearity is significant even when using the first 25% of the signal for the fit. This invites follow-on characterisation to determine the source of the nonlinearity, be it the sense node itself due to its small capacitance, the sense node bias, or just unreliable data.

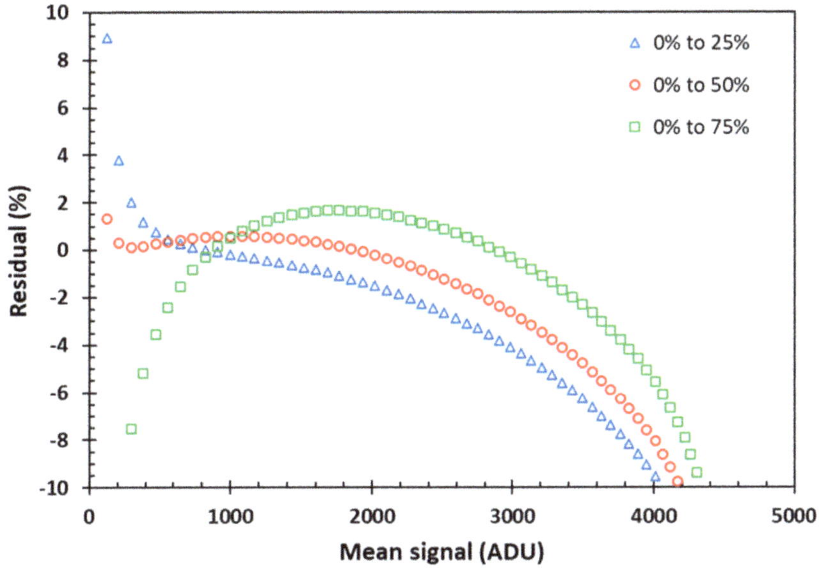

Figure 5.11. Nonlinearity of the photoresponse for three different fit ranges as a percentage of S_{sat}, using the data in figure 5.10(a). The average nonlinearity is 4.1%, 3.1% and 2.5% for the fit ranges 0%–25%, 0%–50% and 0%–75% of S_{sat}, respectively.

Sensor linearity is sensitive to the reset drain voltage V_{RD}, which together with the reset gate determine the sense node potential V_{sn} after reset. Higher V_{sn} helps improve the linearity provided that the signal is kept the same, because the change of the *pn* junction capacitance is lower. Linearity is also sensitive to the transfer gate voltage due to charge spill-back occurring at large signals [8]. Optimising the performance of the sensor requires measuring the linearity at a range of conditions, with some examples shown in figure 5.12. Such plots, together with measurements of the noise and the image lag are the most widely used input for the choice of operating conditions for a sensor.

Once the system gain is known, the sensor output in electrons can be calculated from (5.3). The photoresponse can then be re-plotted together with a linear fit to the data (using the most appropriate of the described methods above) as in figure 5.13. An ideal, linear sensor with the same system gain will have the response shown by the dashed line.

The plot of the signal in electrons versus the signal in ADU is a straight line, by definition, due to the constant system gain (5.3) assumed in linear systems. We can turn this around and ask how many electrons a perfectly linear sensor would register under the same illumination as our nonlinear sensor, provided the system gain is the same. This is effectively re-plotting figure 5.13 with the signal in ADU on the horizontal axis instead of the illumination time [9].

The linear fit, shown in figure 5.14, provides the number of photoelectrons registered by the sensor, and also the number of photons if the QE is known, under the assumption that the photoresponse can be approximated by a straight line.

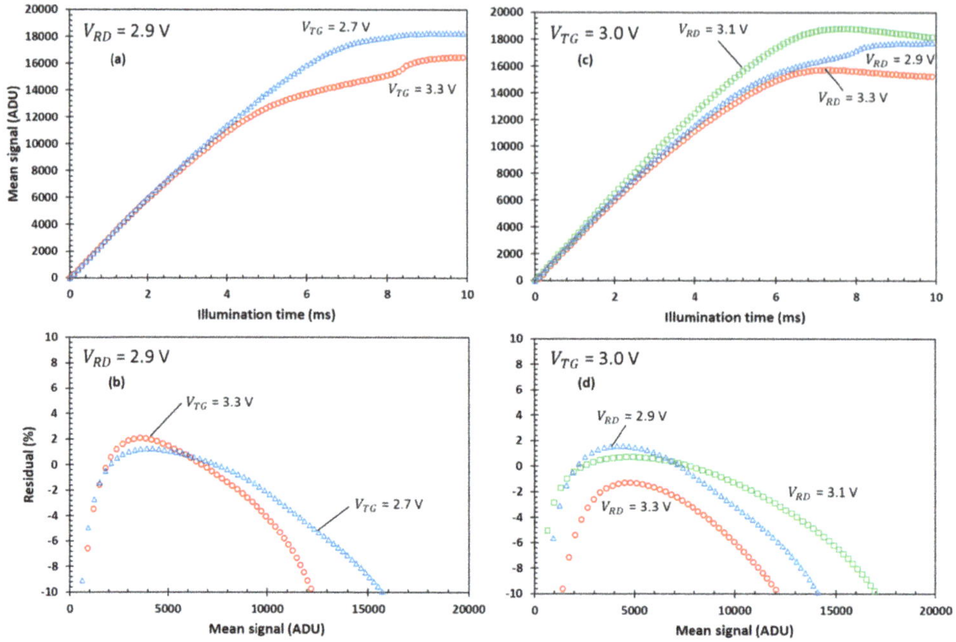

Figure 5.12. Photoresponse (a) and (c) and residuals from a linear fit to the data (b) and (d). For plots (a) and (b) the reset drain voltage V_{RD} is fixed at 2.9 V; for (c) and (d) the transfer gate clock amplitude V_{TG} is 3.0 V and V_{RD} varies. Both linearity plots use non-weighted linear fits between 0% and 50% S_{sat}.

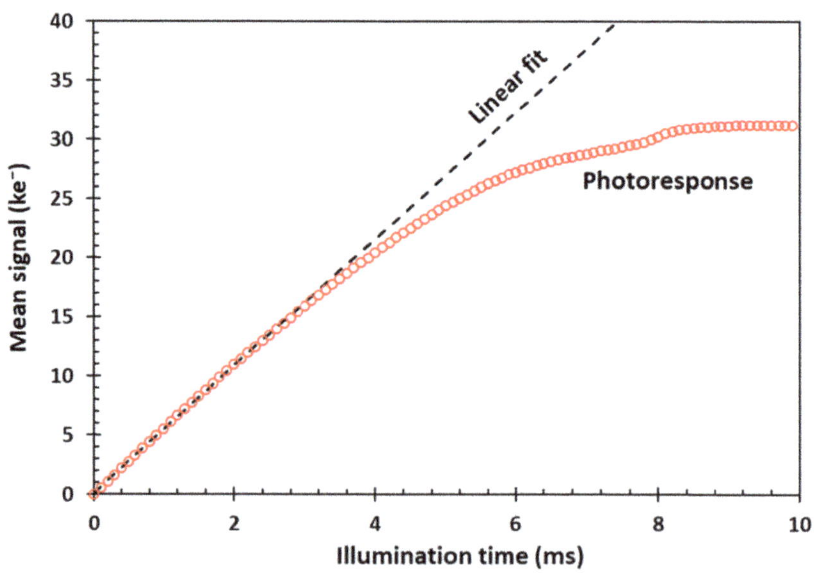

Figure 5.13. Sensor signal in electrons versus the illumination level for the data in figure 5.8. The system gain is $K = 1.76$ e⁻/ADU.

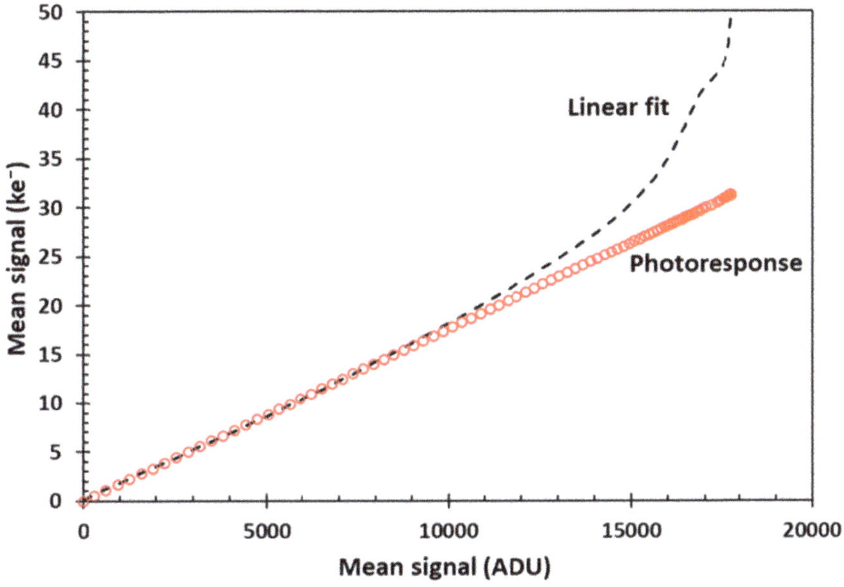

Figure 5.14. Sensor signal in electrons versus the mean signal in ADU. The system gain is $K = 1.76$ e$^-$/ADU.

As the sensor saturates, the relationship between signal and electrons breaks down, and the number of electrons tends to infinity.

5.5 Photon transfer curve

5.5.1 Principles

The photon transfer technique and its variants use that the incoming photons, and therefore the photogenerated electrons exhibit shot noise with standard deviation $\sigma_N = \overline{N_e}^{1/2}$ as per Poisson statistics, where $\overline{N_e}$ is the mean number of collected electrons.

The photon transfer curve (PTC) equation can be obtained from the considerations used in section 5.3. The averaged (mean) signal $\overline{S}$ obtained from a number of pixels and measured in ADU, is proportional to the number of collected electrons $\overline{N_e}$:

$$\overline{S} = \frac{\overline{N_e}}{K} \tag{5.17}$$

The proportionality constant K is the system gain, measured in the dimensionless units of e$^-$/ADU. The signal variance σ_S^2 can be found from (5.17) using the propagation of errors [10] with the addition of the readout noise with standard deviation σ_R

$$\sigma_S^2 = \left(\frac{\partial \overline{S}}{\partial \overline{N_e}}\right)^2 \sigma_N^2 + \left(\frac{\partial \overline{S}}{\partial K}\right)^2 \sigma_K^2 + \sigma_R^2 \tag{5.18}$$

Since K is constant $\sigma_K^2 = 0$ and from Poisson statistics $\sigma_N^2 = \overline{N_e}$, therefore (5.18) becomes

$$\sigma_S^2 = \left(\frac{1}{K}\right)^2 \overline{N_e} + \sigma_R^2 \tag{5.19}$$

After substituting $\overline{N_e}$ from (5.17) in (5.19), we get

$$\sigma_S^2 = \frac{\overline{S}}{K} + \sigma_R^2 \tag{5.20}$$

Equation (5.20) has the measured signal variance σ_S^2 as the sum of two parts: the shot noise component $\overline{S}/K$ and the readout noise variance σ_R^2. This PTC equation lets us obtain the system gain simply by measuring the variance of the signal as a function of its mean, measured in ADU. The system gain K is the inverse of the slope of the straight line fitted to the data. The readout noise is assumed to be purely electronic and signal-independent, therefore constant. Signal data plotted as in equation (5.20) is called *mean-variance curve.*

Figure 5.15 shows a simulated PTC using (5.20) with ideal shot and readout noise, and also adds signal saturation as required for any real system. The system gain is determined from the data points away from signal saturation, and readily yields $K = 2.0$ e$^-$/ADU. The readout noise is determined at $\overline{S} = 0$ and is $\sigma_R = 1.7$ ADU, or 3.4 e$^-$ RMS.

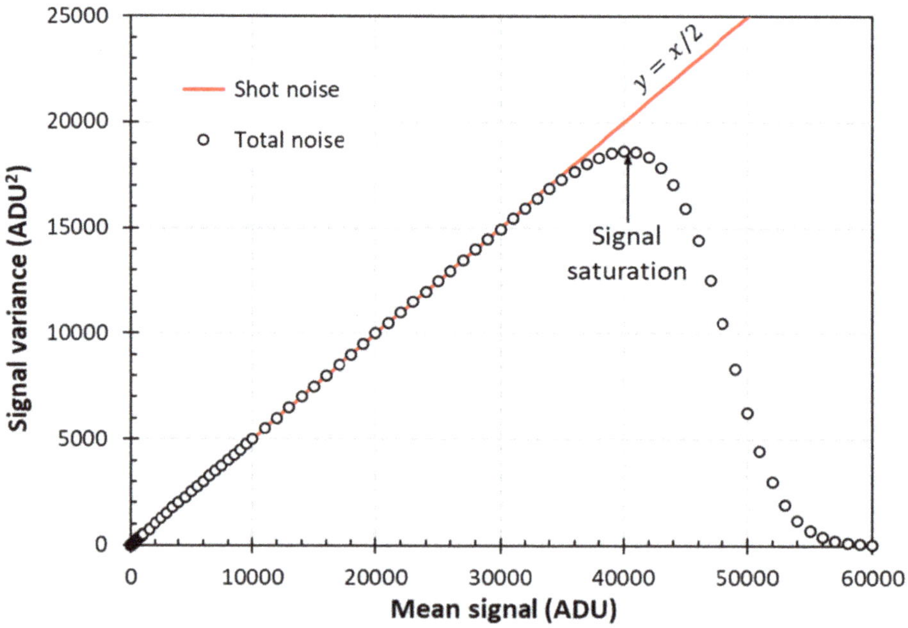

Figure 5.15. Simulated PTC for $\sigma_R = 1.7$ ADU and system gain $K = 2.0$ e$^-$/ADU using (5.20) and pixel saturation. The variance of the readout noise is too small to see on this scale.

Signal saturation occurs when the signal is limited by the pixel's FWC, the sense node voltage span, any amplifiers in the signal chain or the ADC. When the signal is limited its usual statistical fluctuations due to the shot noise are clipped too, therefore the measured variance decreases. Figure 5.15 indicates signal limiting around 40 000 ADU, or 80 ke$^-$. The fall in the signal variance is one of the methods used to estimate the FWC of a pixel, provided there are no other signal-limiting mechanisms.

The photon transfer is a very powerful technique because it does not require any knowledge of the system. The camera is treated as a black box with electrons coming in one end, and digital code out of the other. There is no need to know how the electrons are converted to voltage, what amplifiers are used, their gain, and the scale and the resolution of the ADC.

The electrons do not need to be photogenerated and can be collected from dark current instead, which also follows the Poisson distribution. When operating at ambient temperature the signal consists of both photo- and thermally generated electrons. The PTC technique does not care how the signal electrons are generated as long as they follow the Poisson statistics. This is the main pillar of the technique: if there are significant noise sources in the system that do not originate from the Poisson-distributed number of electrons collected per unit time, the PTC can produce meaningless or spectacularly wrong results. The most important source of error is the fixed pattern noise which can far exceed the shot noise at any signal level. Another source of error could be signal-dependent readout noise, observed in sensors using single-slope ramp ADCs [11]. Since the PTC equation assumes that σ_R is constant, an increase of the readout noise with the signal would increase the variance.

The PTC relies on the signal uniformity across the pixels to calculate the mean signal $\overline{S}$. Uniform illumination, also called *flat field*, is necessary, and thousands of pixels are used so that the statistical variations become small. However, uniform illumination is not needed to calculate the system gain, as we will see in 5.5.6. The PTC technique also works on a single pixel, and naturally there is no issue with uniformity, but very large data sets must be collected to reduce the statistical deviations.

To obtain the classic PTC equation [6] the system gain is expressed from (5.20) and a logarithm with base of 10 is taken on both sides

$$K = \frac{\overline{S}}{\sigma_S^2 - \sigma_R^2} \tag{5.21}$$

$$\log(K) = \log(\overline{S}) - \log(\sigma_S^2 - \sigma_R^2) \tag{5.22}$$

At sufficiently high illumination levels the shot noise is much larger than the readout noise and we have that $\sigma_S^2 \gg \sigma_R^2$, therefore

$$\log(K) \approx \log(\overline{S}) - \log(\sigma_S^2) = \log(\overline{S}) - 2\log(\sigma_S) \tag{5.23}$$

Rearranging (5.23) produces the classic PTC equation:

$$\log(\sigma_S) = \frac{1}{2}\log(\overline{S}) - \frac{1}{2}\log(K) \tag{5.24}$$

It is clear that on a log–log scale the shot noise versus the mean signal has a slope of 1/2. The intercept of the straight line fit to the shot noise part of the PTC with the X-axis at $\sigma_S = 1$ will be at signal $\overline{S} = K$, because then $\log(\sigma_S) = 0$. This is the graphical method for obtaining the system gain K from the PTC.

In addition to the shot and the readout noise, fixed pattern noise (FPN) is present in all pixelated image sensors and can make the PTC look rather differently. FPN is caused by the non-uniformity of the photoresponse of the individual pixels due to slight process variations in the active area, doping, metal coverage, layer thicknesses and transistor parameters. Contrary to the noise sources described in chapter 4, FPN is not temporal, but *spatial noise*. This means that FPN has a predictable pattern, characteristic of each individual sensor, and can even be used for forensic sensor identification [12]. FPN due to photoresponse non-uniformity is, of course, absent in darkness.

The FPN can be the dominant noise source at high illumination levels, and the human eye is extremely sensitive to it, or any other regular patterns in an image. The standard deviation σ_{FPN} is proportional to the average signal level, as we can see from the following.

The signal from each pixel with index i can be written as $S_i = (1 + \delta_i)\overline{S}$, where δ_i is a small random number (positive or negative) describing the variation of the pixel's photosensitivity and $\overline{S}$ is the average signal. The number δ_i is a property of each pixel and is constant. The standard deviation of the signal obtained from n pixels, excluding other noise sources, is

$$\begin{aligned}\sigma_{FPN} &= \sqrt{\frac{1}{n-1}\sum_n (S_i - \overline{S})^2} = \sqrt{\frac{1}{n-1}\sum_n ((1+\delta_i)\overline{S} - \overline{S})^2} \\ &= \sqrt{\frac{1}{n-1}\sum_n (\delta_i\overline{S})^2} = \overline{S}\sqrt{\frac{1}{n-1}\sum_n \delta_i^2}\end{aligned} \tag{5.25}$$

The term multiplying $\overline{S}$ in (5.25) is a proportionality constant P_N, called FPN quality factor [6] or PRNU. Hence, we can write that

$$\sigma_{FPN} = P_N\overline{S} \tag{5.26}$$

which states that the FPN is proportional to the mean signal. The factor P_N is typically around 0.01–0.02 and is not the same as the signal nonuniformity observed in images taken in total darkness, which is due to different DC offsets and dark signal in each pixel.

Figure 5.16 shows a simulated PTC for the same parameters as in figure 5.15, but with added FPN and plotted on a log–log scale as per (5.24). This PTC is typical for CCDs [13]. Based on the dominant noise, four regions can be distinguished:

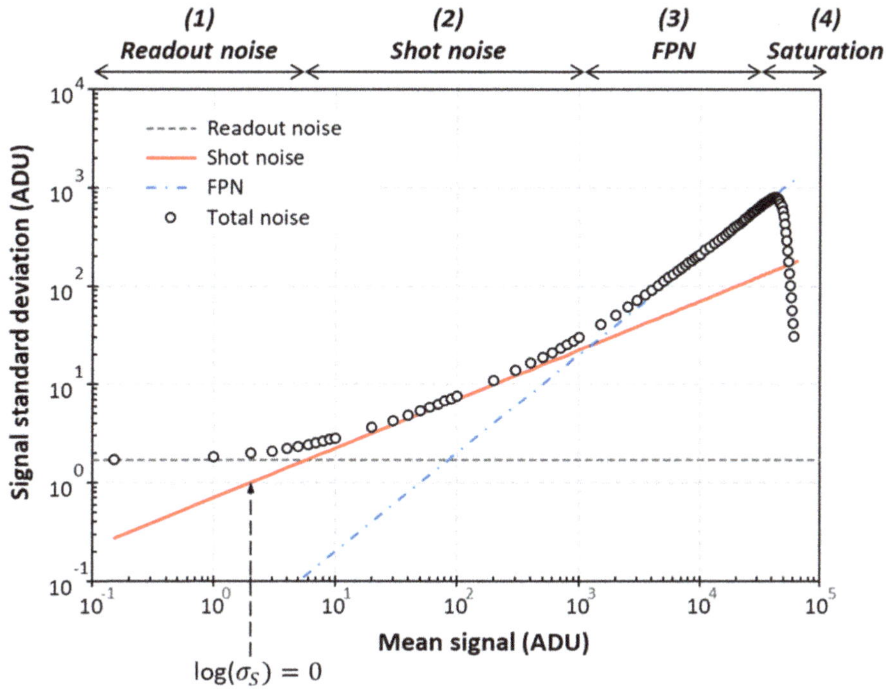

Figure 5.16. Simulated PTC showing photon shot noise, readout noise with σ_R = 1.7 ADU and fixed pattern noise with P_N = 0.02. The system gain (determined at σ_S = 1) is 2.0 e⁻/ADU. Signal saturation occurs around 45 000 ADU.

1. Readout noise dominates at very low signal levels. The standard deviation extrapolated to zero signal gives the readout noise.
2. Shot noise with a slope of 1/2.
3. Fixed pattern noise with a slope of 1 at large signals, because $\sigma_{FPN} = P_N\overline{S}$.
4. Signal saturation indicated by a sharp drop of the standard deviation.

The FPN can greatly complicate the interpretation of the PTC and make the search of the 1/2 slope elusive. In figure 5.16 the shot noise dominates a relatively small part of the data and fitting a straight line to it can be unreliable.

In CIS the FPN can be much larger than in CCDs due to the different DC offsets of each pixel and column. Also, in CIS the FPN can be dominant even at low signal levels, further complicating the data analysis. In the next section we will look into the frame differencing technique for eliminating the FPN, which can greatly help with the use of the photon transfer.

5.5.2 Frame differencing

Frame differencing is a term describing the subtraction of two images taken at the same conditions, and their analysis. FPN caused by pixel non-uniformity and electrical offsets per column is the same in both images, therefore in the differenced image it is removed. The photogenerated signal is also the same and is removed too, and the differenced image has zero mean.

What is not removed are the statistical fluctuations in the images. The two images are taken at the same conditions at different times, and the temporal noise in them is statistically independent. Therefore, the standard deviation of the shot and the readout noise in the differenced image increases by $\sqrt{2}$, and the variance doubles. In this way, the signal variance without FPN can be obtained from a pair of images. Any of the two images (or better, the average of the two) can be used to calculate the signal mean at each illumination level.

Frame differencing is widely used [6] and is indispensable in imagers with large FPN such as CIS. As an example, in the image in figure 5.17 the column FPN

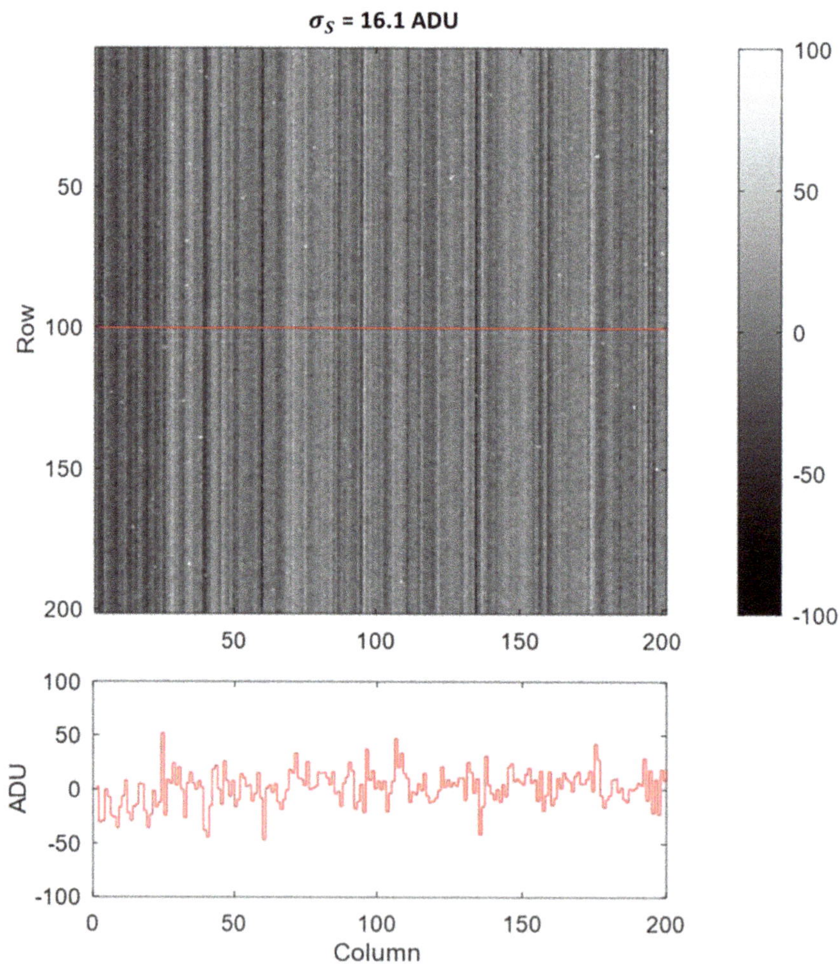

Figure 5.17. Dark image from a 4T CIS showing column FPN and bright pixels in a 200 × 200 pixel area. The mean signal of the whole image has been subtracted. A profile of the signal across the columns at row #100 is shown too. The system gain is 1.74 e⁻/ADU.

completely dominates the standard deviation at zero illumination. If left uncorrected by frame differencing, the readout noise would appear as 16.1 ADU × 1.74 e⁻/ADU = 28.0 e⁻ RMS instead of the actual 5.1 e⁻ RMS. Further on in this chapter all PTCs are mean-variance curves derived by frame differencing, unless stated otherwise.

Figure 5.18(a) shows the frame-differenced mean-variance curve using the same pixel area as in figure 5.17. The signal is generated by LED illumination with

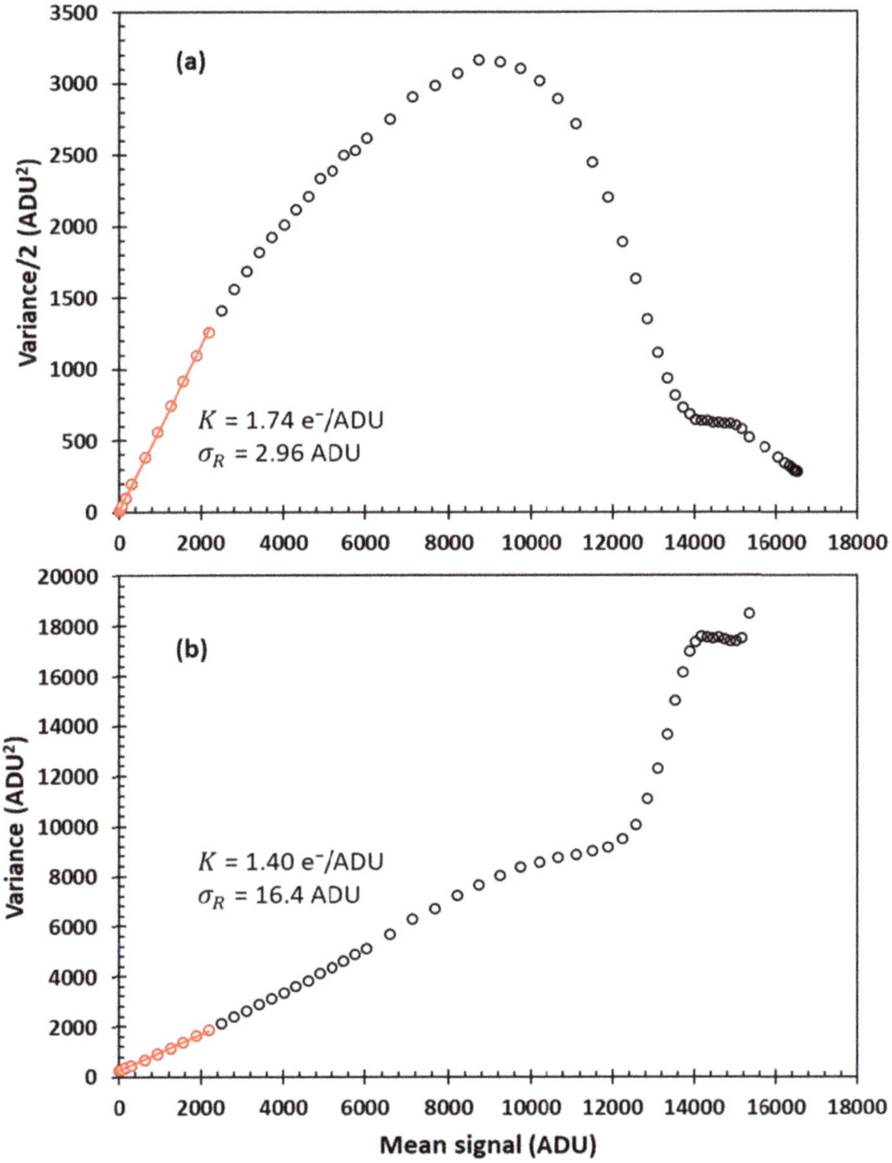

Figure 5.18. Mean-variance curve from a 4T CIS using frame differencing (a); the same data without frame differencing (b), showing large FPN. Data points at the same mean signal (in red) have been used for the linear fit in both figures.

increasing duration, keeping the integration time constant. The signal mean has been offset by the value in darkness, so that $\overline{S} = 0$ at zero illumination time. The data have been acquired using the following pseudocode:

1. Set the illumination time to zero.
2. Take two images one after the other.
3. Subtract the two images pixel by pixel, calculate the variance of the difference and divide the result by two.
4. Add the two images pixel by pixel, calculate the mean of the sum and divide the result by two. Alternatively, use just one image and calculate its mean.
5. Increase the illumination level and go to step 2 until saturation is reached.
6. Plot the signal variance versus the signal mean.

In figure 5.18(b) the same raw image data have been used to calculate the signal variance from a single frame without differencing. The two plots look remarkably different, and even though the variance appears to follow a more linear dependence in figure 5.18(b), the calculated system gain and readout noise are wrong. The variance is overwhelmed by FPN, particularly visible at large signals.

In figure 5.18(a) the variance is distinctly nonlinear before image saturation, unlike the simulated curve in figure 5.15. This could be caused by the sensor's nonlinearity, or by other mechanisms, as we will discuss in section 5.5.4. At low signals we can assume that any nonlinear effects are negligible and can use (5.20) to find the system gain. The standard deviation σ_R obtained from figure 5.18(a) at $\overline{S} = 0$ is higher than the sensor's readout noise due to a dark current of approximately 3 e$^-$/pixel/s, which adds shot noise. To calculate the true readout noise the dark current must be reduced to negligible levels using one of several possible methods.

5.5.3 System gain, CVF and noise

The system gain calculated from the PTC refers to the sense node, where the conversion from charge to voltage happens. The system gain includes all the elements in an imaging system, starting from the sense node and ending with the ADC, and bundles together their contributions.

Very often we want to know the conversion gain of the sensor G_c (also known as CVF—charge-to-voltage factor) because it is an important parameter. Knowing the CVF allows us to calculate the voltage signal in the blocks after the source follower. The CVF is a measure of the sense node capacitance C_{sn} through the relationship derived in chapter 1, describing the change of the sense node voltage ΔV_{sn} from the conversion of one electron:

$$\Delta V_{sn} = G_c = \frac{q}{C_{sn}} \qquad (5.27)$$

The conversion gain is normally given in units of μV/e$^-$ and ranges from few μV/e$^-$ to >100 μV/e$^-$. Calculating the CVF is straightforward if we know the system gain, the gain of all stages in the signal chain leading to the ADC, and the size of the ADU (equal to one least significant bit of the ADC) in volts.

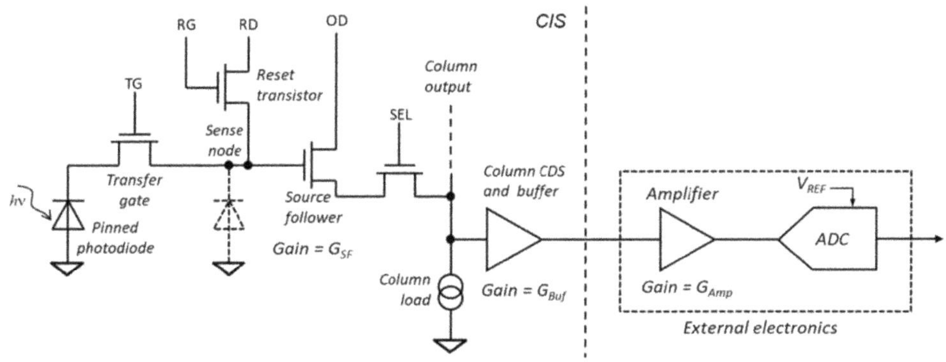

Figure 5.19. System diagram for CVF calculation.

The voltage gain of the signal chain in figure 5.19 is given by the product

$$G_{\mathrm{tot}} = G_{\mathrm{SF}} \times G_{\mathrm{Buf}} \times G_{\mathrm{Amp}} \tag{5.28}$$

and the size of one ADU is

$$V_{\mathrm{ADU}} = \frac{V_{\mathrm{REF}}}{2^n} \tag{5.29}$$

where V_{REF} is the ADC reference voltage and n is the ADC resolution in number of bits. Because the sense node voltage is amplified by G_{tot}, we can think that the size of the ADU *at the sense node* is effectively smaller by the same factor and is $V_{\mathrm{REF}}/G_{\mathrm{tot}}$.

The inverse of the system gain K is the voltage at the sense node per one electron, but expressed in ADU instead of volts. Therefore, the conversion gain is simply $1/K$ multiplied by the ADU size at the sense node:

$$G_{\mathrm{c}} = \frac{1}{K}\frac{V_{\mathrm{ADU}}}{G_{\mathrm{tot}}} \tag{5.30}$$

Normally, V_{ADU} and G_{Amp} are known, but G_{Buf} and G_{SF} may not be. In this case, the product $G_{\mathrm{SF}} \times G_{\mathrm{Buf}}$ can be obtained from an ETF measurement, as explained in section 5.13.

In sensors with digital output both V_{ADU} and G_{tot} could be unknown, and the only way to measure them could be to apply a voltage step through the reset transistor to the sense node mimicking a normal signal, also described in section 5.13.

The readout noise can be calculated from the PTC equation (5.20) in two ways:

1. The intercept of the linear fit to the variance with the Y-axis as zero mean signal ($\overline{S} = 0$);
2. The measured variance at zero signal.

Since the intercept assumes linear variance and depends on the quality of the fit, the two values can be different. If a measurement of the variance at zero signal (which implies negligible dark current) is available, then it would be preferred. If this measurement is not available, then the fit can be used. The following example illustrates the point.

Example 5.2. Calculate the system gain, the readout noise and the CVF from the PTC in figure 5.20, using that $G_{tot} = 1$ and 1 ADU is 305 μV.
Solution: From the linear fit to the variance the system gain is given by (5.20)

$$K = \frac{1}{0.27} = 3.7\ \mathrm{e^-/ADU}$$

The readout noise is the standard deviation at zero mean signal (2.68 ADU), multiplied by the system gain:

$$\sigma_R = \sqrt{2.68} \times 3.7 = 6.06\ \mathrm{e^-\ RMS}$$

If the linear intercept at zero signal were used, the noise would appear much lower, at $\sigma_R = \sqrt{1.10} \times 3.7 = 3.9\ \mathrm{e^-}$ RMS. Even if there is a temptation to use the second value, the direct measurement should be used.

The CVF is calculated from (5.30)

$$G_c = \frac{1}{3.7}\frac{305 \times 10^{-6}}{1} = 82.4\ \mathrm{\mu V/e^-}$$

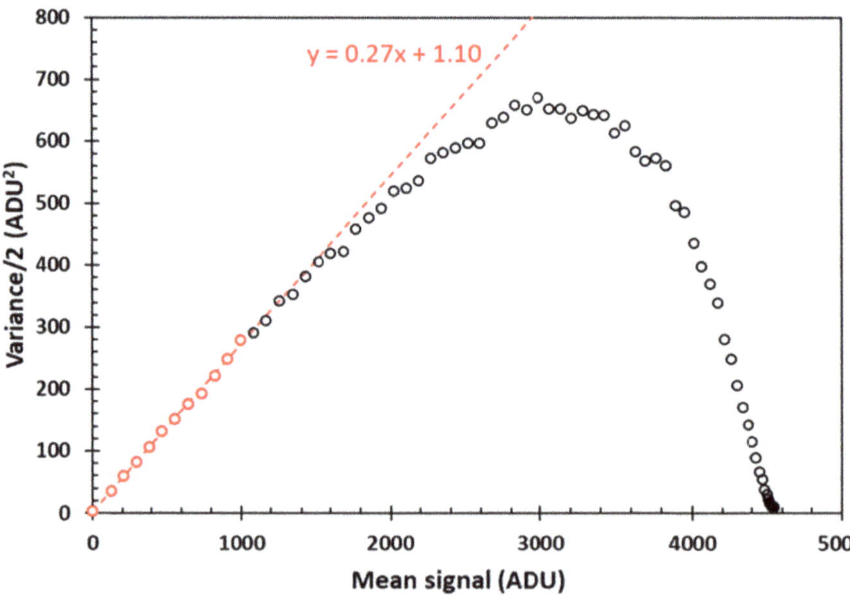

Figure 5.20. PTC of a 10 μm 4T pixel. The variance at zero signal is 2.68 ADU and the dark current is negligible. Only the data points in red have been used for the linear fit.

5.5.4 Nonlinear PTC

In many cases the PTC can be nonlinear before saturation is reached, also illustrated in the example in figure 5.18(a). Since the determination of the system gain assumes linear signal variance, the nonlinearity can be a significant source of error. Near saturation the variance starts to deviate significantly from the Poisson distribution because the statistical fluctuations of the signal are constrained and cannot be used in the PTC.

The nonlinear PTC could be caused by the inherent nonlinearity of the photo-response of the sensor. Indeed, we will see further on that if a sensor is nonlinear then its PTC must be nonlinear too. However, a perfectly linear sensor can still have a nonlinear PTC. Sub-linear variance could be an indicator of change re-distribution, e.g. charge overflowing from one collection site to another, or a signal-dependent correlation in the charge collection mechanism, known as the brighter-fatter effect [14, 15]. This is pointing to the dangers of using the PTC as the only tool for determination of the system gain and sensor characterisation. Whenever possible, other methods such as low energy x-ray illumination should be used too for cross-checks.

The key to understanding the nonlinearity of the PTC is to appreciate that the shot noise is a small signal superimposed on the much larger mean signal. In the language of electronics, the image system responds to shot noise with a *small-signal (noise) gain*, while the overall response is governed by the large-signal gain derived from (5.10). Therefore, the PTC gives the *small-signal parameters* of the system, which match the large-signal parameters only when the system is linear.

The small-signal gain is the change dS of the output signal for a small change dN_e of the number of collected electrons. The output signal is assumed to be a nonlinear, monotonic[3] function $S = f(N_e)$ of the number of the collected electrons. Since the small-signal gain depends on the operating point, it is calculated around the signal N_{e0}

$$g_S(N_{e0}) = \left(\frac{dS}{dN_e}\right)_{N=N_0} \tag{5.31}$$

The standard deviation of the output signal around the mean $\overline{N_0}$, caused by noise at the sense node with standard deviation σ is then

$$\sigma_S = g_S(\overline{N_{e0}})\sigma \tag{5.32}$$

Therefore, the output signal variance from shot noise with mean $\overline{N_e}$ is

$$\sigma_S^2 = g_S^2(\overline{N_e})\overline{N_e} \tag{5.33}$$

Equation (5.33) can also be derived from the expansion of S into Taylor series around $\overline{N_e}$ to second order, as done by Pain and Hancock [16]. Comparing with

[3] The meaning here is that the mean signal never decreases as the sensor collects more electrons. A non-monotonic function would decrease, causing the gain to be negative.

(5.19) we see that the small-signal gain g_S has taken the place of the inverse system gain $1/K$. In a linear sensor g_S does not depend on $\overline{N_e}$ and is constant, therefore (5.33) becomes the familiar PTC equation with $g_S = 1/K$.

In a nonlinear system the signal does not increase linearly with the number of collected electrons. The small-signal gain g_S from (5.31) is therefore not constant, and the signal variance becomes a nonlinear function too. From this it follows that the system gain and the conversion gain are no longer constants, but signal-dependent curves [17].

However, there is one thing that remains linear, and can be useful for sensor characterisation. In the method developed by Pain and Hancock [16] the central assumption is that the number of photogenerated electrons N_e is proportional to the number of photons N_{ph} incident on the sensor

$$N_e = \eta N_{ph} \tag{5.34}$$

The proportionality constant η is the quantum efficiency, which does not depend on the collected signal or the illumination level. The number of photons is calculated from the photon flux multiplied by the illumination time and the pixel area. Using (5.31) and (5.34), we can write

$$\frac{dS}{dN_{ph}} = \frac{dS}{dN_e}\frac{dN_e}{dN_{ph}} = g_S\eta \tag{5.35}$$

and therefore

$$g_S = \frac{1}{\eta}\frac{dS}{dN_{ph}} \tag{5.36}$$

Equation (5.36) lets us calculate the system gain from the rate of signal change with illumination. In contrast, the PTC equation (5.20) does that from the signal and its variance.

The CVF curve of the sensor can be calculated from the system gain [9] and can provide information about any nonlinearities. However, knowing the nonlinear system gain and CVF is not that useful because we need to numerically integrate them to get the information we want—signal versus illumination. What we need much more is the sensor's photoresponse (as in figure 5.8) because it already has this information.

Nevertheless, it is possible to use (5.36) to calculate the nonlinear CVF even if the QE and the photon flux are not known. For this we have to make two reasonable assumptions. The first one is that the number of photons reaching the sensor increases linearly with the illumination time t_i, if the photon flux is constant. This means that we can use the time derivative of the signal instead of differentiating by the number of photons, and express the small-signal gain as

$$g_S = \frac{1}{\eta^*}\frac{dS}{dt_i} \tag{5.37}$$

In (5.37) the constant η^* combines the QE and photon flux information. To find η^* we make the second assumption—that at low signals any nonlinearities are negligible and the classic PTC can be used [18], therefore $g_S = 1/K$. From (5.30) we know that $G_c \propto 1/K$, therefore $G_c \propto dS/dt_i$.

As an example, figure 5.21(a) shows the signal-dependent CVF using the same data as in figure 5.20. The data have been normalised to the small-signal CVF $G_c(0)$ calculated in example 5.2 (which is equivalent to finding η^*) using the formula

$$G_c(t_i) = \frac{G_c(0)}{\left(\frac{dS}{dt_i}\right)_{t_i=0}} \left(\frac{dS}{dt_i}\right) \tag{5.38}$$

The normalisation term multiplying dS/dt_i is a constant, ensuring that the derived CVF equals $G_c(0)$ at small signals. The CVF shows a gradual decrease to 60 μV/e$^-$, a nearly 30% fall from the small-signal CVF, before decreasing more rapidly as the signal approaches saturation. To put it in perspective, the photoresponse in figure 5.21(b) is plotted on the same horizontal scale. If the mean-variance curve in figure 5.20 were used to calculate the signal-dependent CVF the result would be completely wrong because the CVF would be negative for signals above 3000 ADU, where the variance peaks.

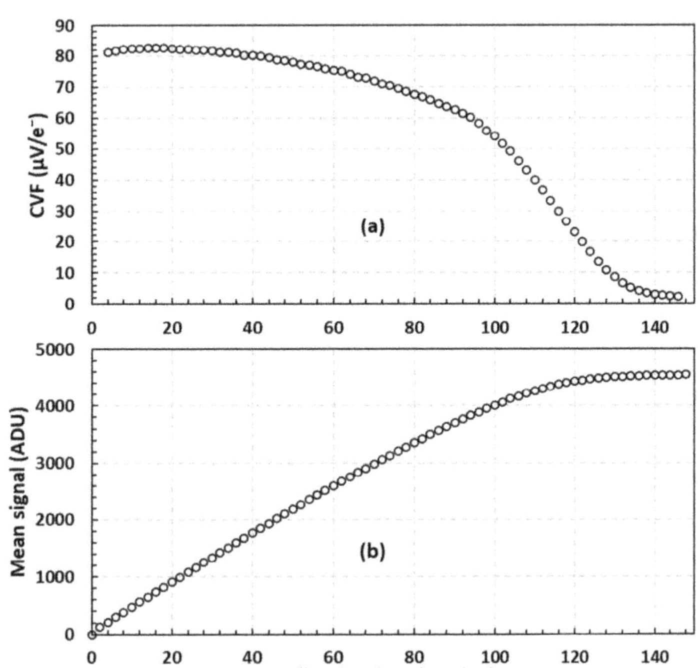

Figure 5.21. CVF versus the illumination time (a), calculated from the nonlinear system gain (5.38) and (5.37), and output signal versus the illumination time (b) of a 4T sensor.

Assuming that a sensor is linear for small signals may not always be true because the zero-signal DC bias point could be in a nonlinear part of the transfer characteristic. For example, if the reset level happens to be in the upper part of an S-shaped transfer characteristic, then small signals could also suffer from nonlinearity.

5.5.5 PTC from dark current

Since the dark current has Poisson distribution, it too can be used to obtain a PTC. A notable advantage of this method is that it does not require a light source or any other equipment. However, there are several difficulties, arising mostly because the dark current is very low. At few electrons per second it would take a very long time to accumulate a sizeable signal or reach saturation. If the goal is just to calculate the system gain, then saturation is not needed, and the dark current PTC can be useful.

The dark current is non-uniform, therefore averaging over a sufficiently large number of pixels does not work well. Instead, each pixel must be treated individually, with its own signal and variance. This increases the collected data and the computational burden considerably, and the mean-variance becomes a large scatter plot.

CIS almost invariably contain hot pixels with dark current much above the average. There could be hundreds of pixels with dark current above few hundred electrons per second (an example is in figure 5.35), which can provide the larger signals needed to obtain a PTC. However, with most pixels having small signals the statistical error in obtaining the system gain can be large.

A dark current PTC can be built using two sets of images using the following pseudocode:

1. Take many images at short integration time, containing minimum dark signal.
2. Take an equal number of images at long integration time, ensuring large dark signal.
3. Average the two sets of images pixel by pixel and subtract the two averages. This gives the average dark current signal for each pixel.
4. Subtract the images at long integration times in pairs, e.g. subtract image 2 from image 1, image 3 from image 2 and so on.
5. Calculate the variance for each pixel from the differenced images and divide the result by two.
6. Plot a scatter plot of the signal variance versus the signal mean and fit it with a straight line.

Figure 5.22 shows the result from a 4T CIS. The method can tolerate considerable dark current in the images taken at short integration times. This offsets the calculated mean signal but does not change the slope, which is the only thing we are interested in.

The calculated system gain in figure 5.22 is very close to the one obtained with light-induced signal, but the statistical error here is considerable. Repeating this analysis on the same data but with a different set of selected pixels indicates that the error is at least 15%. The error is large because the PTC relies on the 'nice' behaviour of the dark current and its variance, but in hot pixels this is by no means guaranteed.

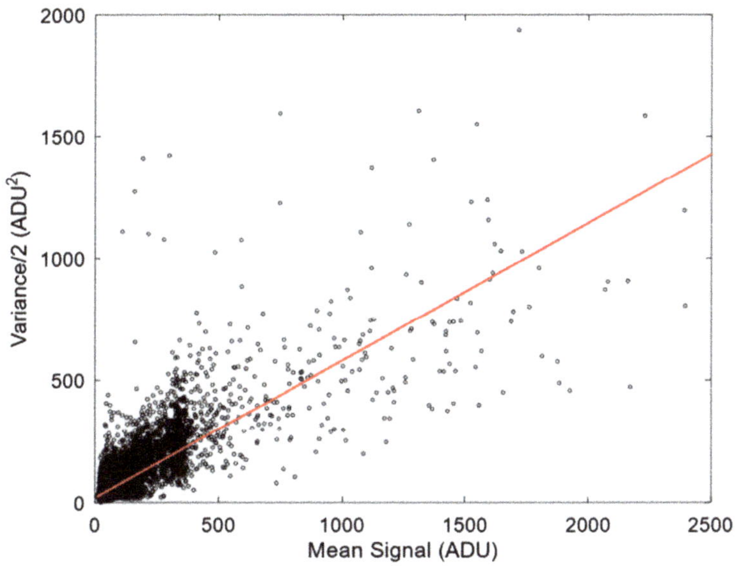

Figure 5.22. Mean-variance PTC from dark current at room temperature using a 500×500 pixel area and 20 images each at 1 and 4 s integration times. The system gain is calculated at 1.78 e$^-$/ADU, while 1.74 e$^-$/ADU was calculated from the light-generated PTC in figure 5.18(a).

The system gain could be underestimated due to hot pixels exhibiting RTS and higher variance.

5.5.6 Practical tips for the PTC

The PTC is a great technique which relies on the Poisson distribution of the collected charge and on uniform device illumination. The data needed to obtain the PTC is the same as that for the photoresponse, the PRNU and the nonlinearity—a series of images taken at increasing illumination, spanning from zero to saturation. Due to the frame differencing, the PTC requires at least two images at each illumination level, and often many more are taken to reduce the statistical uncertainty.

A good idea is to use more than two images, and to average the calculated variance from the image pairs. This is particularly useful when the image area and the number of pixels is small. Frame differencing can be used on any two image pairs—for example with three images, there are three possible combinations: 1–2, 2–3 and 3–1, and with n images the number of combinations is $n(n-1)/2$. Using multiple images in this way saves time and disk space.

A small part of the image area in a large sensor can be used when the goal is to obtain a photoresponse and a PTC without the PRNU. A smaller number of pixels can be much more uniformly illuminated than the whole of a large image area. The question here is: how small can the selected part be?

Figure 5.23 shows the effect of increasing the size of the image area on the statistical uncertainty, when only two images per illumination level are used. The area of 100×100 pixels appears sufficiently 'clean' to allow the calculation of the system gain at small

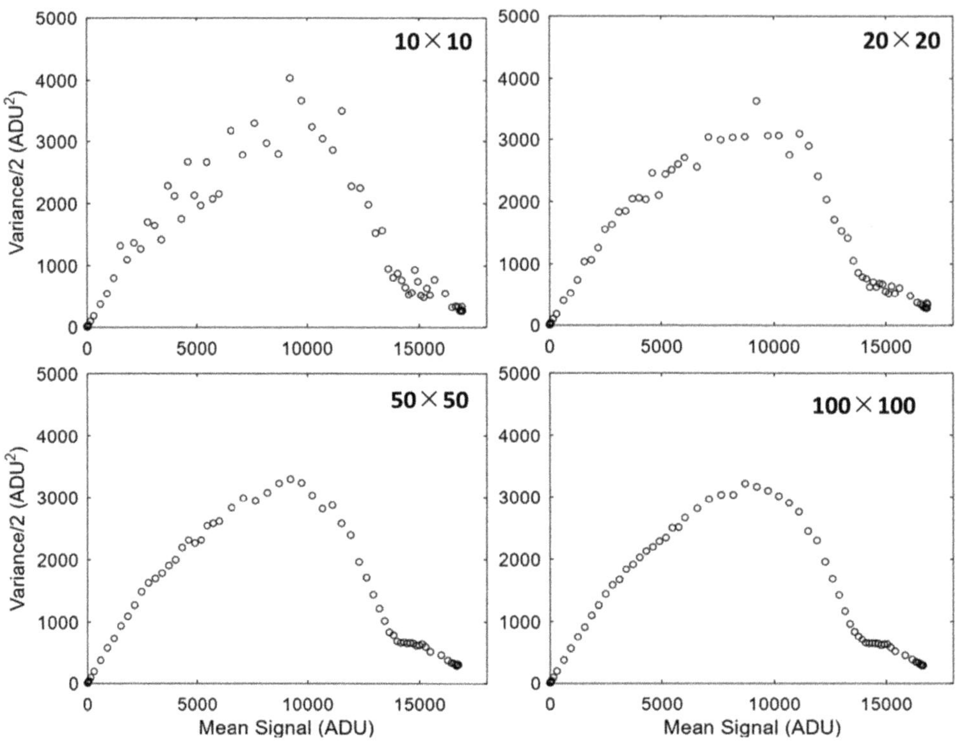

Figure 5.23. Mean-variance curves from two images per signal level, using the same data but with different number of selected pixels.

signals, and there is not much to be gained by increasing it. The uncertainty is expected to fall off as $1/\sqrt{N_{\mathrm{pix}}}$, where N_{pix} is the number of pixels used, therefore, an image region of 100×100 pixels has one tenth the statistical uncertainty of 10×10 pixels. Following the method in [6], the statistical variance in the system gain K is

$$\sigma_K^2 = \frac{4K^2}{N_{\mathrm{pix}}} \tag{5.39}$$

When frame differencing is used, the variance increases by a factor of two. The relative standard deviation in percent is then

$$\frac{\sigma_K}{K} = 100 \times \sqrt{\frac{8}{N_{\mathrm{pix}}}} \tag{5.40}$$

From (5.40) it follows that for a <1% uncertainty in K the number of pixels must be greater than 80 000. For a region of 100×100 pixels the statistical error is 2.8%, provided that only two images are used for the frame differencing.

With sufficient averaging, very small image regions can be used to obtain the system gain. However, it is much better to choose as large as possible a region, provided it is uniformly illuminated. This helps with averaging out local non-uniformities such as hot pixels and shadows from dust on the sensor's window, and speeds up the data taking and analysis.

A piece of dust or other obstruction can affect the PTC because the signal and the photon shot noise in that region are lower than in the rest of the device. Figure 5.24 shows an extreme example of a device in which a piece of tape shades a large part of the image area, blocking the otherwise uniform illumination. Two ROIs with identical number of pixels (200×200) are studied: (1) is uniformly illuminated, and (2) is partially shaded.

The photoresponse of ROI 2 in figure 5.25(a) is simply a scaled down version of the uniformly illuminated ROI 1 by approximately two thirds. Similarly, the signal variance in figure 5.25(b) has also been shrunk by the same amount.

Because both the signal and the variance have been scaled down by the same factor, the slope of the variance at small signals is identical for the two ROIs. This means that the calculated system gain is the same, despite one third of the second ROI not receiving any light at all. This may look surprising, but determining the system gain from the PTC does not require uniform illumination [6], as we can see from the following considerations.

Let us suppose that under uniform illumination a pixel region receives on average $\overline{N_e}$ electrons per readout in each pixel. The average is determined in the usual way of summing the signals from all pixels and dividing by the number of pixels. In a non-uniformly illuminated pixel region the average number of electrons per pixel is $a\overline{N_e}$, where $a \neq 1$ is a constant determined by the light distribution, shading, dust etc. The signal in the non-uniform region is $S = a\overline{N_e}/K$ and the variance is $\sigma^2 = a\overline{N_e}$, therefore the PTC equation (5.20) does not change. Despite arriving at the same

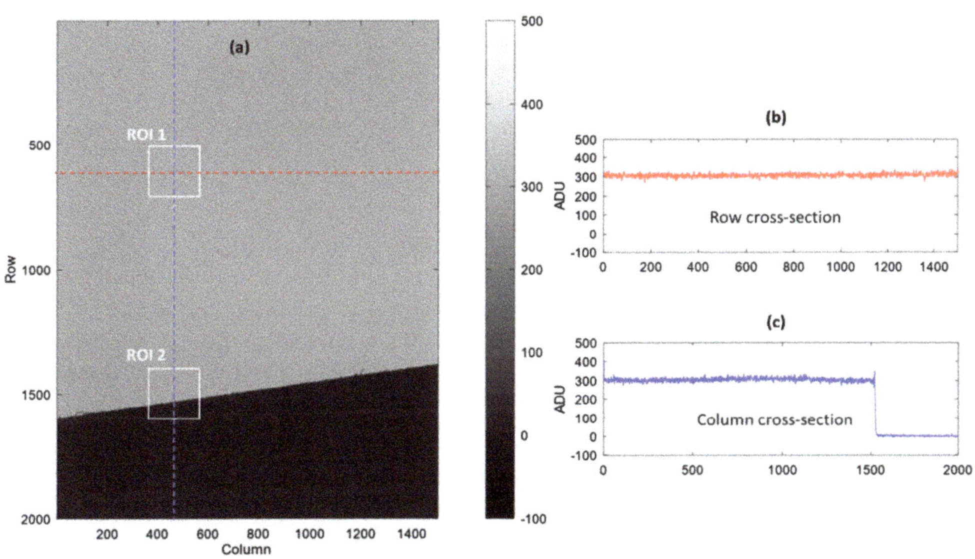

Figure 5.24. Image from a CIS with partially shaded image area (a), with two ROIs marked with white squares. The red and the blue dashed lines are at row 600 and column 480, respectively, and mark the positions of the row (b) and the column (c) cross-sections.

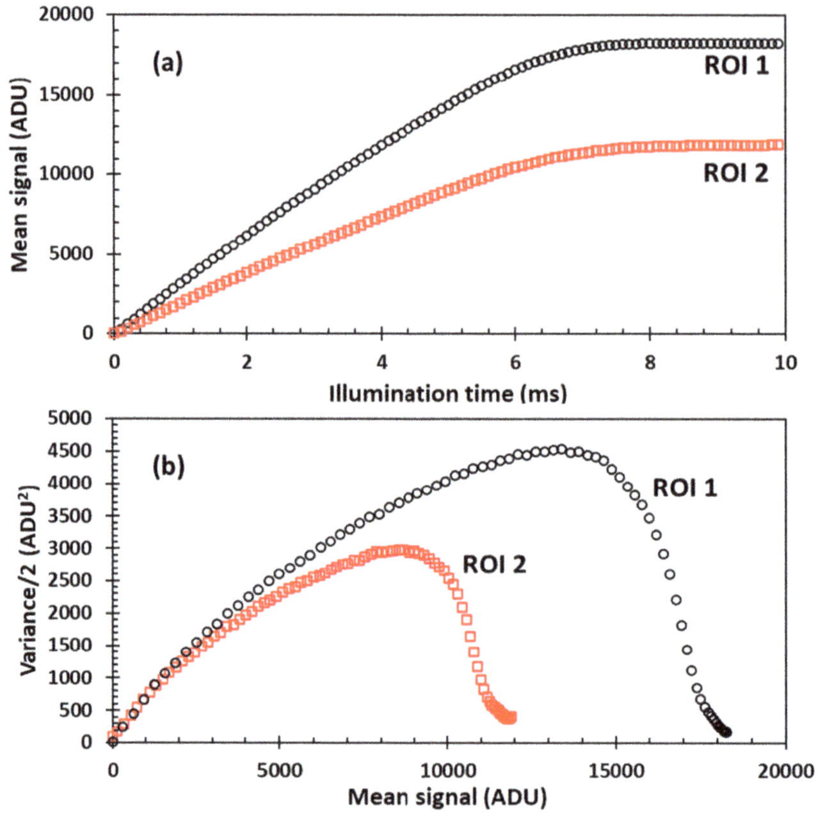

Figure 5.25. Photoresponse (a) and a mean-variance PTC (b) for the two regions in figure 5.24.

system gain, the PTC from the non-uniformly illuminated ROI 2 is still wrong because it gives a lower FWC.

5.5.7 The PTC as a diagnostic tool

The PTC can be sensitive to subtle effects which may be difficult to observe with other methods. Interpreting the features of the PTC is not straightforward. Being a statistical method, the PTC tells us that something is happening, but the cause of the effect is rarely obvious.

Any 'kinks' or 'dips' in the PTC are indicative of either a reduction of the normal statistical fluctuations or a correlation between the charge carriers, with both resulting in sub-Poisson variance. An example of correlation during charge collection is the brighter-fatter effect mentioned in section 5.5.4, which causes a curvature of the PTC in otherwise very linear sensors. Nonlinearity in the PTC is more often caused by a nonlinear conversion gain, however this is much more straightforward to establish from the photoresponse.

'Kinks' in the PTC have been attributed to incorrect clock amplitude which limits the pixel capacity in CCDs by degrading the charge transfer efficiency at large signals [6]. 'Dips' in the PTC are often observed in 4T CIS [9] and have been explained by charge spill-back, causing image lag at large signals [19]. Figure 5.26 illustrates this in a 4T sensor in which the transfer gate voltage V_{TG} has been varied. As V_{TG} decreases, the onset of spill-back is pushed towards higher signals [20] and the dip in the variance, barely visible at V_{TG} = 3.3 V, deepens and follows it.

In this example it would be very difficult to find out what causes the dip without measuring the image lag at the same conditions.

Figure 5.27 is an example of a PTC indicating an underlying problem, in this case a poor charge transfer inefficiency (CTI) in a specialised time delayed integration

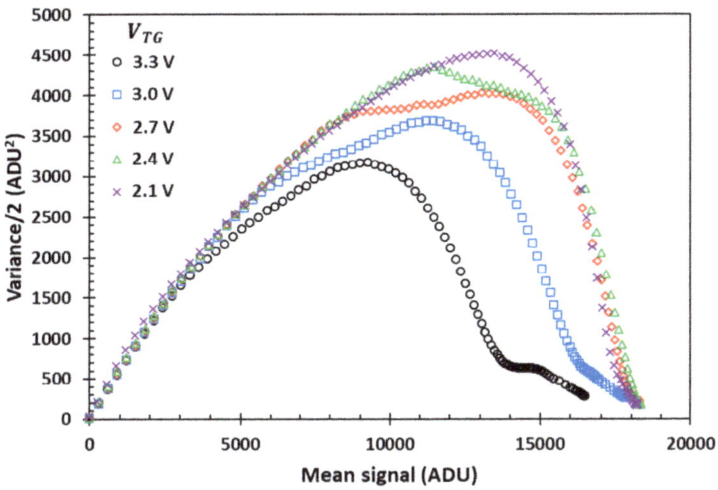

Figure 5.26. Mean-variance curves of a 4T CIS taken at different transfer gate voltages, with all other parameters fixed.

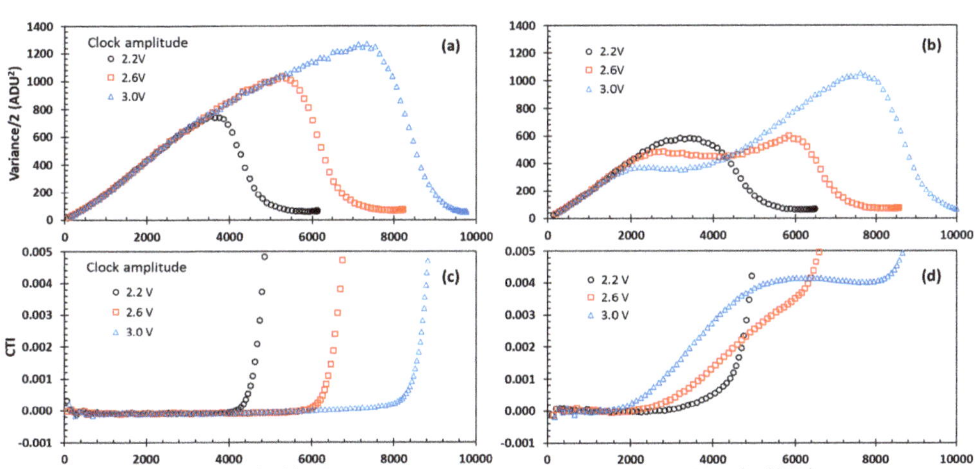

Figure 5.27. Mean-variance curves (a)(b) and CTI (c)(d) in a CMOS TDI pixel arrays. Figure pairs (a) and (c), and (b) and (d) correspond to pixel variants with small differences in their designs.

(TDI) imager. In figure 5.27(a) the PTC does not arouse any suspicions, and the corresponding CTI in (c) is very good, up to the point when the potential wells begin to overflow. The PTC in figure 5.27(b) has pronounced dips which match the signal at which the CTI in (d) begins to increase, at much smaller signals than in the 'good pixel' (c). Since there was only one difference between the pixel designs, the problem was identified as a manufacturing issue which caused an excessive charge trapping in the 'bad pixel' (with plots in figure 5.27(b) and (d)).

5.6 X-ray calibration

As an alternative to the PTC, calibration with x-rays provides another way to determine the system gain. This method is preferred in cases when an optical PTC cannot be taken, for example if there are metal layers covering the photodiode intended for backside illumination (BSI) operation, or the device is optically blind, as required for most x-ray sensors. Also, this method is useful when it is not possible to adequately eliminate the FPN, or the photon shot noise cannot be cleanly isolated.

In principle, the system gain can be determined by introducing a well-known packet of charge in the pixel. Such charge can be created via the photoeffect by monoenergetic soft x-rays (<10 keV). Provided that the charge is well above the noise floor and is collected in a single pixel, this is a quick, effective, and accurate way to measure the system gain.

One downside of x-ray calibration is that it is usable only with sensors without a protective glass window. The absorption length for 10 keV x-rays in glass is less than 100 μm, so they cannot penetrate the window, which can be up to 1 mm thick.

The x-ray source can be a radioactive element, such as the ^{55}Fe isotope discussed in chapter 1. ^{55}Fe is widely used due to the convenient energy of its characteristic manganese x-rays: 5.89 keV (Mn-K_α) and 6.49 keV (Mn-K_β), generating charge of 1614 e^- and 1778 e^- at 300 K, correspondingly. The ionisation energy in silicon is E_w = 3.65 eV at 300 K and increases at lower temperatures; therefore, the device temperature must be known to precisely calculate the amount of charge [21].

Figure 5.28 shows a typical ^{55}Fe spectrum obtained with a CMOS image sensor. As with optical characterisation, an averaged dark image should be stored and subtracted from all x-ray images to eliminate spatial FPN and dark current. A histogram of the baseline-corrected images is then calculated and plotted. There should be a large peak at ≈0 ADU corresponding to pixels which do not have x-ray signal. Due to the baseline subtraction their value is around zero, which leaves the system noise to define the width of the peak. The system gain is calculated simply by dividing the charge in electrons corresponding to Mn-K_α or Mn-K_β by their measured value in ADU. A good number of x-ray hits is needed to form a peak with easily measurable position, and normally only the Mn-K_α line is used because of its better signal statistics.

Only x-ray charge which has been collected in a single pixel produces the two distinctive peaks in figure 5.28 and can be used for calibration. For most x-ray hits

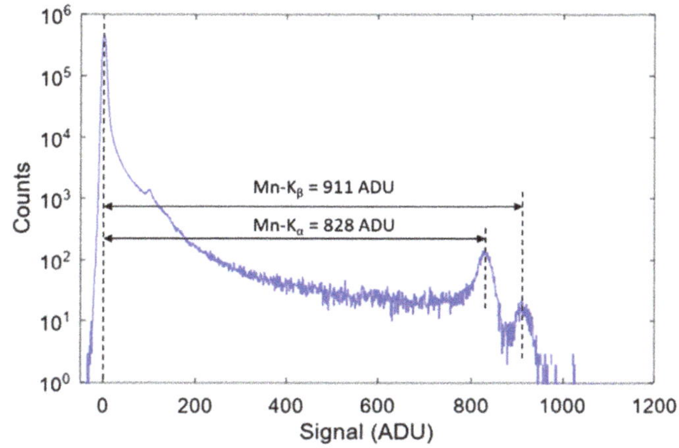

Figure 5.28. ^{55}Fe spectrum from a 4T CIS with 10 μm pixels, read at 5 fps at 25 °C. The baseline has been obtained by averaging of 10 dark images and then subtracted from all the images containing x-rays.

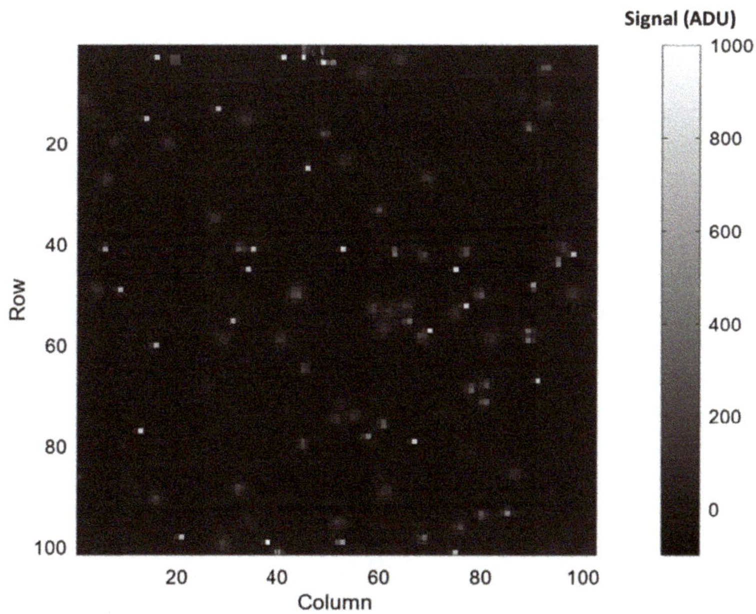

Figure 5.29. ^{55}Fe x-ray image from a 4T CIS with 10 μm pixels, built on a 24 μm-thick epitaxial layer with 1000 Ω.cm resistivity.

the charge is shared between several pixels, which is seen as the 'shoulder' between the noise and the Mn-K_α peak. This is also visible in the images—the fuzzy 'blobs' in figure 5.29 are the shared charge, while the sharp isolated pixels correspond to the two peaks. X-ray calibration works better for large pixels because of the larger probability of containing the charge in a single pixel.

Example 5.3. Calculate the system gain from the spectrum in figure 5.28, taken at 300 K.
Solution: The peak of the Mn-K_α line is at 828 ADU, therefore the system gain is

$$K = \frac{1614}{828} = 1.95e^{-}/\mathrm{ADU}$$

For the Mn-K_β line the result is

$$K = \frac{1778}{911} = 1.95e^{-}/\mathrm{ADU}$$

Ideally, the dark images should be taken with the x-ray source in a shielded position, followed by moving the source in front of the sensor, as shown in figure 5.6. If the source cannot be blocked or moved, which is more difficult in vacuum, data analysis techniques can help obtain a baseline image which is very close to a pure dark one, even if the sensor is constantly illuminated by x-rays.

The simplest way to do this is to use the median or the mode signal value instead of the mean. If the x-ray hit occupancy is low, this provides a good estimation of the baseline signal because the large, but rare x-ray signals are excluded. In comparison, the signal from just one Mn-K_α x-ray, present in only one out of 10 images, would increase the mean for the hit pixel from $\approx$0 ADU to about 83 ADU for the data in figure 5.28. More sophisticated methods using thresholding for detection of x-ray events and their subsequent removal for the baseline averaging are possible and could deliver better results, but are usually slower.

The characteristic K_α and K_β x-rays from a radioactive source provide only a couple of closely spaced calibration points and cannot say much about the sensor's linearity, unlike the optical response. The bremsstrahlung from an x-ray tube can be used to excite fluorescence emission x-rays from a number of target chemical elements and provide multiple data points. Commonly used elements are aluminium (K_α = 1.49 keV), titanium (K_α = 4.51 keV), copper (K_α = 8.05 keV), and of course manganese. The charge they generate provides K_α calibration points in the range from 407 e^- (Al) to 2205 e^- (Cu). Silicon rapidly becomes transparent for x-rays above 10 keV (the absorption length grows beyond 100 μm), and therefore elements with higher atomic number and higher K_α energy become less effective. A typical setup using fluorescence x-rays for device characterisation could look like figure 5.30. The x-ray energy is selected by rotating the target wheel to the appropriate position.

Using an x-ray tube instead of a radioactive source has many advantages despite the increased complexity. The tube can be turned on and off under computer control and be synchronised to the sensor readout, and the x-ray intensity can be adjusted so that it is as low or as high as necessary. Besides, there is no radiation when the tube is powered off, and there is no radioactive decay to contend with, which makes the ^{55}Fe source (with half-life of 2.74 years) rather short lived.

And finally—whatever x-ray source you use, don't forget to remove the glass cover from your sensor, or you will get no x-ray spectrum at all.

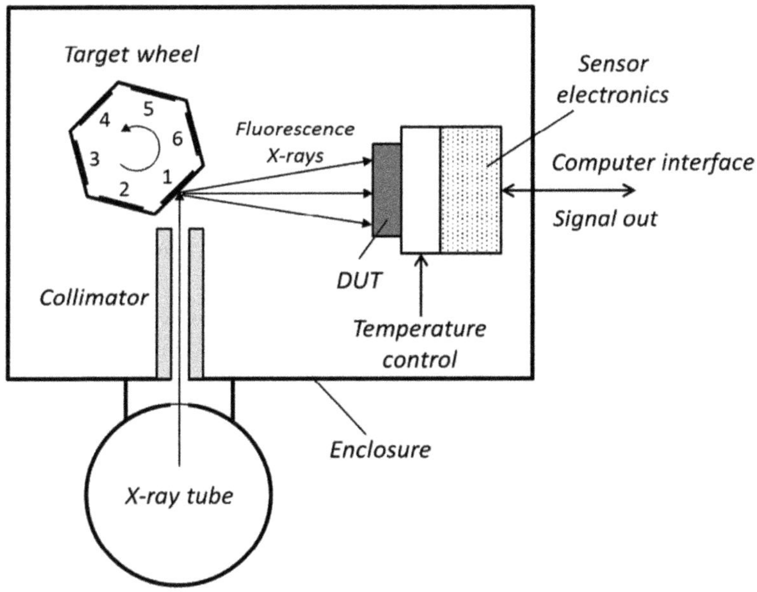

Figure 5.30. Characterisation setup using fluorescence x-rays. Each side of the target wheel is made of a different material.

5.7 Full well capacity and dynamic range

The maximum charge that can be stored in a pixel is called full well capacity (FWC)[4] and is an important design parameter. Together with the readout noise, it is used to calculate the dynamic range (DR) of a sensor:

$$\mathrm{DR} = 20\log\left(\frac{\mathrm{FWC}}{\sigma_{\mathrm{R}}}\right) \tag{5.41}$$

Measuring the FWC is not straightforward and is not always possible because the output signal can be limited by the voltage swing at the sense node or the following electronics, and not the capacity of the photodiode. A practical limit on the output signal is often the power supply; with the usual 3.3 V analogue supply the optical output can rarely exceed 2 V. On the other hand, the typical area storage capacity in a PPD is between 2 and 4 ke$^-$ μm^{-2}. A 5 μm pixel can store up to 80 ke$^-$ (if the PPD occupies 80% of the pixel), and a modest CVF of 50 μV/e$^-$ would create a signal of 4 V. The signal gets smaller for smaller pixels because the signal capacity decreases proportionally to the square of the pixel pitch, while the CVF increases much more slowly.

There are three ways to define the FWC, and they give very different answers:

1. The output signal at saturation S_{sat};
2. The output signal S_{lin} at which the nonlinearity exceeds a limit, normally 5%;
3. The signal S_{PTC} corresponding to the peak of the variance in the mean-variance curve.

[4] Also known as charge handling capacity (CHC), or charge storage capacity. The term CHC is more appropriate to 3T pixels which do not have potential wells.

A good starting point is a technology CAD (TCAD) simulation of the charge collection in the pixel, since the sense node capacitance and the readout electronics play no role. The true FWC can be measured in pixels where the CVF has been intentionally reduced by adding a capacitor in parallel with the sense node.

In most 4T sensors signal saturation occurs before the PPD's FWC is reached, therefore due care should be exercised to understand what is limiting the output signal. Regardless of where the limiting occurs, the characterisation methods are the same, and for simplicity we can consider them as a 'FWC measurement' even though it may not be the photodiode doing the limiting.

Figure 5.31 shows the photoresponse of a 4T sensor, used to determine S_{sat} and S_{lin} at 5% nonlinearity, and the PTC derived from the same image data. The pixel in this example has design FWC of 45 ke$^-$ and relatively low CVF. The saturation signal is approximately 0.8 V ($S_{\mathrm{sat}} \approx$ 2600 ADU), however, the output is still limited by the on-chip readout and not by the pixel's well capacity.

The saturation signal S_{sat} can be expressed in electrons using the system gain (derived from the small-signal part of the PTC) and would be approximately 22 ke$^-$. This may look OK, but remember figure 5.14—in saturation the signal corresponds to an infinite number of electrons because the system calibration is no longer valid. Therefore, S_{sat} does not provide very good, or even adequate measure of the FWC. The saturation signal in volts, however, is a very useful characteristic that helps with the design of the camera system.

The signal at which the linearity exceeds 5% in figure 5.31(b) is 18.4 ke$^-$. This is a much more robust estimate of the FWC because it is better defined and corresponds to a finite charge. It also makes practical sense because it links to the linearity, and has been adopted by some manufacturers. On the flip side, S_{lin} depends on how the linear fit to the photoresponse is performed, and on the threshold used (here 5%).

The third method, based on the PTC, relies on the shot noise of the collected charge. As the signal approaches the FWC, the normal statistical fluctuations begin to be clipped and the signal variance starts to fall off. Therefore, the variance has a peak, and the signal corresponding to it is taken as the FWC.

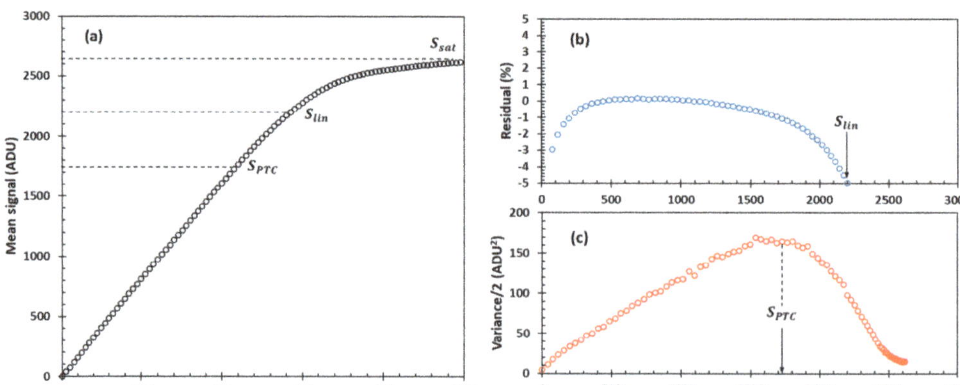

Figure 5.31. Photoresponse (a), residual linearity error (b) and a mean-variance curve (c) from a 5.4 μm 4T CIS. The system gain is $K = 8.35$ e$^-$/ADU and the CVF = 36.5 μV/e$^-$.

The signal variance is much more sensitive to signal limiting than the linearity. In figure 5.31(c) the peak of the variance is at S_{PTC} = 14.6 ke$^-$, which is nearly 4 ke$^-$ lower than S_{lin}. From Gaussian statistics, the probability of the shot noise fluctuations being within three standard deviations from the mean is 99.7%. For a signal of 14.6 ke$^-$ this makes only ±362 e$^-$ and is very far from reaching S_{lin}.

Because of the high sensitivity, the drop of the signal variance is the first indicator of signal limiting. It may look like a conservative estimate of the FWC because it is about 20% lower than S_{lin}, but has been adopted by the EMVA 1288 Standard [1].

5.8 Dark current and DSNU

The dark current is one of the easiest things to measure, but can be the hardest to explain. Naturally the device must be in complete darkness, and the device temperature must be known due to its strong influence on the dark current. No additional equipment except the readout electronics is required.

Measuring the dark current is quite straightforward—the signal is recorded at different integration times and the data fitted with a straight line. This signal is either the average from many pixels (from a ROI) or from a single pixel. For each integration time many images are averaged to reduce the readout and the shot noise. The signal from a ROI is therefore averaged twice: once over time (averaging images on a pixel-by-pixel basis) and once spatially (averaging over all the pixels in an ROI). The slope of the line is the dark current per pixel, measured either in electrons per second (e$^-$ s^{-1}) or in amperes (A) at a *given device temperature*. Stating the dark current without the temperature at which it was measured is not very useful.

Larger pixels tend to have higher dark current than smaller pixels in the same technology because of the larger pixel volume and Si–SiO_2 interface area. To compare the dark current between devices with different pixel sizes and technologies, we use the *dark current density*, defined as the dark current per pixel divided by the pixel area. The dark current density can easily span three orders of magnitude between the pixel technologies. In non-pinned imagers, such as in 3T pixels and TDI CMOS the density can be >1 nA cm^{-2} at 20 °C, while in high quality 4T pixels it can be below 1 pA cm^{-2}.

In figure 5.32 the averaged dark signal from 250 000 4T pixels has been measured at four integration times. The data have been offset by the minimum signal so that it equals zero at the shortest time; this is OK because we are only interested in the slope.

Example 5.4. Calculate the dark current in e$^-$ s^{-1} from the data in figure 5.32 using system gain K = 1.73 e$^-$/ADU. Calculate the dark current density in units of pA cm^{-2}, considering that the pixel is 7 μm square.

Solution: The dark current is the slope of the dark signal multiplied by the system gain. The time is in milliseconds, therefore the slope must be multiplied by 1000:

$$I_d = K \times \frac{dS}{dt_{int}} = 1.73 \times 0.0106 \times 1000 = 18.3 \text{ e}^-\text{s}^{-1}$$

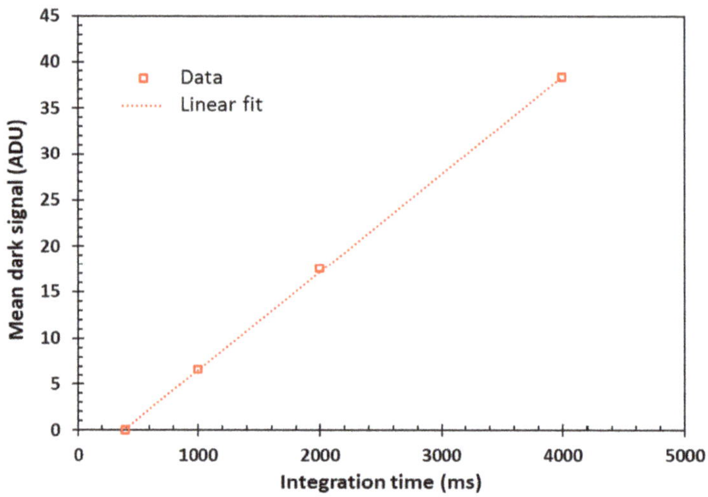

Figure 5.32. Averaged dark signal at 25 °C versus the integration time from a 500×500 pixel ROI and a linear fit to the data.

The dark current density is the dark current in pA, divided by the pixel area. The dark current is converted to pA by multiplying by the elementary charge q and by 10^{12}:

$$\frac{I_d}{A_{pix}} = \frac{I_d \times q \times 10^{12}}{A_{pix}} = \frac{18.3 \times 1.6 \times 10^{-19} \times 10^{12}}{(7 \times 10^{-4})^2} = 6.0 \text{ pA cm}^{-2}$$

The general assumption is that the dark signal increases linearly with time, and the data in figure 5.35 does not show any exceptions. However, this is not always true. If the dark signal is large, the nonlinear sense node capacitance would make the rate of increase nonlinear too [6]. Another, more subtle and difficult to spot effect occurs due to the movement of the depletion edge and the increase of charge density in the vicinity of a trap, which also shows as a nonlinear dark signal increase [22].

The average signal in figure 5.32 hides the behaviour of individual pixels. Most pixels have dark current around the average, but for a sizeable fraction, called 'hot pixels', it can be much higher. Figure 5.33 shows the increase of the dark signal with the integration time for a subsample of 100 pixels from the 250 000 pixels used in figure 5.32. The large number of parallel lines correspond to the 'normal' pixels; they are spread vertically due to the spatial FPN but have very similar slopes, and therefore similar dark current. Only six pixels deviate significantly from the average dark current.

The dark signal non-uniformity (DSNU) is the parameter used to describe the spread of the dark signal. This can be confusing unless we make a clear distinction between dark signal and dark current.

Dark signal is the output of a sensor in darkness at a particular integration time, and includes both the DC offsets and the dark current signal collected over that

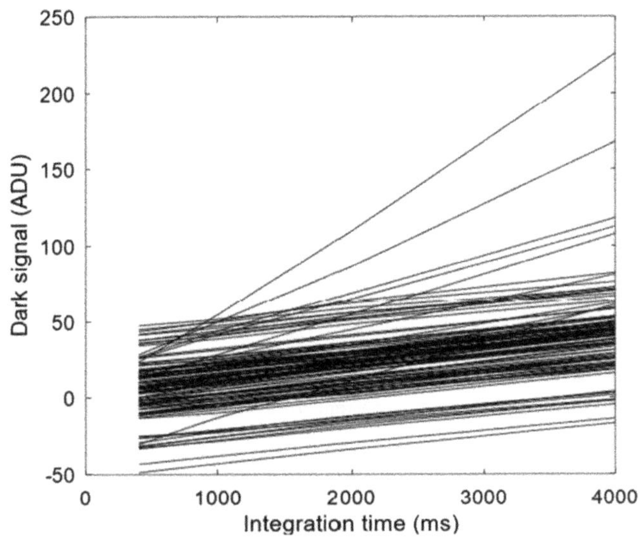

Figure 5.33. Dark signal versus the integration time for 100 pixels (10×10 pixel area), taken from the data used in figure 5.32.

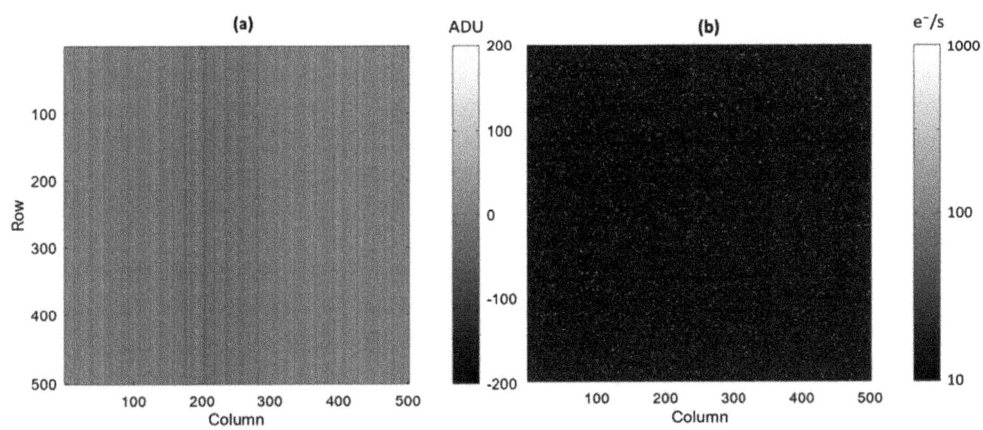

Figure 5.34. Dark signal (a) and dark current image (b) from the same data used in figure 5.32. The mean dark signal from the whole ROI in (a) has been subtracted, and column FPN is clearly visible. The dark current in (b) is in a logarithmic scale.

time. Even in a sensor having zero dark current there could be a large spread in the dark signal, due to pixel and column offsets.

The dark current is the rate of increase of the dark signal, and is therefore insensitive to any DC offsets. The dark signal is measured in ADU or electrons, while the dark current is measured in $e^{-}\ s^{-1}$.

Very often, DSNU is taken to mean *dark current non-uniformity* (DCNU) even though they are different things, as the images in figure 5.34 demonstrate. In figure 5.34(a) the dark signal is clearly dominated by the spread of column and pixel

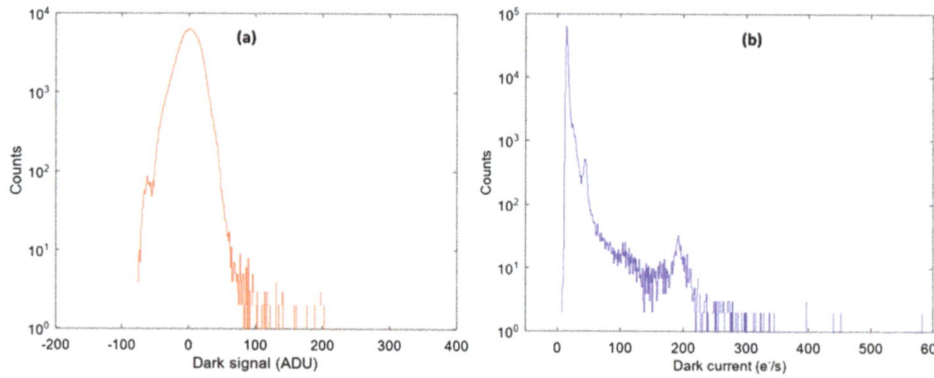

Figure 5.35. Dark signal (a) and dark current (b) histograms from the data in figure 5.34.

DC offsets giving rise to spatial FPN, while the image in figure 5.34(b) has a uniform baseline, peppered with pixels having much higher dark currents.

The corresponding distributions are given in figure 5.35(a) and figure 5.35(b), respectively. The DSNU is calculated as the standard deviation of the dark signal, and the same can be applied to the dark current. The mean dark current in figure 5.35(b) is 18.3 $e^- s^{-1}$ and its standard deviation is 26.2 $e^- s^{-1}$ due to the number of hot pixels.

The dark current distribution in figure 5.35(b) shows some interesting characteristics. Many pixels have dark current far above the average, and some exceed it by more than 30 times. The peaks at around 50 and 200 $e^- s^{-1}$ indicate that the dark current is quantised and can point towards the type of trap responsible. Traps can be identified by their activation energy using dark current spectroscopy [23], leading to the development of many mitigation techniques [24].

Normally, the biggest contributor to the DCNU are the hot pixels, therefore the measurement temperature for this parameter should be specified. As the sensor is cooled down and the dark current vanishes, the non-uniformity will disappear too.

Hot pixels are not the only reason for DSNU and DCNU—there are also defective pixels (dark or 'dead' pixels), rows and columns which produce very little or no signal under illumination. The usual practice is to exclude those defects from the calculations.

5.9 Noise measurement

The readout noise is one of the main sensor parameters and is invariably given in the datasheets. The term 'readout noise' is taken to mean only the electronic noise in the readout chain and excludes the shot noise from any optical signal or the dark current. Noise measurements are conducted without illumination, and the dark current should be removed as well.

One method to eliminate the dark current is to cool down the sensor. This is excellent but is not always practical or possible because a cooling setup is not something that can be easily assembled. Often the temperature must be below the dew point of ambient air, therefore a vacuum system, air displacement or de-humidifying must be used to prevent condensation on the sensor, which adds complexity.

If cooling is not an option, a useful alternative is to read the sensor at several integration times t_{int}, corresponding to different shot noise from the dark current, and to extrapolate the data to zero integration time, when the dark current becomes zero too. The measured signal variance σ_S^2 is the sum of the readout noise variance σ_R^2 (at zero dark current) and the variance of the dark current shot noise, equal to the number of dark current electrons N_{dark}. Since $N_{\mathrm{dark}} = I_{\mathrm{DC}} t_{\mathrm{int}}$, where I_{DC} is the dark current in units of electrons per second, we can write

$$\sigma_S^2 = \sigma_R^2 + I_{\mathrm{DC}} t_{\mathrm{int}} \tag{5.42}$$

Figure 5.36 demonstrates the method, using dark image pairs at each integration time to remove the FPN. The intercept of the extrapolated data at $t_{\mathrm{int}} = 0$ gives the square of the readout noise σ_R. Since the signal variance is measured in ADU, it has been converted to electrons (squared) by multiplying by the square of the system gain.

This method essentially emulates a sensor with infinitely short readout time, read continuously. In some sensors it may be possible to achieve similar effect by continuously reading a small ROI at sufficiently high rates, so that the dark current becomes negligible.

Another way to measure the readout noise in 4T sensors is to eliminate the dark current by not transferring charge to the sense node. Signal sampling is done in the usual manner, but the TG pulse is either not applied, or the transfer gate voltage is set to zero. Figure 5.37 shows the timing diagram for this 'Zero TG' measurement technique. Dark current continues to accumulate in the PPD and the charge will eventually start overflowing, but the reset of the sense node should take care of the

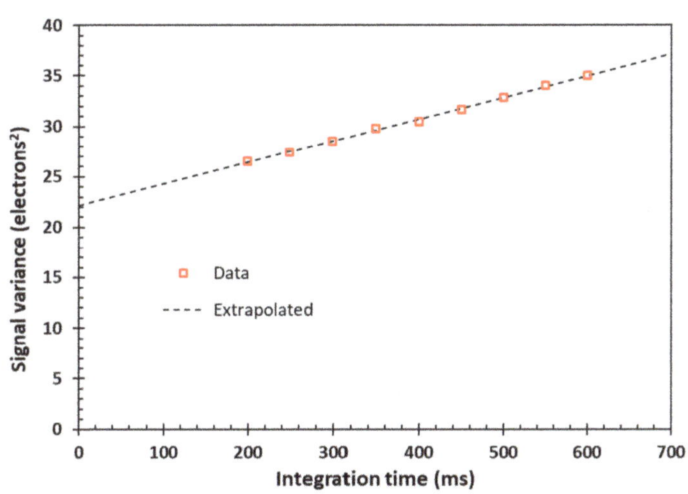

Figure 5.36. Signal variance in darkness, frame-differenced, from a 200×200-pixel area as a function of the integration time. The readout noise of this sensor is calculated at 4.7 e⁻ RMS.

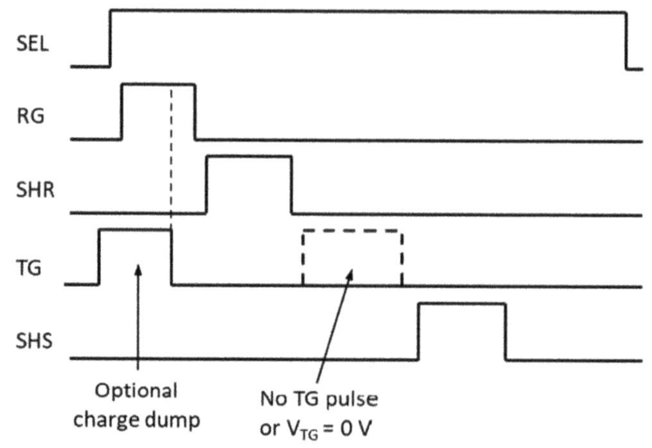

Figure 5.37. Timing diagram for elimination of the dark current in 4T sensors.

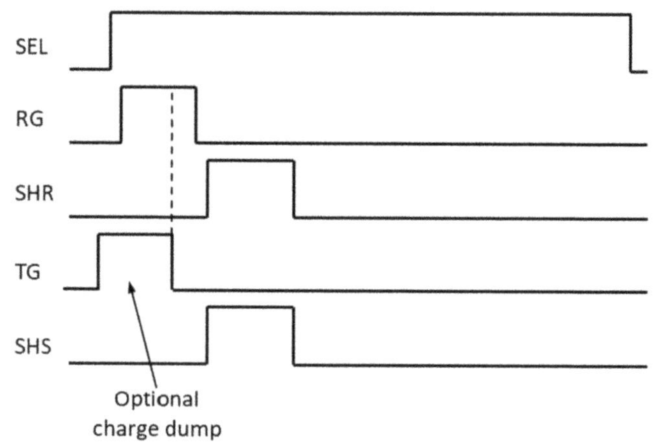

Figure 5.38. Timing diagram with overlapping reset and signal sampling for noise measurement of the CDS circuit and the output amplifier.

leaking charge. To be certain that the PPD is held in depletion, an optional charge dump can be performed while the sense node is being reset.

In practice this technique is most likely to be used with $V_{\mathrm{TG}} = 0$ because modifying the sensor timing is more difficult. It works well and gives noise values very close to the extrapolation method; for the sensor in the example in figure 5.36 the 'Zero TG' technique gives 4.4 e^- RMS. A word of warning—the 'Zero TG' image cannot be used as a dark reference for images taken with the TG applied. This is because the TG pulse couples to the sense node and changes the DC offsets, therefore the FPN would be very different.

A further extension of this noise measurement technique is shown in figure 5.38. The overlapping SHS and SHR store the same signal, therefore the noise of the preceding blocks—source follower and column amplifier, is cancelled. What remains is the noise of the CDS and the output circuits.

When doing a noise measurement, it is important to prove that the result is due to the sensor and not to the external parts of the system, such as amplifiers and ADCs. A well-designed system should add negligibly small noise, but it is still a good idea to verify that the external electronics is quiet.

One way to measure this is to ground the inputs of the signal amplifier as in figure 5.39(b) and take an image. The standard deviation of this image σ_{ext} is due only to the amplifier and the ADC. The sensor readout noise σ_{R} is obtained from the subtraction of σ_{ext} in quadrature from the system noise σ_{sys}

$$\sigma_{\mathrm{R}} = \sqrt{\sigma_{\mathrm{sys}}^2 - \sigma_{\mathrm{ext}}^2} \tag{5.43}$$

The system noise is the noise of a normally operated image sensor, using any of the methods for dark current elimination described previously.

A distinct characteristic of CMOS image sensors is the spread of the readout noise. This is because each pixel is different, having a separate source follower serving it, and each transistor has its own noise characteristics. Therefore, we need to measure the noise of each pixel. This is simply done by taking many dark images (with the dark current removed), calculating the standard deviation per pixel, and plotting it as a histogram in figure 5.40(a). Frame differencing is not required because fixed pattern noise does not exist for a single pixel.

While most pixels sit around the average, many exhibit much higher noise, commonly attributed to 1/*f* and RTS noise [18, 25]. It is those pixels that form the long tail in the noise distribution. Tight noise spread is of course preferred, and many improvements to the manufacturing process [26] have been developed to reduce RTS noise.

The noise distributions are commonly characterised by their mean and median values. In figure 5.40(a) both have been calculated for two designs with the same layout, but with different gate oxide treatment for the source follower. The noise distribution of the pixels with the 'low noise' oxide clearly has smaller tail, and the

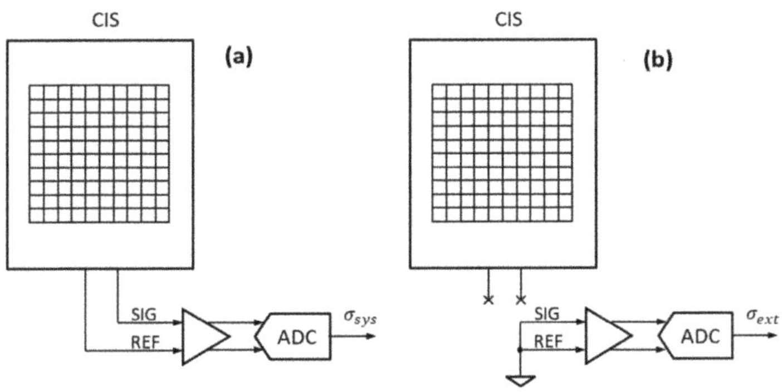

Figure 5.39. Normal operation of a CIS with differential outputs (a); connections for the noise measurement of the external electronics (b).

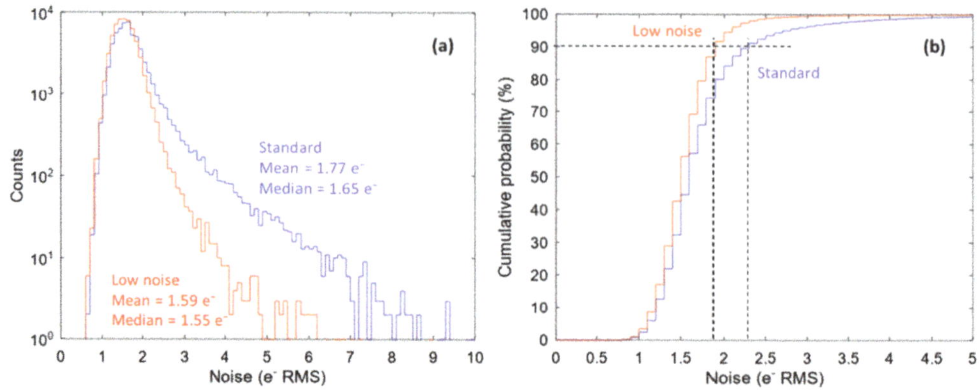

Figure 5.40. Noise distribution in 4T pixels (a) and cumulative probability for the noise to be below a certain level (b). The pixel with plots in red has been made with a 'low RTS' gate oxide process and is otherwise identical to the pixel in blue. The data were taken using the 'Zero TG' technique.

mean and the median closely match. A tell-tale sign of a pixel with high RTS noise is when the mean noise is much higher than the median because the median suppresses the contribution of the large noise outliers.

The cumulative noise distributions (figure 5.40(b)), derived from the summation of the histogram bins, are also frequently used to quantify the noise performance. The proportion of pixels below a certain noise level is a convenient metric for noise characterisation. For example, figure 5.40(b) shows that 90% of the pixels with the 'low noise' oxide have noise below 1.9 e^- RMS, while for the standard process the figure is 2.3 e^-.

5.10 Image lag

PPD-based image sensors suffer from incomplete charge transfer, called image lag, and the mechanisms causing it were described in chapter 2. Image lag is usually seen as charge remaining in the pixel in dark images following several bright ones, in which case it is called trailing edge, or discharging lag. A matching effect happens on the rising edge too—a bright image following a string of dark ones has less signal than the following bright images and is naturally called leading edge, or charging lag. Both lags are connected because charge is conserved—whatever charge has not been transferred in the leading edge must come out in the trailing edge.

In addition to the size, structure, and the doping profiles of the PPD, the image lag depends on the amplitude and the duration of the transfer gate pulse, the bias of the sense node, the size of the signal, and the operating temperature. This makes it a rich area for experimental investigation. Lag is most often characterised as a function of the signal and is measured either in absolute units (electrons) or relative to the signal in steady-state illumination (in percent).

Image lag in good quality PPDs is much better that 1%. Because of that, usually only the first dark image in the trailing edge has lag signal that can be measured, provided that it is above the noise. Therefore, the normal practice is to characterise the trailing edge lag in a dark image following a bright one, called 'first image lag'. Figure 5.41 gives an example of a timing diagram for first image lag measurement. The sensor is read out twice after each optical signal, created by a light pulse from an LED. A separate averaged dark image, obtained after the sensor has been read out for sufficiently long time in darkness, should be subtracted from all images to eliminate the dark signal. The fixed integration time guarantees the same dark signal in all images.

Lag at signal levels increasing from zero to saturation is the most frequently required measurement. Figure 5.42 shows this in a 10 μm pitch, 4T pixel, for three different transfer gate voltages. The figure illustrates the effect on the image lag at low and high signals, and how choosing the optimal V_{TG} can improve it. At $V_{TG} = 3.6$ V the onset of charge spill-back is clearly seen around 10 ke$^-$.

More sophisticated charge transfer methods have been developed to study the different contributing factors to the observed lag [8, 27]. When the lag is substantially larger, one pair of bright and dark images may not be enough to see the full picture. In this case a series of bright and dark images are needed to characterise the lag beyond the first image, and the timing diagram in figure 5.43 can be used.

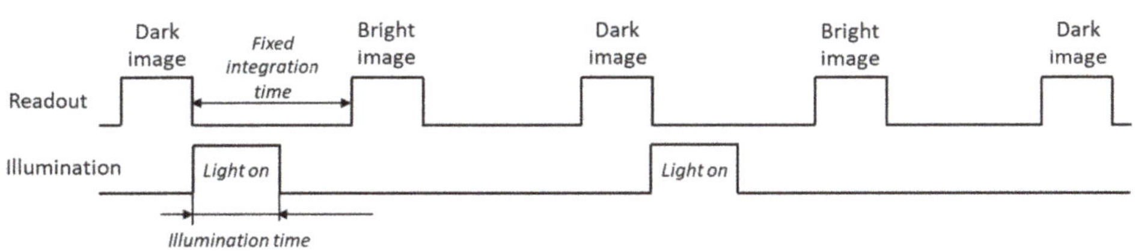

Figure 5.41. Timing diagram for measurement of trailing edge lag in the first image.

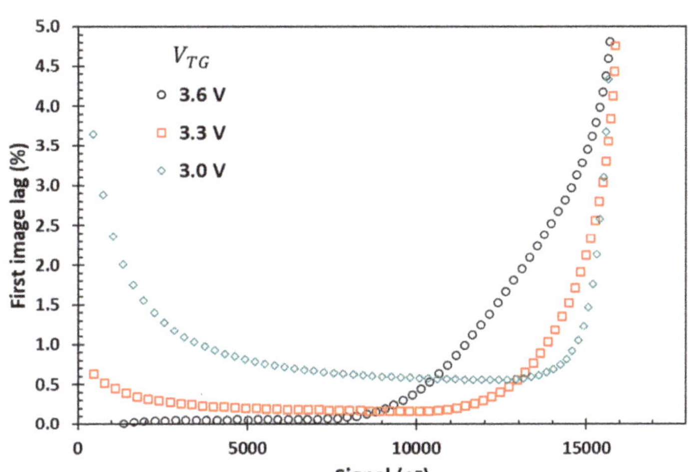

Figure 5.42. Image lag in the first dark image in a 10 μm 4T CIS at different transfer gate voltages. The low level of V_{TG} is zero.

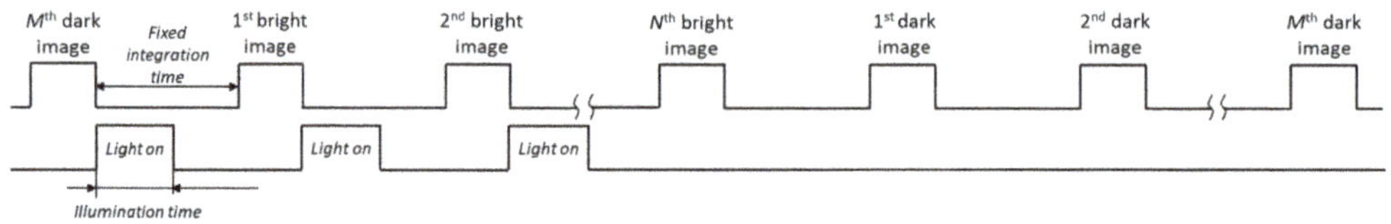

Figure 5.43. Timing diagram for generation of N bright and M dark images for the measurement of image lag.

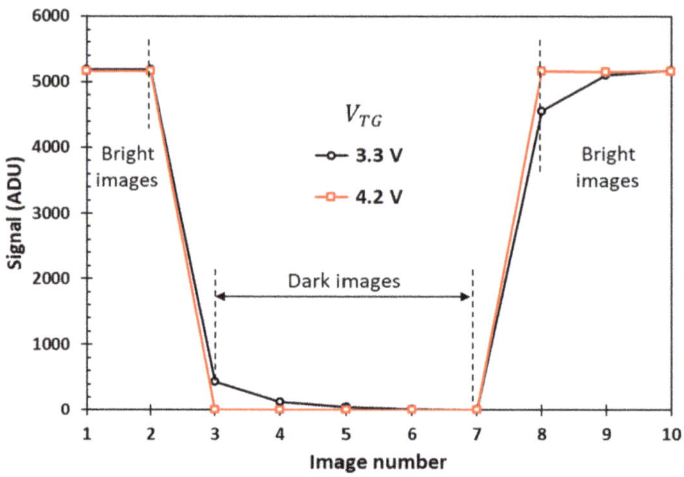

Figure 5.44. Signal from a 4T sensor taken with two different transfer gate voltages V_{TG}. The illumination is a series of five dark images following five bright ones. The system gain is 1.13 e$^-$/ADU.

Using five bright and dark images, figure 5.44 shows the output of a 4T sensor having lag problems even at relatively high transfer gate voltages, with visible lag signal beyond the first dark frame. At $V_{TG} = 3.3$ V, the first image trailing lag is 8.5% (500 e$^-$) and drops to 2.4% (140 e$^-$) in the second. Increasing V_{TG} to 4.2 V makes the first image lag fall below 0.1%.

5.11 Quantum efficiency

5.11.1 Principles

The quantum efficiency describes the effectiveness of the conversion of photons into charge. An ideal sensor fully absorbs light at all wavelengths without any loss and converts each photon into a photoelectron (in the visible range) without charge loss. Such sensor would have QE of unity, or 100%.

Quantum efficiency depends on the wavelength of light and the temperature but is practically independent on the signal. The wavelength dependence comes through light reflection, photon absorption length and charge losses. The temperature dependence is due to the increase of the bandgap at lower temperatures, leading to longer absorption length and lower QE at NIR wavelengths.

Almost universally, the QE is measured by comparing the photoresponse of the sensor (DUT) to a reference (calibrated) photodiode receiving the same light power per unit area, as shown in figure 5.45. The wavelength dependence of the QE is of primary interest; therefore, the light is normally coming from a monochromator as in figure 5.6. The experimental setup can be very complex [3] or quite simple [28].

In essence, the QE η is calculated as the ratio of the photogenerated current per unit area j_{DUT} in the DUT over the current in the reference photodiode j_{REF}, which has known QE η_{REF} [13]. This can be written as:

$$\eta_{\mathrm{DUT}} = \eta_{\mathrm{REF}} \frac{j_{\mathrm{DUT}}}{j_{\mathrm{REF}}} \tag{5.44}$$

Because η_{REF} is known in absolute terms from a measurement in a reference lab, equation (5.44) gives the absolute QE of the sensor under test.

The key to a successful QE measurement is to accurately measure the two currents while making sure that the illumination of the DUT and the reference is either the same, or if not, their ratio is well known from geometry and other factors. For large sensors it could be difficult to generate uniform illumination over their entire area. In such cases, a smaller part of the image sensor can be used, and the photodiode can be moved in and out of the illuminated area. This requires full understanding of the influence of the changing geometry on the light flux for both devices.

The mean signal from the sensor $\overline{S}$ can be converted to photogenerated charge per pixel by multiplying by the system gain and the elementary charge

$$\overline{Q} = qK\overline{S} \tag{5.45}$$

The current density is the charge $\overline{Q}$ divided by the pixel area A_{pix} and the exposure (integration) time t_{int}

$$j_{\mathrm{DUT}} = \frac{qK\overline{S}}{A_{\mathrm{pix}} t_{\mathrm{int}}} \tag{5.46}$$

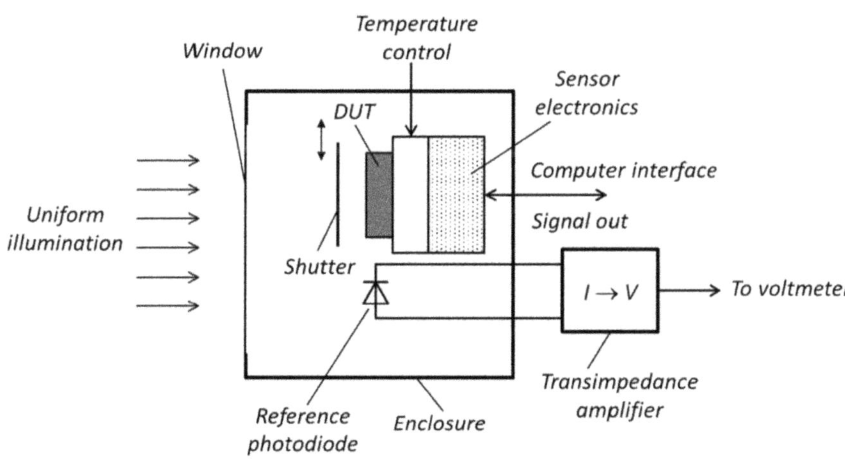

Figure 5.45. Principles of the QE measurement.

The current in the reference photodiode is converted to voltage in the transimpedance amplifier and is measured at each wavelength. The current density is simply the current divided by the photodiode area A_{REF}

$$j_{\mathrm{REF}} = \frac{I_{\mathrm{REF}}}{A_{\mathrm{REF}}} \tag{5.47}$$

Substituting (5.46) and (5.47) into (5.44) gives the QE formula

$$\eta_{\mathrm{DUT}} = \eta_{\mathrm{REF}} \frac{A_{\mathrm{REF}}}{A_{\mathrm{pix}}} \frac{qK\overline{S}}{t_{\mathrm{int}} I_{\mathrm{REF}}} \tag{5.48}$$

Very often the QE of the photodiode is not available directly, but the photosensitivity (responsivity) S_{REF} (in units of A/W) is given instead. In this case η_{REF} can be found from the ratio of S_{REF} over the theoretical maximum photosensitivity $S_{\mathrm{opt}}^{\mathrm{max}}$, corresponding to 100% QE. As derived in chapter 1, $S_{\mathrm{opt}}^{\mathrm{max}} = \lambda/1240$, where λ is the wavelength in nanometres. Therefore,

$$\eta_{\mathrm{REF}} = \frac{S_{\mathrm{REF}}}{S_{\mathrm{opt}}^{\mathrm{max}}} = \frac{1240}{\lambda} S_{\mathrm{REF}} \tag{5.49}$$

Using (5.49) in (5.48) gives the QE of the sensor under test using the photosensitivity of the reference. A typical photosensitivity curve of a reference silicon photodiode is shown in figure 5.46.

As with most other measurements, the dark signals in the sensor and in the photodiode must be accounted for by recording and subtracting them from the data. The QE at wavelengths longer than about 700 nm falls with temperature due to the increasing bandgap. Therefore, the operating temperatures of the DUT and reference photodiode, and the temperature at which the photodiode was calibrated must be known and used for any corrections.

Example 5.5. Calculate the QE of a BSI sensor for the following conditions: $\lambda = 650$ nm, $S_{\mathrm{REF}} = 0.335$ A/W, $A_{\mathrm{REF}} = 13$ mm^2, $I_{\mathrm{REF}} = 10$ nA, pixel size of 10 µm, $K = 2.0$ e$^-$/ADU, $\overline{S} = 15\,000$ ADU, $t_{\mathrm{int}} = 0.05$ s.
Solution: First, the QE of the reference photodiode is found from (5.49):

$$\eta_{\mathrm{REF}} = \frac{1240}{650} \times 0.335 = 0.639$$

Converting the diode and the pixel area to cm^2 gives $A_{\mathrm{REF}} = 0.13$ cm^2 and $A_{\mathrm{pix}} = 10^{-6}$ cm^2. Substituting everything into (5.48) gives

$$\eta_{\mathrm{DUT}} = 0.639 \times \frac{0.13}{10^{-6}} \times \frac{1.6 \times 10^{-19} \times 2.0 \times 15000}{0.05 \times 10^{-8}} = 0.798$$

Figure 5.47 shows the measured QE of a 5T CIS in both front and backside illuminated variants, demonstrating the huge increase of the QE resulting from BSI processing.

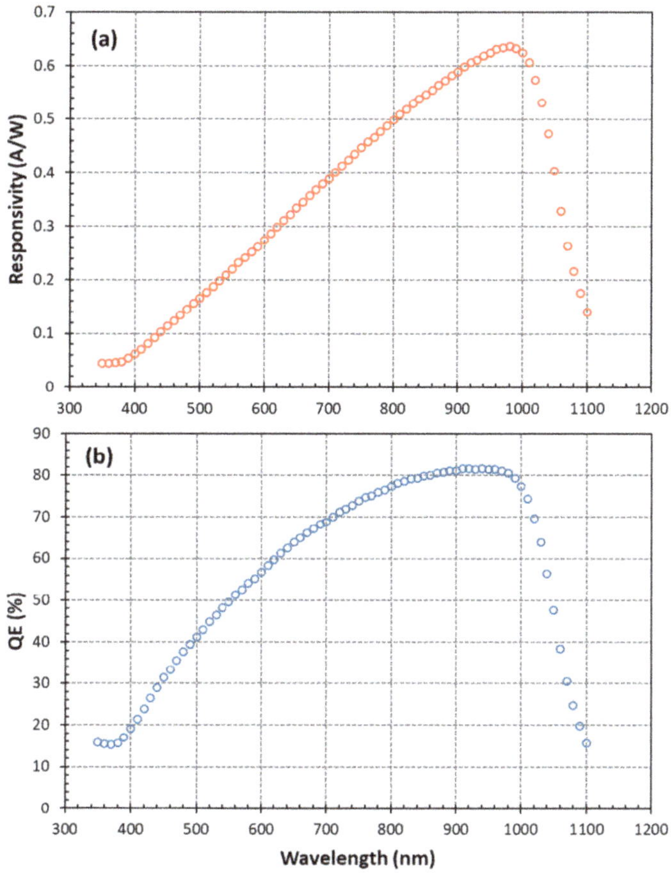

Figure 5.46. Photosensitivity (a) and QE calculated with (5.49) (b) of a silicon photodiode with NIST traceable calibration at 25 °C (Thorlabs model FDS100-CAL, www.thorlabs.com). Reprinted with permission from Thorlabs Inc.

QE measurements can be challenging and require careful consideration of all sources or error and their elimination. The calibration uncertainty of the reference can be as good as ±1%, but it can increase above ±5% at UV and NIR wavelengths. This is the ultimate limit for the accuracy of the measurement, but in practice other factors—geometry, dark current and temperature can add higher systematic errors.

Very often it is not possible to have both the DUT and the reference illuminated at the same time. This requires that the measurements are performed consecutively by moving both in and out of the light beam, and this can add errors due to instabilities and tolerances in the geometry of the setup. Monitoring the light source with a separate photodiode (diode #1 in figure 5.6) can help provide a correction for its fluctuations. Temperature fluctuations in all components can add to the systematic error, especially in the dark current of the reference and the DUT at low illumination levels.

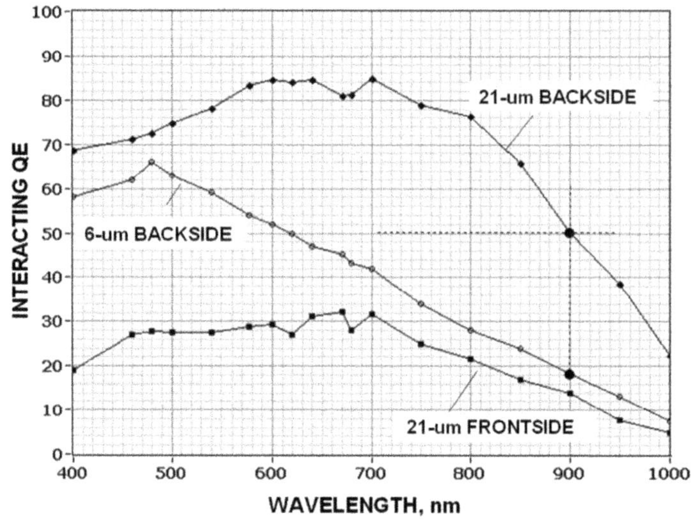

Figure 5.47. QE of a 5T CIS in FSI and BSI variants. Reprinted with permission from [29].

5.11.2 Pain–Hancock method

From equation (5.46) we can see that the calculated QE is sensitive to the system gain, and therefore to nonlinearities in the conversion gain of the sensor. Since the conversion gain can change considerably with the signal, as demonstrated in figure 5.21, the calculated QE will also change. However, we know that the QE must be independent of the signal, therefore the nonlinearity of the CVF should not affect it.

The Pain–Hancock method [16] has been developed to eliminate the nonlinearity of the CVF and produce more accurate QE measurements for large signals, and is particularly useful for CIS. As described in section 5.5.4, the method introduces another way to analyse the signal variance, where the light flux (or the equivalent—illumination time) is the primary variable, and not the signal as in the classic PTC. It also makes the central assumption that the number of collected electrons is proportional to the number of incoming photons. Sensor linearity is not assumed, and the concept of small-signal gain g_S (5.31) around an operating point of $\overline{N_e}$ collected electrons is used.

Adding the readout noise variance to (5.33), we can write the following for a nonlinear system with signal-dependent small-signal gain g_S:

$$\sigma_S^2 = g_S^2(\overline{N_e})\overline{N_e} + \sigma_R^2 \tag{5.50}$$

Substituting for $\overline{N_e}$ from (5.34) and g_S from (5.36) in (5.50), we arrive at the main result of the Pain–Hancock method:

$$\sigma_S^2 = \left(\frac{1}{\eta}\frac{dS}{dN_{ph}}\right)^2 \eta N_{ph} + \sigma_R^2 \tag{5.51}$$

This gives the formula for the quantum efficiency:

$$\eta = \frac{N_{\rm ph}}{\sigma_{\rm S}^2 - \sigma_{\rm R}^2}\left(\frac{dS}{dN_{\rm ph}}\right)^2 \tag{5.52}$$

Often, we prefer to obtain the QE from a linear fit to the data because it is more accurate than a calculation at a single point and allows us to spot any irregularities. In this case, the QE can be calculated from the linear slope of the signal variance $\sigma_{\rm S}^2$ versus $N_{\rm ph}(dS/dN_{\rm ph})^2$, as in the linear relationship $y = (1/\eta)x + b$

$$\sigma_{\rm S}^2 = \frac{1}{\eta}\left[N_{\rm ph}\left(\frac{dS}{dN_{\rm ph}}\right)^2\right] + \sigma_{\rm R}^2 \tag{5.53}$$

The number of photons $N_{\rm ph}$ is proportional to the illumination time t_i, therefore the term multiplying $1/\eta$ can be expressed as

$$N_{\rm ph}\left(\frac{dS}{dN_{\rm ph}}\right)^2 = \text{const} \times t_i\left(\frac{dS}{dt_i}\right)^2 \tag{5.54}$$

Equation (5.54) allows us to work with the illumination time as the primary variable instead of the number of photons, just as we did in section 5.5.4 for the nonlinear CVF. Using the raw data in figure 5.21, equation (5.53) and the signal variance against the illumination time are plotted in figure 5.48. The derivative dS/dt_i for the point with index n has been calculated numerically from the data as

$$\left(\frac{dS}{dt_i}\right)_{(n)} = \frac{S_{(n+1)} - S_{(n-1)}}{t_{i(n+1)} - t_{i(n-1)}} \tag{5.55}$$

An important feature of the data is that $t_i(dS/dt_i)^2$ is nearly zero at low signals because $t_i \approx 0$, and also at high signals because in saturation $dS/dt_i \approx 0$. This creates the interesting behaviour shown in figure 5.48(a): the variance increases linearly with $t_i(dS/dt_i)^2$ up to its peak at around $t_i = 70$ μs (figure 5.48(b)), and then the data loop back to the origin of the coordinate system. The rising section is the linear part used for the QE calculation. Figure 5.48(a) indicates that the Pain–Hancock method encounters a limit around the peak of the signal variance, despite theoretically being able to cope with any conversion gain nonlinearities.

It is worth looking into the signal dependence of the QE of a nonlinear sensor calculated by the simple method based on the assumption of constant CVF and system gain. Bearing in mind the QE definition (5.34) and that $N_{\rm e} = KS$ and $N_{\rm ph} \propto t_i$, we can write the following for a system where a constant K is assumed:

$$\eta = \frac{dN_{\rm e}}{dN_{\rm ph}} \propto K\frac{dS}{dt_i} \tag{5.56}$$

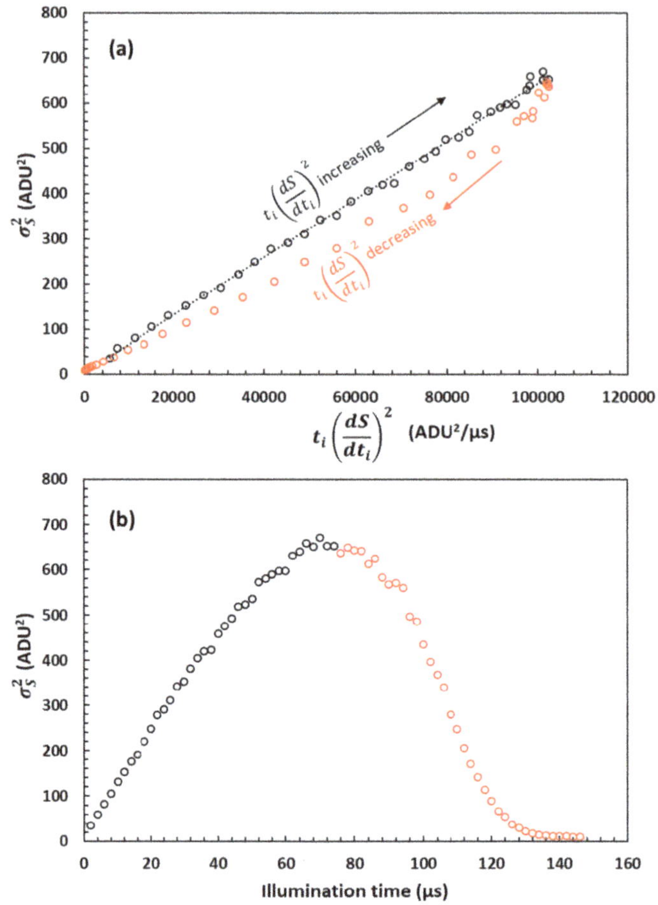

Figure 5.48. Signal variance σ_S^2 against $t_i(dS/dt_i)^2$ and a linear fit (a); and the signal variance against the illumination time (b). The data points before and after the variance peak for both plots are drawn in black and red, correspondingly.

In such system K has been derived from the PTC at low signals. The derivative dS/dt_i is a measure of the sensor's linearity, and with K fixed, can be used to estimate the change of the calculated QE.

Figure 5.49(a) shows that dS/dt_i falls significantly over the same signal range where the Pain–Hancock method produces the very linear response in figure 5.48(a). This will under-estimate the QE by 15% at the end of the range, given as 71% S_{sat} in figure 5.49(b), compared to the QE at much smaller signals.

As we know, the correct method for QE measurement should be signal-independent. One obvious question is this—why try to measure the QE at large signals and suffer from nonlinearities, when sensors are generally much more linear at small signals? The first part of the answer is straightforward—because the relative errors due to dark current and noise become larger at small optical signals. The second part comes from the most widely used measurement of the QE as a function

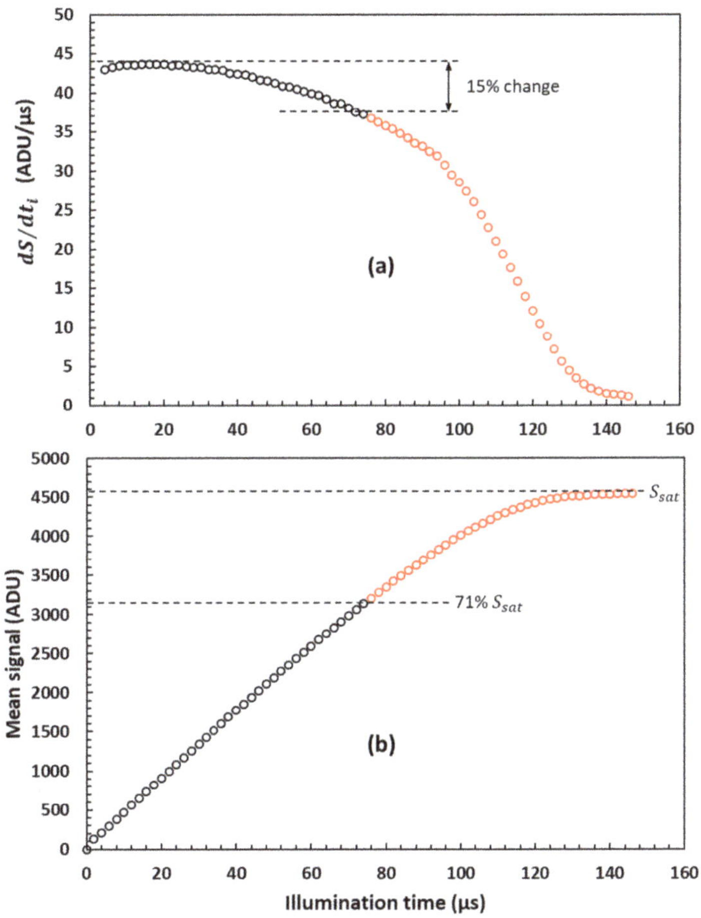

Figure 5.49. Signal derivative (a) and photoresponse (b) for the data in figure 5.48.

of wavelength, using a monochromator. The light flux can vary by an order of magnitude at different wavelengths due to the spectrum of the light source, therefore the QE method should be able to work well with both small and large signals. The Pain–Hancock method is very useful in such situations but is not infallible and must be used within its limitations, as figure 5.48(a) demonstrates.

5.11.3 Modulation transfer function

The modulation transfer function (MTF) is a measure of the spatial resolution of a sensor, which is its ability to resolve detail in an image. The classical way to visualise and measure the response of the sensor is to illuminate it with a sinusoidally modulated light, generated by passing it through a target. The intensity of the input light changes between $I_{\max}$ and $I_{\min}$, as shown in figure 5.50, and for a perfect target $I_{\min}$ would be zero.

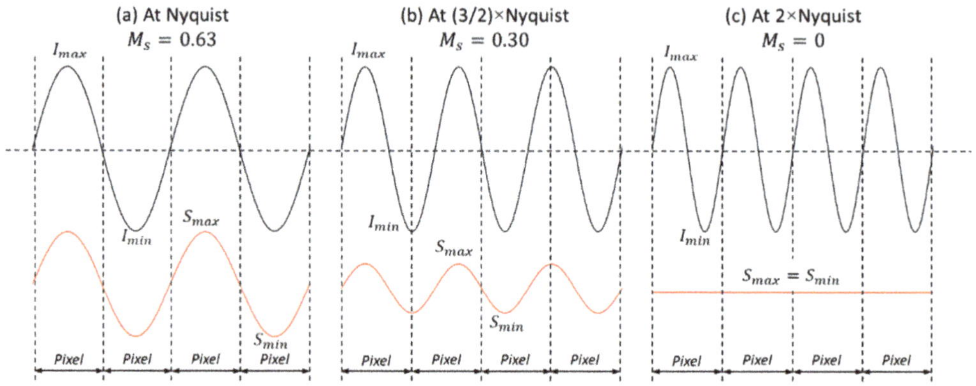

Figure 5.50. Input projected sinewave illumination (black line) and the sensor response (red line) at three spatial frequencies.

The modulation of the projected light equals the target modulation M_{T}, and is defined as:

$$M_{\mathrm{T}} = \frac{I_{\max} - I_{\min}}{I_{\max} + I_{\min}} \tag{5.57}$$

The modulation recorded by the sensor is

$$M_{\mathrm{S}} = \frac{S_{\max} - S_{\min}}{S_{\max} + S_{\min}} \tag{5.58}$$

where $S_{\max}$ and $S_{\min}$ are the maximum and the minimum signals in figure 5.50. Due to the finite pixel size, charge diffusion and optical effects the light modulation registered by the sensor M_{S} cannot be higher than the projected modulation. The ratio of the registered to the projected light modulation is the MTF:

$$\mathrm{MTF} = \frac{M_{\mathrm{S}}}{M_{\mathrm{T}}} \tag{5.59}$$

If the target is perfect $I_{\min} = 0$ and $M_{\mathrm{T}} = 1$, therefore the sensor MTF becomes:

$$\mathrm{MTF} = M_{\mathrm{S}} = \frac{S_{\max} - S_{\min}}{S_{\max} + S_{\min}} \tag{5.60}$$

It is obvious that the MTF depends on the ratio between the period of the sinewave illumination and the pixel pitch p. The MTF will be high when the illumination period is much larger than the pitch because the sinewave spreads across many pixels and the captured image is closer to the original. When the illumination period decreases the MTF decreases too; when both periods match as in figure 5.50(c) the captured image is flat and the MTF is zero.

Instead of periods we use the term *spatial frequency*: the spatial frequency of the illumination f is simply the inverse of its period; similarly, the spatial frequency of the pixel array is $1/p$. The spatial frequency is usually given in units of lines cm^{-1} (cm^{-1}); for a sensor with pixel pitch $p = 10$ μm the spatial frequency is $f_p = 1000$ cm^{-1} because it has 1000 pixels per cm.

Due to the pixelated structure of image sensors the captured image is discrete and subject to the Nyquist sampling theorem [30]. As in electrical signal theory, sampling an analogue sinewave using sampling frequency f_s results in a faithful capture only if the frequency of the sinewave is below $f_s/2$, known at the Nyquist frequency f_N. Applied to image sensors, this means that a sensor with pixel pitch p samples the signal using spatial frequency of $f_p = 1/p$, and the Nyquist frequency is

$$f_N = \frac{1}{2p} \tag{5.61}$$

Spatial frequencies slightly above f_N would appear as frequencies slightly above zero as captured by the sensor, or they 'alias' to lower frequencies.

In an ideal sensor each pixel collects 100% of the photons reaching it, all the image area is photosensitive and there is no charge sharing between the pixels. The MTF of such ideal sensor has a theoretical maximum, called the integration MTF:

$$\text{MTF}_\text{I} = \frac{\sin\left(\frac{\pi f}{2f_N}\right)}{\frac{\pi f}{2f_N}} = \text{sinc}\left(\frac{\pi f}{2f_N}\right) \tag{5.62}$$

At the Nyquist frequency ($f = f_N$) the modulation MTF is $\text{sinc}(\pi/2) = 0.637$. The MTF of real devices can be close, but not higher than (5.62). The main reason for MTF degradation is diffusion of photogenerated charge away from the pixel where it has been generated, described in detail in [13]. Another cause could be incomplete charge transfer caused by image lag. Both effects reduce the signal modulation registered by the sensor.

The classic method to measure the MTF is to project many sinewave-modulated images at different spatial frequencies onto a sensor and to calculate the MTF using (5.59) for each frequency. A data point is generated from each image, and for an ideal sensor the points will coincide with the modulation MTF as in figure 5.51.

No matter how straightforward it looks, sinewave projection is rarely used because the measurement is time consuming and requires multiple targets, precise focusing and large sensor area. Besides, manufacturing targets with smooth, sinusoidal change in transmittance is quite difficult. Binary targets with alternating 0% and 100% transmission are much easier to make and are preferred instead. They take the form of various bar configurations, slanted edges and Siemens star targets. Since the light input is no longer sinusoidal and contains higher spatial harmonics, some computation is required to obtain the MTF [31].

The slanted bar is one of the most popular and easy to use techniques because it allows one to build the edge spread function (ESF) from the rows it covers from just one image [30]. The ESF is a measure of the sensor's response to an ideal light-dark step illumination. Using the slanted bar, the ESF becomes *oversampled* because the

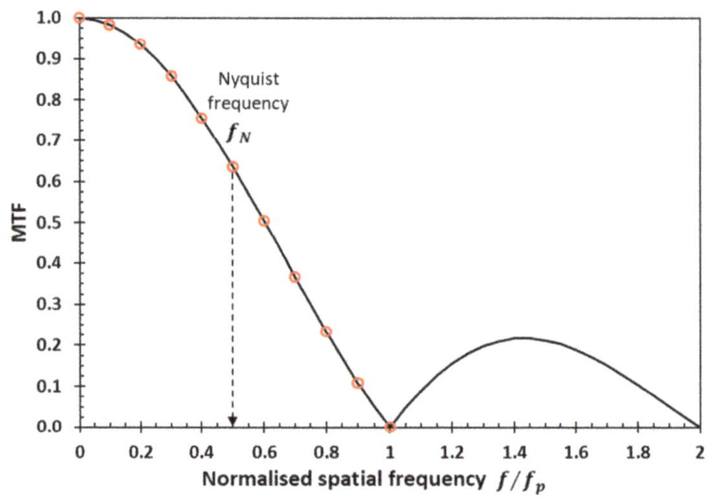

Figure 5.51. Ideal integration MTF calculated with (5.62) (black line), and data points from an ideal measurement.

light-dark transition is spread over many pixels, and each pixel sees a different phase of the transition. The oversampled ESF can also be built with a bar parallel to the pixels by taking a series of images as the bar moves in sub-pixel sized steps, but this is more difficult to do experimentally and is time consuming. Differentiating the ESF produces the line spread function (LSF), which is a measure of the response to an infinitely narrow line source, described by a one-dimensional delta function. Finally, the normalised magnitude of the Fourier transform of the LSF gives the MTF [30].

All methods using optical projection of targets must account for the MTF of the lenses and mirrors used in the system because the optical MTF multiplies the measured MTF. Also, the measurement can be degraded due to imperfect image focus and alignment, or due to reflections. Far better is to not use target projection at all, but to deposit the target directly on the surface of the sensor. For example, a slanted bar can be made with the top metal layer in a CIS [32, 33] and this makes the optical setup extremely simple—all that is needed is collimated light.

When using the embedded slanted bar technique, the edge can be aligned with excellent precision to the pixels underneath. The edge is designed to create a full light-dark transition cycle over integer number of pixels, called an oversampling ratio (OSR). An example with OSR = 8, forming a right triangle with sides ratio of 1:8, is shown in the image in figure 5.52. The angle of the edge is $\theta = \arctan(1/\text{OSR})$ and is important for capturing the transition with sufficient resolution. It has been shown that the angle must be below 10° for good MTF measurement [33].

The ESF is built from the pixels in the transition area covering a number of rows equal to the OSR. Using the following pseudocode, the ESF is extracted from the image in figure 5.52 and plotted in figure 5.53(a) together with the corresponding LSF:

1. Choose an image area of $N \times N$ pixels covering the edge, such as the 32 × 32 pixels in figure 5.52. Choose a starting row.

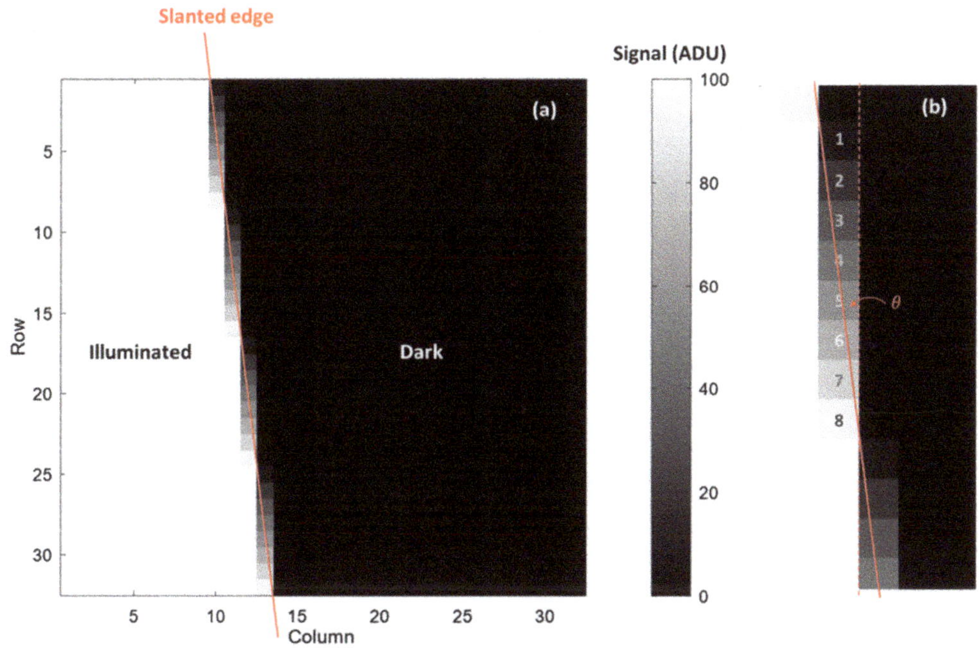

Figure 5.52. (a) Simulated image of a slanted edge with sides ratio of 1:8 ($\theta = 7.125°$) and signal spanning between 0 and 100 ADU; (b) zoom-in on the simulated image around the slanted edge.

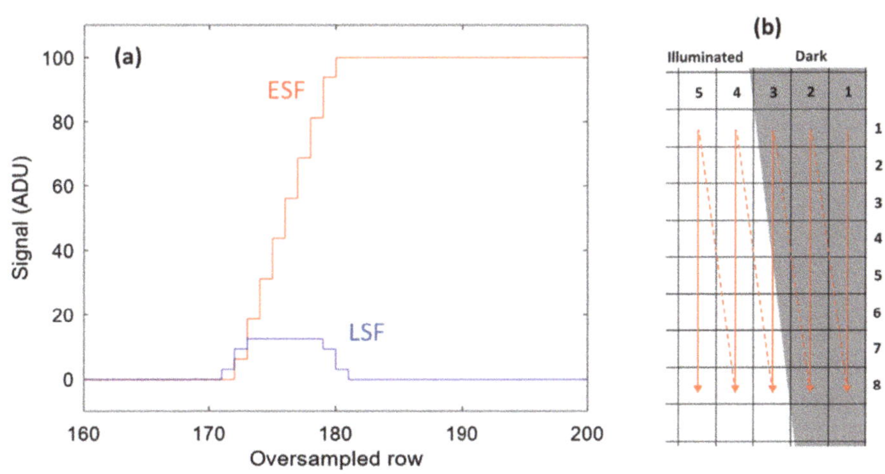

Figure 5.53. (a) ESF and LSF derived from the image in figure 5.52; (b) diagram showing the pixel order for the build-up of the ESF.

2. Starting from the rightmost non-illuminated column of the image, build the ESF as a one-dimensional vector by moving down along the column and adding the signal of the next row as the next element of the ESF.
3. Once OSR pixels have been added, move back to the staring row and to the column to the left. Repeat (3) and (2) until the last column is reached, as

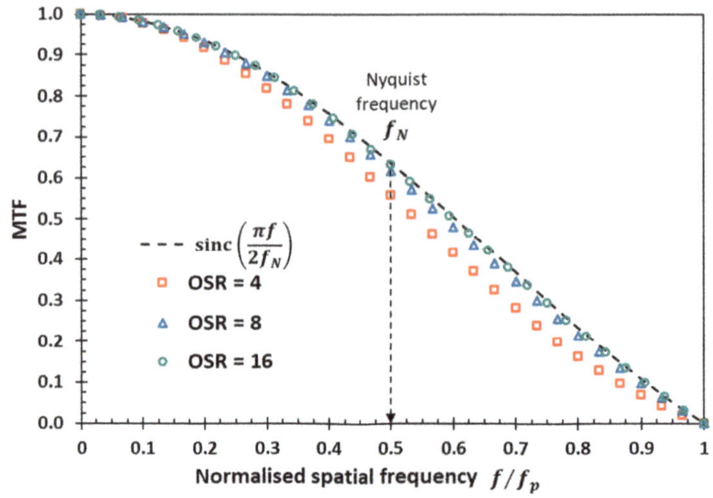

Figure 5.54. The integration MTF (dashed black line) and calculated MTF with the slanted edge method using OSR = 4 ($\theta = 14.036°$), OSR = 8 ($\theta = 7.125°$) and OSR = 16 ($\theta = 3.576°$).

illustrated in figure 5.53(b). The number of elements in the ESF vector is now $N \times \text{OSR}$.

4. Calculate the LSF by numerically differentiating the ESF using the three-point method (for example, *gradient* in MATLAB and Octave).
5. Perform a Fourier transform of the LSF and take the absolute values. Normalise by dividing by the sum of all elements in the LSF vector to create the Y-values of the MTF plot.
6. Create the normalised values of X-axis of the MTF plot from the integers between 0 and $(N \times \text{OSR}) - 1$, divided by N.

Figure 5.54 shows the calculated MTF for three different oversampling ratios from simulated images like figure 5.52. For OSR $\geqslant$ 8 the calculated MTF is very close to the theoretical integration MTF. Since the image simulates an ideal sensor, this is a good indicator that the analysis works well. When applied to real sensors, the image must be free from FPN, and the influence of the readout noise and dark current should be reduced by averaging many images.

5.12 Electrical transfer function

Despite the reliability of CAD simulations, it is often desirable to experimentally characterise the on-chip circuitry serving the pixel array. In many CIS with analogue outputs, it is possible to measure the signal gain from the sense node to the output by taking the electrical transfer function (ETF). This includes the gain of the source follower and the subsequent amplifiers which are needed to calculate the CVF.

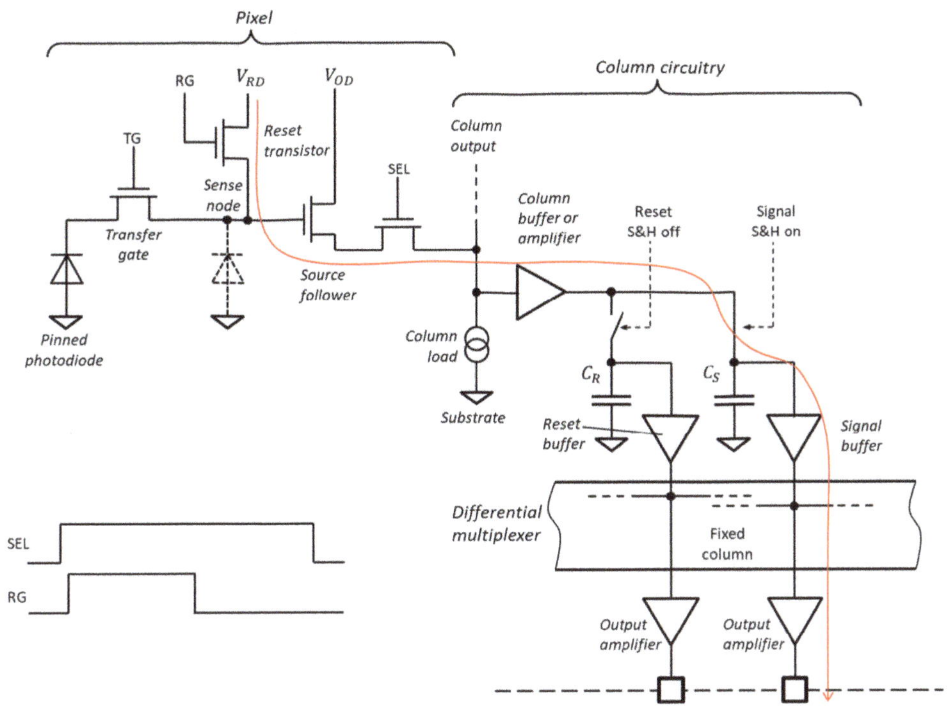

Figure 5.55. Signal path for ETF measurement in a sensor with differential output. TG and the Reset S&H switch are off; RG, SEL and Signal S&H are on.

To measure the ETF, a single pixel is selected by fixing the row and column addresses, and at least one CDS sample and hold (S&H) switch is turned on, as shown in figure 5.55. With the reset transistor on, the sense node voltage equals the reset drain voltage V_{RD}, provided that RG is biased sufficiently high. By scanning V_{RD} and measuring the output, the transfer characteristic of the signal chain can be determined. If both S&H switches are on, the signal will appear on both outputs. This allows one to measure the gain difference between the two halves of the differential signal path.

Measuring the ETF can be either static, with the RG and SEL permanently on, or dynamic. The latter is shown experimentally in figure 5.56 with a timing diagram corresponding to the inset in figure 5.55. Recording the output voltage taken from the scope trace versus the DC bias of the reset transistor drain V_{RD} produces the plot in figure 5.57. By fitting a straight line to the data between V_{RD} = 1.0 to 3.2 V the signal gain from the sense node to the output is calculated from the slope as 0.877. The fit also tells us that the output is linear to better than 1% between 1.3 and 3.2 V.

The ETF measurement gives a wealth of information but relies on full external control of the readout chain which may not be possible in many sensors. A different way to measure the gain is to apply a step change ΔV_{RD} to the reset drain voltage at

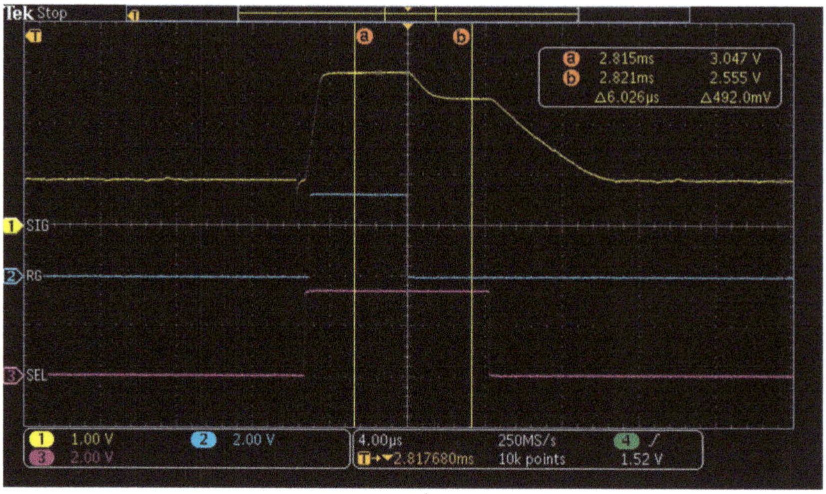

Figure 5.56. Oscilloscope trace showing the output (SIG) measured during reset at marker position (a) for V_{RD} = 3.0 V. RG and SEL are the control signals to the sensor with 3.3 V CMOS logic levels. In this setup V_{OD} = 3.3 V and the voltage applied to the gate of the reset transistor is V_{RG} = 3.6 V.

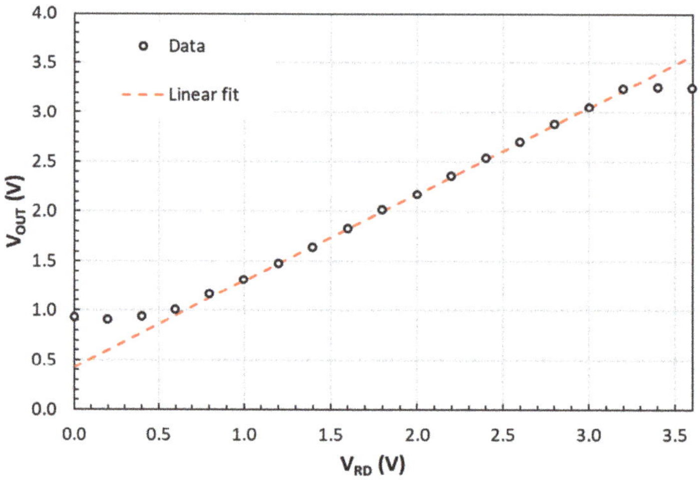

Figure 5.57. Measurement of the electrical transfer function. The slope of the linear part of the data corresponds to a gain of 0.877.

the place where the transfer gate pulse TG normally is, as shown in the timing diagram in figure 5.58. The voltage step is captured by the CDS and becomes the output signal. To do this, the reset transistor must be on for the duration of time covering both SHS and SHR, and the voltage step on RD should be synchronised to start shortly before SHS.

Of course, this method can also be applied to a sensor which allows full external control of its operation, as in figure 5.55.

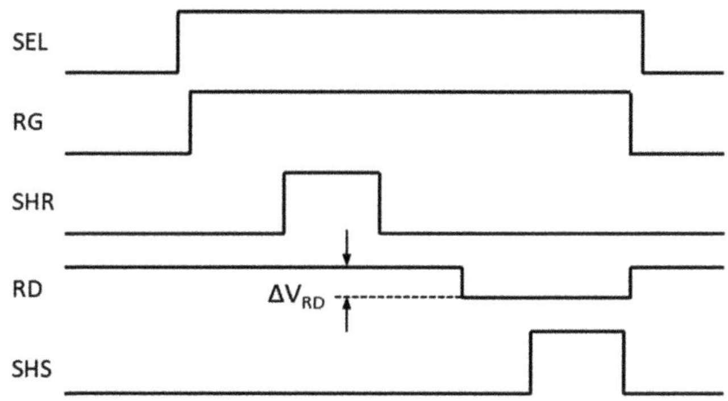

Figure 5.58. Timing diagram for ETF measurement using a voltage step input to the sense node.

Chapter summary

1. The system gain is a proportionality constant between the output digital signal and the number of collected electrons in a pixel, and is measured in e^{-}/ADU.
2. The conversion and the electronic gain of a sensor are nonlinear, leading to overall nonlinear photoresponse. Normally, the conversion gain nonlinearity dominates.
3. The deviation from a linear fit to the data is used to estimate the nonlinearity of the photoresponse, and the result depends on the chosen data range.
4. The photon transfer curve is a powerful method for image sensor characterisation. It is used for the calculation of the system gain, noise, conversion gain, full well capacity, fixed pattern noise, and for system diagnostics.
5. The PTC relies on the Poisson statistics describing the standard deviation of the collected electrons. Deviations from Poisson statistics can make the PTC nonlinear but do not mean that the sensor is nonlinear too.
6. Frame differencing is used to remove the fixed pattern noise from images, leaving only shot and readout noise for subsequent use in a PTC.
7. The mean-variance curve is one of the most convenient methods for PTC representation. The system gain is determined from the inverse slope of the signal variance plotted against the mean signal.
8. The system gain is generally nonlinear, and noise and signal gain should be distinguished. The PTC gives the small-signal gain, which in linear sensors is the same as the large-signal gain.
9. The well-known charge created by photoeffect from characteristic x-rays can be used for the measurement of the system gain. X-ray calibration does not rely on the properties of the readout noise and is more robust than the PTC.
10. The readout noise of a sensor should be measured with zero dark current to eliminate its shot noise.

11. The Pain–Hancock method can be used to remove the sensor nonlinearity for the measurement of the QE. This method uses the number of received photons as the primary variable, unlike the PTC, which uses the sensor's signal.
12. The ETF provides a method to trace a signal from the sense node to the output and is used for characterising the full signal chain.

References

[1] EMVA Standard 1288 Standard for Characterization of Image Sensors and Cameras, European Machine Vision Association, 15 March 2021 (www.emva.org)

[2] International Organization for Standardization (https://www.iso.org/home.html)

[3] Jacquot B C, Monacos S P, Hoenk M E, Greer F, Jones T J and Nikzad S 2011 A system and methodologies for absolute quantum efficiency measurements from the vacuum ultraviolet through the near infrared *Rev. Sci. Instrum.* **82** 043102

[4] Spivak A, Belenky A, Fish A and Yadid-Pecht O 2009 Wide-dynamic-range CMOS imagesensors—comparative performance analysis *IEEE Trans. Electron Devices* **56** 2446–61

[5] Fernandez F, Steward B, Gross K and Hawks M 2019 Implementation of a non-linear CMOS and CCD focal plane array model in ASSET *Proceedings of SPIE, 110010C (Baltimore)*

[6] Janesick J 2007 *Photon Transfer DN* $\rightarrow \lambda$ (Bellingham, WA: SPIE Press)

[7] Wang F and Theuwissen A J P 2017 Linearity analysis of a CMOS image sensor IS&T Inter. Symp. on Electronic Imaging *(San Francisco)*

[8] Bonjour L, Blanc N and Kayal M 2012 Experimental analysis of lag sources in pinned photodiodes *IEEE Electron Device Lett.* **33** 1735–7

[9] Soman M, Stefanov K, Weatherill D, Holland A, Gow J and Leese M 2015 Non-linear responsivity characterisation of a CMOS active pixel sensor for high resolution imaging of the Jovian system *J. Instrum.* **10** C02012

[10] Berendsen H J C 2011 *A Student's Guide to Data and Error Analysis* (Cambridge: Cambridge University Press)

[11] Levski D, Wäny M and Choubey B 2021 Compensation of signal-dependent readout noise in photon transfer curve characterisation of CMOS image sensors *IEEE Trans. Circuits Syst. II: Express Br.* **68** 102–5

[12] Okura S *et al* 2019 A 2-Mpixel CMOS image sensor with device authentication and encryption key generation based on physically unclonable function *Int. Image Sensor Workshop (Snowbird, UT)*

[13] Janesick J 2001 *Scientific Charge-Coupled Devices* (Bellingham, WA: SPIE Press)

[14] Downing M, Baade D, Sinclaire P, Deiries S and Christen F 2006 CCD riddle: (a) signal vs time: linear; (b) signal vs variance: non-linear *Proc. of the SPIE 627609 (Orlando, FL)*

[15] Stefanov K D 2013 A statistical model for signal-dependent charge sharing in image sensors *IEEE Trans. Electron Devices* **61** 110–5

[16] Pain B and Hancock B R 2003 Accurate estimation of conversion gain and quantum efficiency in CMOS imagers *Proc. of SPIE 5017 (Santa Clara, CA)*

[17] Bohndiek S E *et al* 2008 Comparison of methods for estimating the conversion gain of CMOS active pixel sensors *IEEE Sensors J.* **8** 1734–44

[18] Janesick J, Andrews J and Elliott T 2006 Fundamental performance differences between CMOS and CCD imagers: part I *Proc. of SPIE 6276, 62760M (Orlando, FL)*

[19] Michelot J, de Ipanema Moreira A, Monsinjon P and Caranhac S 2016 Effects of transfer gate spill back in low light high performances CMOS image sensors *Photon Counting, Low Flux and High Dynamic Range Optoelectronic Detectors Workshop (Toulouse)*

[20] Ivory J, Stefanov K D and Holland A D 2020 Mitigating charge spill-back induced image lag with a multi-level transfer gate pulse in PPD image sensors *Proc. of SPIE 11454*

[21] Lowe B and Sareen R 2007 A measurement of the electron–hole pair creation energy and the Fano factor in silicon for 5.9 keV X-rays and their temperature dependence in the range 80–270 K *Nuclear Instrum. Methods Phys. Res.* A **576** 367–70

[22] Widenhorn R, Dunlap J C and B E 2010 Exposure time dependence of dark current in CCD imagers *IEEE Trans. Electron Devices* **57** 581–7

[23] McGrath D, Tobin S, Goiffon V, Magnan P and Le Roch A 2018 Dark current limiting mechanisms in CMOS image sensors IS&T *Int. Symp. on Electronic Imaging 2018, Image Sensors and Imaging Systems (Burlingame, CA)*

[24] Teranishi N 2013 Dark current and white blemish in image sensors *2013 Int. Symp. on VLSI Technology, Systems and Application (VLSI-TSA) (Hsinchu, Taiwan)*

[25] Chao C Y-P *et al* 2019 Identifying the sources of random telegraph noises in pixels of CMOS image sensors *Int. Image Sensor Workshop (Snowbird, UT)*

[26] Kwon H-M *et al* 2013 Effects of high-pressure annealing on random telegraph signal noise characteristic of source follower block in CMOS image sensor *IEEE Electron Device Lett.* **34** 190–2

[27] Gao W, Guidash M, Li N, Ispasoiu R, Ailuri P R, Palaniappan N, Tekleab D and Rahman M 2017 Photodiode barrier induced lag characterization using a new lag versus idle time methodology *Int. Image Sensor Workshop (Hiroshima, Japan)*

[28] Crews C, Soman M, Allanwood E A, Stefanov K, Leese M, Turner P and Holland A 2020 Quantum efficiency of the CIS115 in a radiation environment *Proc. of SPIE, 114540E*

[29] Janesick J, Andrews J, Tower J, Grygon M, Elliott T, Cheng J, Lesser M and Pinter J 2007 Fundamental performance differences between CMOS and CCD imagers: part II *Proc. of SPIE 6690, 669003 (San Diego, CA)*

[30] Boreman G D 2021 Point-, line-, and edge-spread function measurement of MTF *Modulation Transfer Function in Optical and Electro-Optical Systems* 2nd edn (Bellingham, WA: SPIE Press) pp 67–84

[31] Zhang X, Kashti T, Kella D, Frank T, Shaked D, Ulichney R, Fischer M and Allebach J P 2012 Measuring the modulation transfer function of image capture devices: what do the numbers really mean? *Proc. of SPIE 8293 (829307) (Burlingame, CA)*

[32] Janesick J, Pinter J, Potter R, Elliott T, Andrews J, Tower J, Grygon M and Keller D 2010 Fundamental performance differences between CMOS and CCD imagers: part IV *Proc. of SPIE 7742, 77420B (San Diego, CA)*

[33] Estribeau M and Magnan P 2004 Fast MTF measurement of CMOS imagers at the chip level using ISO 12233 slanted-edge methodology *Proc. of SPIE 5570 (Maspalomas)*

Chapter 6

Electronics

6.1 On-chip electronics

6.1.1 Architecture

In addition to the pixels, a CIS contains numerous other circuits for row and column selection, row drivers, CDS, signal multiplexers, amplifiers and ADCs. This on-chip electronics, normally the domain of the IC designer, is essential for the operation and the performance of an image sensor.

In sensors with analogue outputs the additional electronics is rather simple and could look like the block diagram in figure 6.1. The row and column address decoders are straightforward digital circuits, and the row drivers and the output amplifiers are also easy to design. The column readout and the CDS are much more critical as they determine the performance of the sensor, together with the pixel.

Sensors with digital outputs are much more complex and contain typically one ADC per column, high speed timing and ramp generators, control over SPI or I^2C interface, and even image processing and compression. The outputs use high speed drivers with physical standards such as low voltage differential signalling (LVDS), scalable low voltage signalling (SLVS), current mode logic (CML) or mobile industry processor interface (MIPI).

Certainly, it is much easier to study and understand a sensor with analogue outputs. It is even possible to follow the schematic of the complete readout chain from sense node to output for one selected pixel. Figure 6.2 shows a simplified analogue schematic of the complete readout chain, with only the pixel shown at transistor level. The different buffers and amplifiers are normally simple circuits containing a handful of transistors, and the switches are just a NMOS–PMOS pair. Even at transistor level the whole signal path schematic (excluding current and voltage bias circuits) could fit on a single sheet of A4 paper.

Being able to follow the signal from the sense node to the chip's outputs can give us a greater understanding of the workings of an image sensor. In the following sections we are going to explore the blocks and circuits in figures 6.1 and 6.2 in more detail.

doi:10.1088/978-0-7503-3235-4ch6

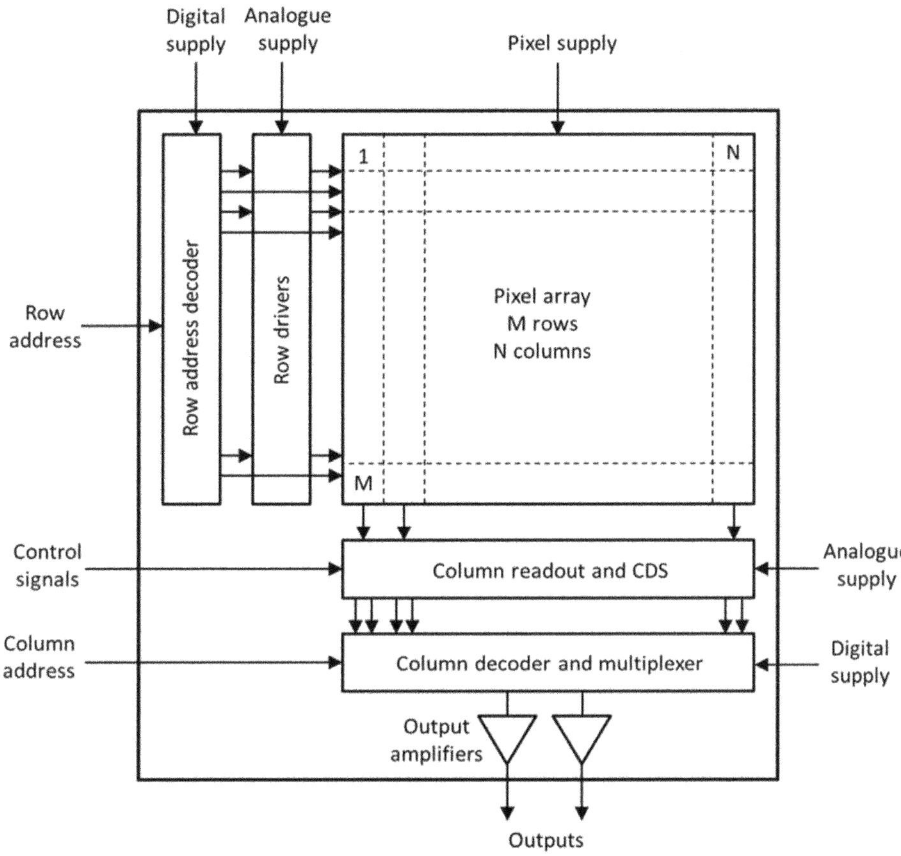

Figure 6.1. Simplified block diagram of an image sensor with analogue output.

6.1.2 Column buffers

The in-pixel source follower must drive the following column circuitry, but due to its small size and low bias current it may struggle with the load, and the settling time can be too long. The simplest way to improve this is to buffer with another source follower running at much higher bias current and capable of driving heavier loads.

The in-pixel NMOS follower can be buffered by another NMOS follower, as in figure 6.3(a). The voltage at the source of M1 is $V_{G1} - V_{GS1}$, where V_{GS1} is its threshold voltage, and this is approximately the maximum output voltage swing to ground, sparing few hundred millivolts for M2. If we take that the threshold voltages of M1 and M3 are the same, then the output voltage is approximately $V_O = V_{G1} - V_{GS1} - V_{GS3} \approx V_{G1} - 2V_T$, therefore the output swing is reduced by one transistor threshold compared to the output of M1. This could be severely limiting in a sensor supplied with only 3.3 V.

To avoid this, a common practice is to buffer an NMOS with a PMOS source follower, as in figure 6.3(b), and vice versa. For this circuit $V_O = V_{G1} - V_{GS1} + V_{GS3} \approx V_{G1}$ and the output swing of $V_{G1} - V_{GS1}$ is not reduced. The other advantage of the configuration in

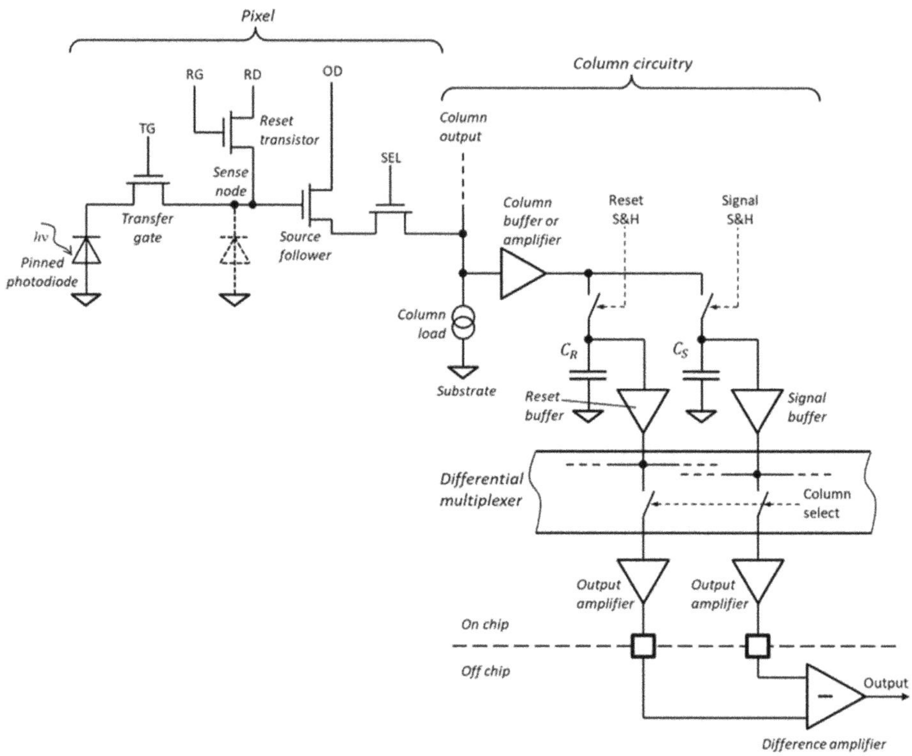

Figure 6.2. Simplified analogue schematic of a pixel and its readout.

figure 6.3(b) is the near-unity gain with the PMOS source follower due to the avoidance of the body effect by the connection of its *n*-well to the source, therefore the overall gain of the circuit increases.

Figure 6.4 illustrates the signal gains and the limitations in the output swing of the two source follower configurations in figure 6.3. The first stage is typical of an in-pixel source follower, having a gain of 0.834 and a gate–source threshold approaching one volt (at high V_{G1}) due to the large body effect. After another threshold drop in the second NMOS source follower the usable output swing is reduced to only 1.2 V and the gain to 0.68. With a PMOS second stage, however, the threshold loss is largely recovered, the output swing is about 2 V and the gain increases to 0.828.

6.1.3 Column amplifiers

Very often the pixel signal must be amplified before it is passed on to the sensor's outputs or is digitised by an ADC. Multiple selectable gains can be used to boost the signal to a level suitable for further processing under variable light conditions. Since a buffer has a gain of slightly less than one, we need an amplifier for each column.

The column amplifier should be simple and physically small because it must fit within the pixel pitch. All column amplifiers are enabled and working at the same time when a row is selected; therefore, their power dissipation cannot be high.

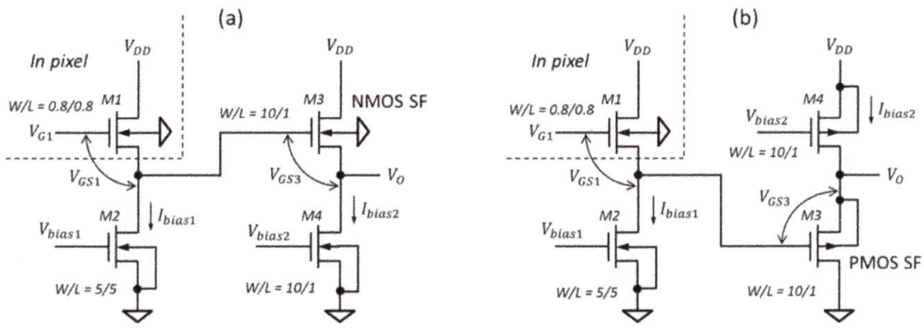

Figure 6.3. NMOS–NMOS (a) and NMOS–PMOS (b) consecutive source followers.

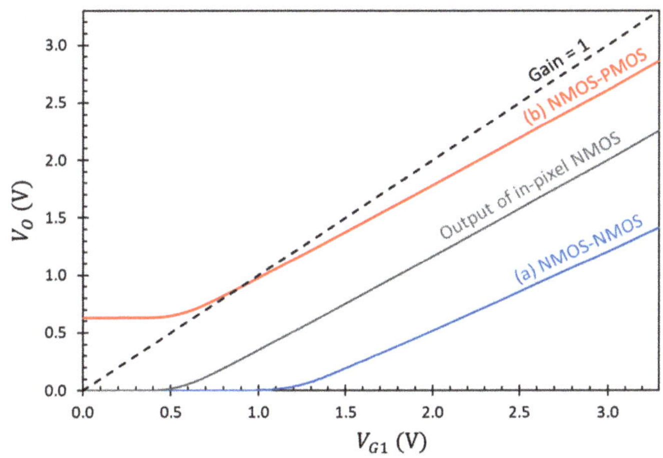

Figure 6.4. SPICE simulation of the consecutive source followers in figure 6.3 for $I_{\text{bias1}} = 2\ \mu\text{A}$, $I_{\text{bias2}} = 10\ \mu\text{A}$ and $V_{\text{DD}} = 3.3$ V. The dashed line shows the response of an ideal buffer with a gain of one and zero threshold voltage.

For a sensor with typical >1000 columns the current consumption per amplifier should not be much higher than about 10 μA—this would make the total current >10 mA, or >33 mW power dissipation at 3.3 V supply. The column amplifiers rarely have gains larger than 4, but they are required to have large output swing to maximise the dynamic range under low supply voltages.

These requirements mean that single-ended, simple transistor circuits such as the capacitive feedback amplifier in figure 6.5 are the most widely used [1]. The amplifier is AC-coupled and uses a reset switch, controlled by the signal *RST*, to set the output to a defined state before the signal appears at the input. This operation is well suited for the signals generated from 3T and 4T pixels, with their distinctive reset and signal time intervals. The amplifier in figure 6.5 works by transferring the charge in the input capacitor C_{in}, generated by the change in the input signal, to the much smaller feedback capacitor C_{fb}, thus creating a voltage gain.

With the reset switch closed, the output voltage V_{O} is equal to the gate voltage of M1, which is simply the transistor threshold voltage for the bias current I_{bias}. During

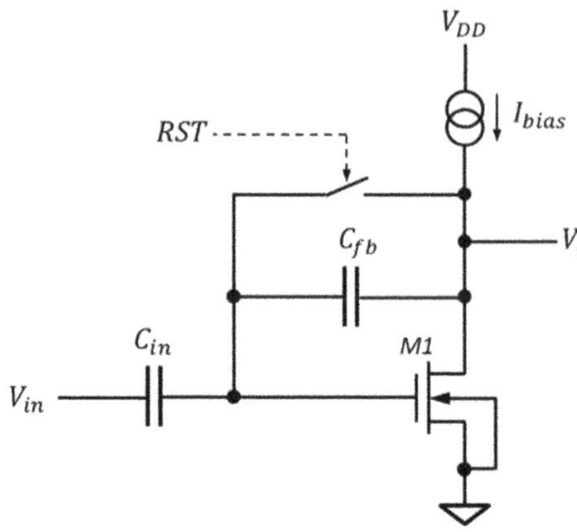

Figure 6.5. Single-stage capacitive feedback amplifier.

reset, the feedback capacitor C_{fb} is discharged and the voltage across it is zero. When the reset switch opens the output is free to take higher potential, but the common point of the capacitors C_{in} and C_{fb} remains at a nearly constant potential, equal to the threshold voltage of M1.

With the reset switch open, the charge Q passing through the input capacitor C_{in} is shared between the feedback capacitor C_{fb} and the much smaller input capacitance of M1, therefore most of the input charge will end up on C_{fb}. The voltage across C_{fb} becomes $V_{fb} = Q/C_{fb}$ and since the common point of C_{in} and C_{fb} is at a fixed potential, the output voltage V_O changes by $\Delta V_O = V_{fb}$.

If the input voltage decreases, the charge in C_{in} would decrease too by the amount $Q = \Delta V_{in} C_{in}$. Since charge is conserved and transferred to C_{fb}, the voltage across C_{fb} would increase, therefore the amplifier is *inverting*. The gain of the circuit is $G = -\Delta V_O/\Delta V_{in}$, where $\Delta V_O = Q/C_{fb}$ and $\Delta V_{in} = Q/C_{in}$, or simply

$$G = -\frac{C_{in}}{C_{fb}} \tag{6.1}$$

Therefore, the gain is determined by the two capacitors. Multiple gains can be realised by connecting a different feedback capacitor through additional analogue switches [2].

An example of a single-stage capacitive feedback amplifier, based on [1], is shown in figure 6.6. Here M5 is the main amplification element, M6 is connected in a cascode configuration for higher bandwidth, and the PMOS transistor M8 is an active load. The output voltage is always equal to or higher than the gate voltage of M5, therefore the simple NMOS switch M7 is adequate. The feedback and the input capacitors have relatively small values and are typically realised as a metal–insulator–metal (MIM) element.

The SPICE simulation in figure 6.7 illustrates the operation of the capacitive feedback amplifier. The gate of M2, simulating a sense node in a pixel, is reset to V_{RD} while M7 keeps the amplifier in reset. After *RST* is released, at 4 μs the current sink I_Q connected to the gate of M2 is turned on to generate 10 ke$^-$ signal, which results in

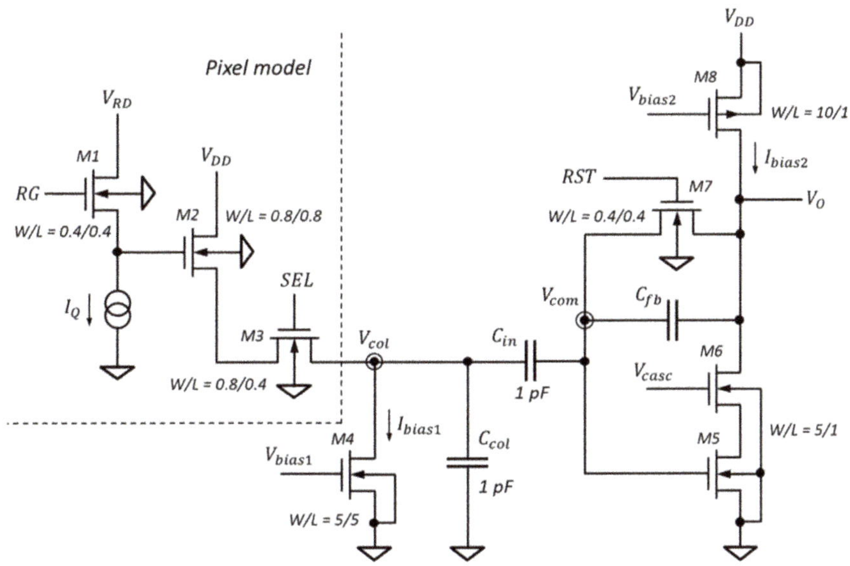

Figure 6.6. Example schematic of a single-stage capacitive feedback amplifier with input from a pixel model.

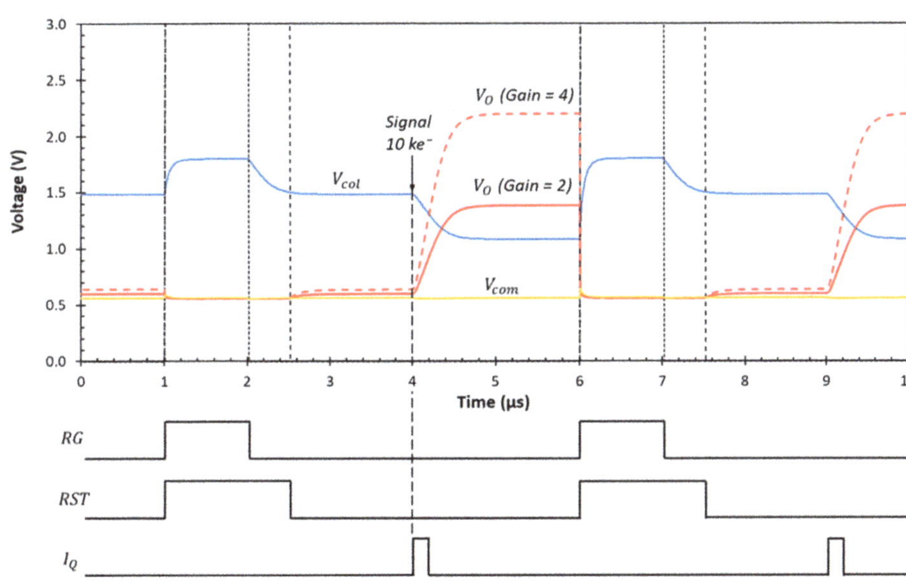

Figure 6.7. SPICE simulation of the column amplifier in figure 6.6 for $I_{\text{bias1}} = 2$ µA, $I_{\text{bias2}} = 10$ µA, $V_{\text{casc}} = 1.8$ V, $V_{\text{DD}} = 3.3$ V, $V_{\text{RD}} = 2.8$ V, $SEL = V_{\text{DD}}$. The feedback capacitor C_{fb} is 0.5 pF for a gain of 1.96 and 0.25 pF for a gain of 3.91. A 16 nA current sink operating over 100 ns generates 1.6 fC signal charge (10 ke^-).

a 400 mV step in the column voltage V_{col}. This voltage change is amplified with a gain deviating by about 2% from the expected from formula (6.1). The voltage V_{com} at the common point of C_{in} and C_{fb} stays nearly constant as explained earlier. The output can be sampled after it has settled sufficiently—approximately between 3 and 4 µs for the reset level and between 5 and 6 µs for the signal level.

6.1.4 CDS circuits

The PPD allows CDS to be implemented by storing the reset and the signal samples in a circuit serving a whole column. What is more, this can be done *in parallel* for thousands of pixels in a selected row. Due to this large parallelism the signal storage can be done at relatively relaxed timescales, in the few microseconds range (i.e. bandwidth in the 100 kHz range), using low noise circuits. The stored signals can then be read out sequentially at very high speed (tens of MHz) while maintaining good noise performance by using a small number of dedicated circuits. This parallel operation is made possible by the ability to implement complex circuits in CMOS technology and brings about major performance advantages.

Different variants of column readout circuitry are found in practically all CIS, regardless of whether they have analogue or digital output. Figure 6.8 shows a typical implementation of a differential sample and hold CDS and its timing diagram. With the select transistor on, the reset level at the gate of the source follower is sampled and stored on C_R after the electronic switch SHR is closed for few microseconds. Similarly, the signal level is stored on C_S after the charge has been transferred to the sense node by pulsing the transfer gate TG. At the end of this operation the two switches are open and the two capacitors hold the two stored voltages. Due to the high impedance buffers connected to them there is little self-discharge and the voltages can be

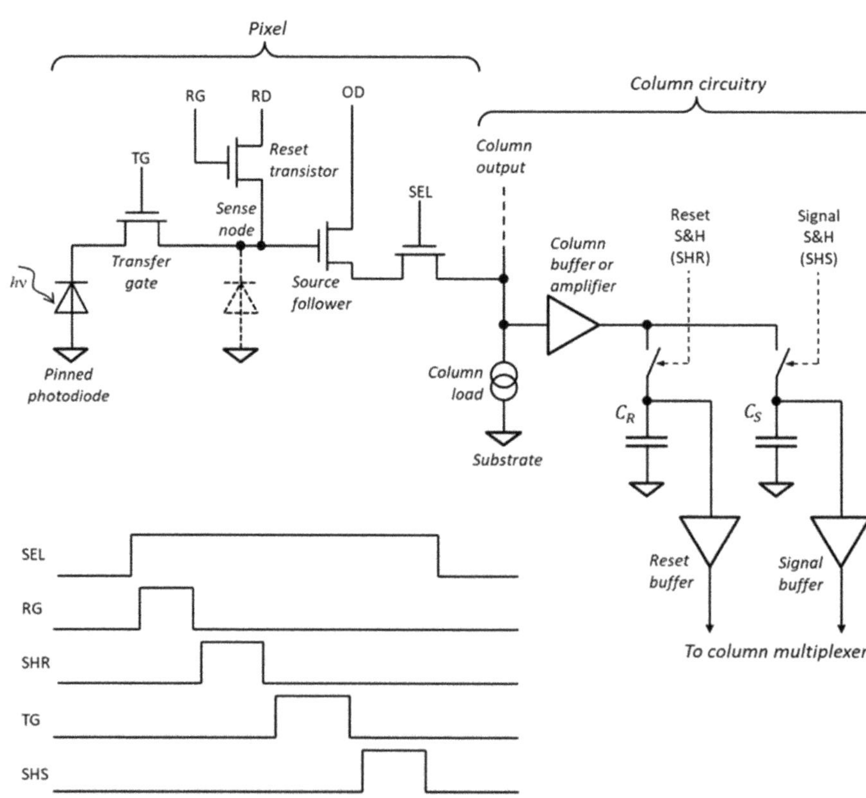

Figure 6.8. Column circuitry and timing diagram of a differential sample-and-hold CDS column circuit.

preserved for a long time. The buffers also allow the two stored voltages to be fed to the following circuits such as multiplexers, amplifiers and ADCs. The buffer can be a simple source follower or a more complex buffer-amplifier.

Using a difference amplifier (usually external) the output voltage from the pixel becomes the difference between the reset and the signal voltage levels. This performs CDS using the double sampling method discussed in chapter 4. More advanced CIS would include an ADC at this point and could output the digitised difference instead of the analogue signal.

The storage capacitors C_R and C_S are typically in the range of several picofarads in order to reduce the reset noise below the noise level of the in-pixel source follower.

Example 6.1. Calculate the thermal noise in electrons RMS generated in a 1 pF signal storage capacitor for a sensor with CVF = 50 μV/e⁻ at 20 °C.
Solution: The RMS noise voltage is

$$\overline{V_n} = \sqrt{\frac{kT}{C}} = \sqrt{\frac{1.38 \times 10^{-23} \times 293}{10^{-12}}} = 63.6\ \mu\mathrm{V}$$

Referred to the input this corresponds to ENC of 63.6/50 = 1.3 e⁻ RMS. For the circuit in figure 6.8 the noise will be $\sqrt{2}$ times higher, or 1.8 e⁻ RMS, because it is differential.

This example shows that even a 1 pF storage capacitor may be too small for low noise applications. Such high capacitance is practical only as a MOS capacitor, either NMOS or PMOS, due to the large area capacitance of the gate oxide. However, MOS capacitance is nonlinear and equals the gate oxide capacitance when the channel is in inversion, therefore care should be exercised to always bias the gate well above threshold.

Figure 6.9 shows the schematic of a differential sample-and-hold CDS circuit with input from a pixel model. The signal storage branch is identical to the reset branch, but its switch is controlled by the signal sample and hold (SHS). No buffer or amplifier is used between the column and the CDS circuit for simplicity; this is acceptable for sensors with moderate frame rate.

The bidirectional switch M9–M10 is used here to accommodate large signal amplitudes. If only the NMOS M9 were used, the column output could be clipped at high column voltages due to the large threshold of M9 (which suffers from body effect); this could mean that the reset signal level is not captured properly.

As the photogenerated signal goes up, the output column voltage decreases and could be near zero at saturation; therefore, an NMOS storage capacitor would not work well here because its gate–source voltage could fall below threshold[1]. Instead, a PMOS capacitor is used, implemented by M5 with its body connected to the supply V_{DD}. Since the reset voltage at V_{col} (the highest voltage the gate of M5 will see) is at

[1] An NMOS storage capacitor could be used if the column signal is inverted.

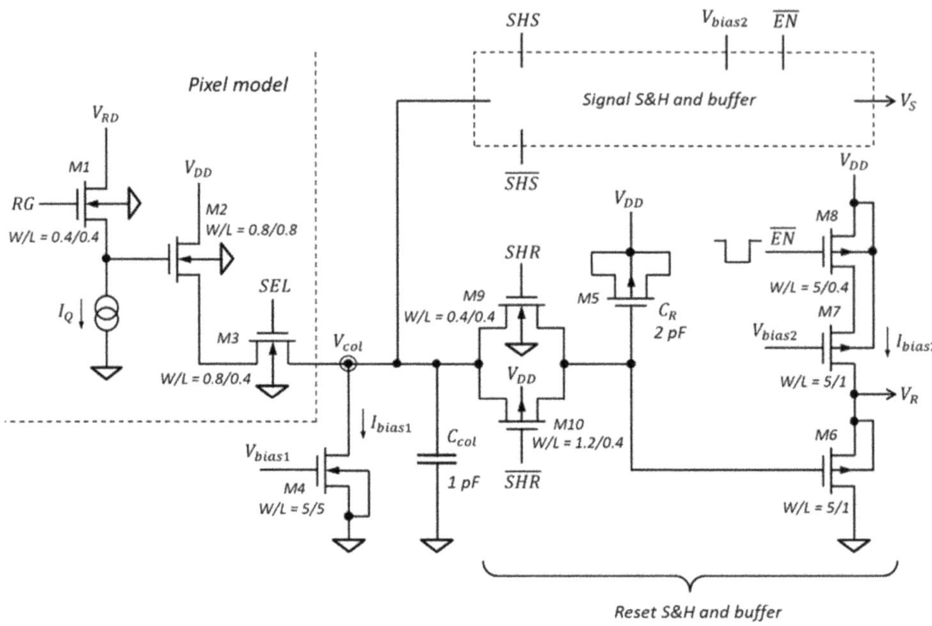

Figure 6.9. Transistor level schematic of one half of a differential sample-and-hold type CDS circuit using a PMOS signal storage capacitor and a PMOS source follower buffer with enable. The signal and the reset S&H and buffer are identical.

least a volt below V_{DD}, the channel of M5 is always in inversion and its gate capacitance is the maximum oxide capacitance C_{ox}. Connecting the storage capacitor to the substrate (as an NMOS transistor would be connected) or to the supply voltage (for a PMOS transistor) is identical because both are AC grounds. The substrate and the supply should be clean for good noise performance; however, even if they are not very quiet the differential storage and readout can suppress most of the low frequency interference. Storage capacitance of 2 pF means that M5 is very large and normally not a single device but several transistors in parallel. For example, with $C_{ox} = 5$ fF μm^{-2} the gate area of M5 would be 400 μm^2.

Once the signal is stored in a capacitor, it can remain there regardless of whether the PMOS source follower M6 is biased or not. M6 is needed only when the column is addressed and can be turned off during the rest of the time, resulting in substantial power savings because the current I_{bias2} is normally much higher than I_{bias1}. The transistor M8, acting as a switch, has been added in series with the current load M7 of the source follower, so that it can turn off its bias current I_{bias2} when not addressed.

Figure 6.10 shows the SPICE simulation of the circuit in figure 6.9. With the storage capacitors C_R and C_S disconnected and the S&H switches turned on, the output from both branches would look like the dashed purple line—it follows V_{col} with an offset equal to the threshold of M6. This output serves as our reference to help verify that the sampled reset and the signal levels are identical to those without CDS.

After closing the SHR switch during the reset period and the SHS switch during the signal period, both voltages remain stored in their corresponding capacitors.

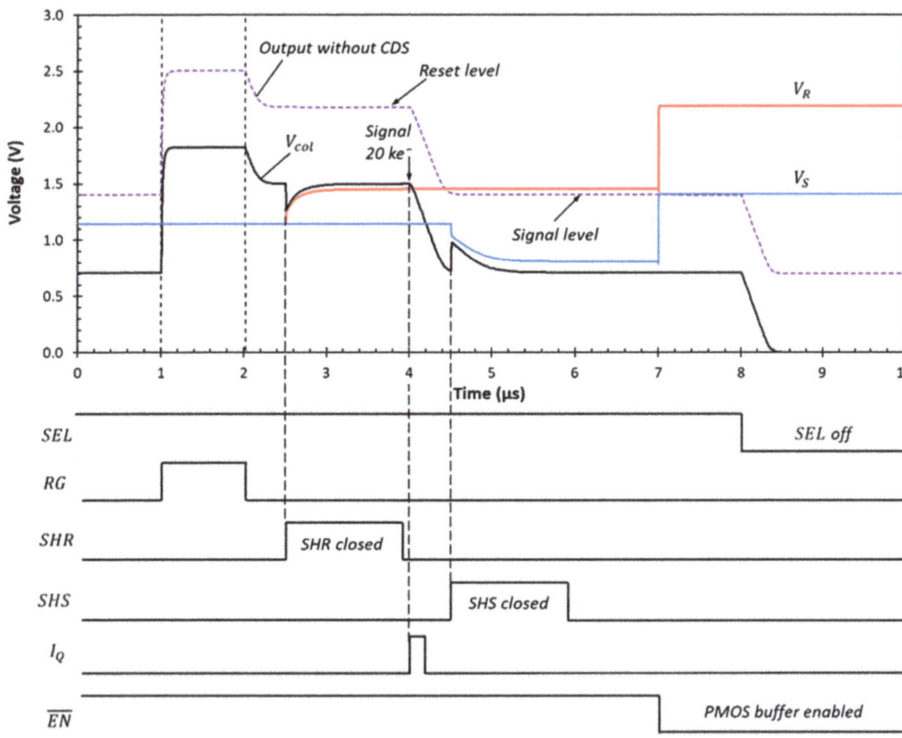

Figure 6.10. SPICE simulation of the circuit in figure 6.9 for $I_{\text{bias1}} = 2$ µA, $I_{\text{bias2}} = 10$ µA, $C_R = C_S = 2$ pF, $V_{DD} = 3.3$ V, $V_{RD} = 2.8$ V. A 16 nA current sink operating over 200 ns generates 3.2 fC signal charge (20 ke⁻). The dashed purple line shows the response with the S&H circuits bypassed and $\overline{EN} = 0$ V.

When the buffers are enabled ($\overline{EN} = 0$ V at 7 µs) both stored signals appear at the outputs with their expected voltages. When *SEL* is off, the column output discharges to zero, and this could happen either before or after the buffers are enabled. We can see that the maximum voltage at the gate of M5 is $V_{\text{col}} \cong 1.8$ V, therefore the gate–source voltage is always higher than 1.5 V and is more than necessary to keep M5 in inversion.

While the differential CDS is popular, it has the inherent disadvantage that the readout noise increases by $\sqrt{2}$ compared to a single-ended circuit, which could also be simpler. Figure 6.11 shows a single-ended CDS using the 'clamp and sample' technique familiar from CCDs. The subtraction of the reset level from the signal level is accomplished by the capacitive coupling between the column output and the following Buffer 1, and its operation is more subtle compared to the differential CDS.

After reset, the CLAMP switch is turned on and connects the floating end of the clamp capacitor C_C to the stable voltage VREF. The voltage across C_C becomes the difference between VREF and the reset voltage, which also captures the instantaneous reset noise voltage. After the CLAMP switch is turned off, the signal charge is transferred to the sense node, and the signal appears on the column output. The coupling via C_C transfers the *difference* between the old and the new voltage at the column output, i.e. the input of Buffer 1 will see a step change equal to the signal change at the column. Therefore, the reset noise is removed together with the reset

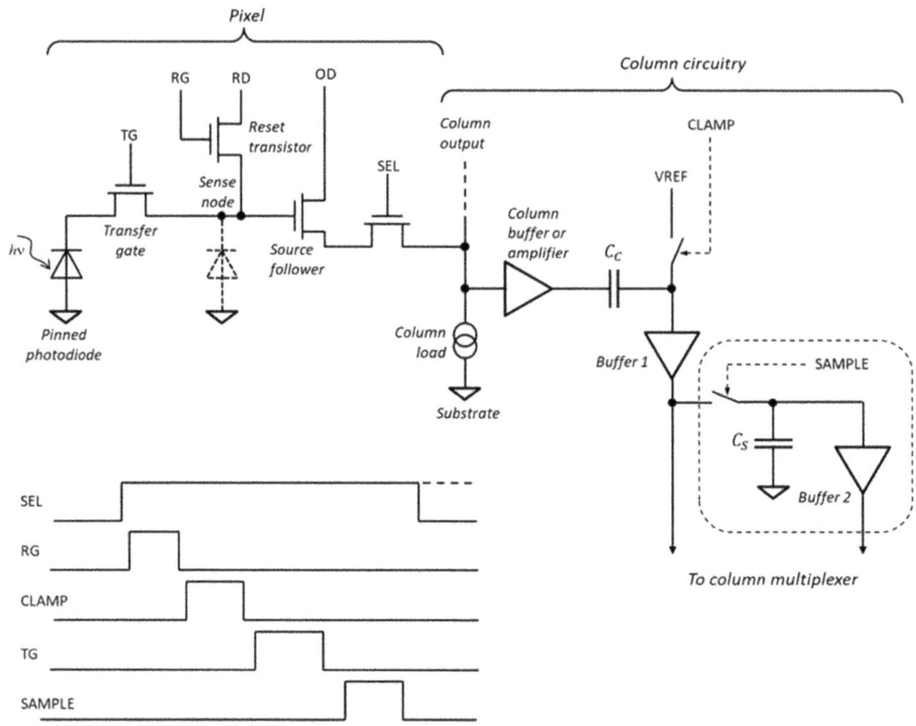

Figure 6.11. Single-ended CDS using the 'clamp and sample' method.

voltage level. The buffered output has the signal voltage change, now referenced to VREF, and offset by a transistor threshold if a source follower is used for the Buffer 1.

Provided that the column output does not change any further, the differenced voltage at the output of Buffer 1 stays stable and can be digitised. This means that SEL must be kept high throughout, as shown by the dashed line in figure 6.11. If SEL is turned off, the column output will float and that will change the output. In this case, a second capacitor C_S and another buffer must be added. Sampling the output of Buffer 1 by the SAMPLE signal makes sure that the CDS output voltage is safely stored regardless of what the column output is doing.

Figure 6.12 shows what a single-ended CDS schematic could look like. Similarly to the differential CDS, the clamp capacitor is realised with a PMOS transistor due to the column output voltage approaching ground. The reference voltage V_{REF} must be sufficiently high so that M5 is always in inversion, and this makes the clamp voltage V_C at least one transistor threshold higher than the column voltage. This higher voltage suits an NMOS source follower better than a PMOS one.

There is some flexibility in choosing V_{REF} and this could be a useful feature because the output can be re-biased to a different DC voltage. However, with two buffers and two capacitors, the single-ended CDS is ultimately not simpler than the differential CDS. It also needs a low noise, stable reference voltage, which the differential CDS does not.

Both the differential and the clamp and sample CDS circuits described here are of the 'double sample' type and have a transfer function given in chapter 4.

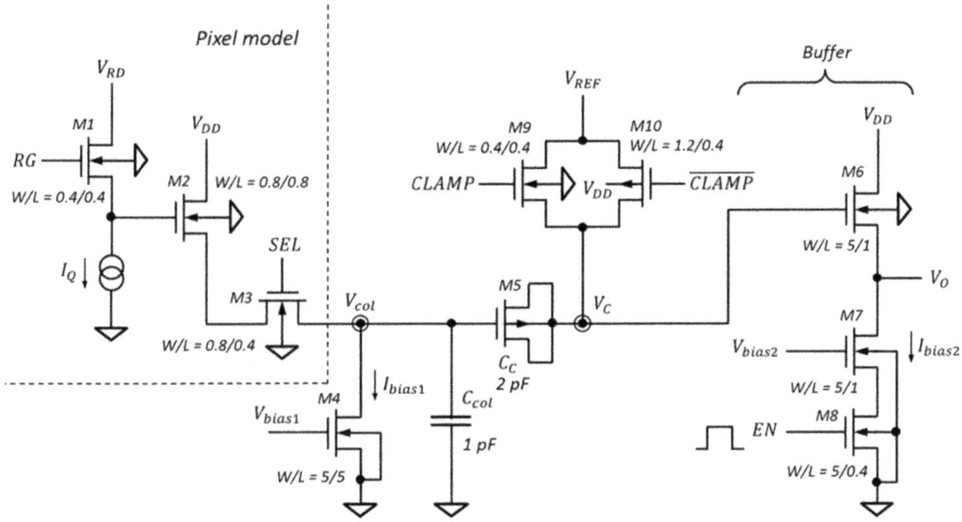

Figure 6.12. Simulation schematic of a single-ended CDS as in figure 6.11 without a sample capacitor and the second buffer.

6.1.5 Row drivers

Row drivers generate the control signals to the pixels such as the row select, transfer, and reset gate pulses. The voltage levels for each of them are generally different; the row select can swing between the rails of the digital supply (e.g. 0–3.3 V), while the transfer gate may need lower voltage and the reset gate higher. This calls for separate supplies for all row drivers that need to generate voltages different from the digital supply, and also for appropriate level shifting. Furthermore, transfer gate drivers may need to switch between three levels instead of two [3], or to have much longer fall time than the rise time in order to reduce charge spill-back.

Figure 6.13 shows a simple row driver built with two CMOS inverters and capable of level shifting. The purpose of the first inverter (M1 and M2) is to drive the second stage; it uses small transistors and is driven with input voltage V_{in} swinging between ground and V_{DD1}. The second inverter (M3 and M4) uses transistors with much larger aspect ratio W/L so that it can drive the capacitance of the row line, here taken as 10 pF. The circuit as a whole is non-inverting.

The supply to the second inverter V_{DD2} can be lower or higher than V_{DD1} and this is what accomplishes the level shifting, but there are limits to how different the supplies can be. The input voltage to the second stage V_{G2} swings between ground and V_{DD1} and so does the gate–source voltage of M4, but the gate–source voltage of M3 swings between $V_{DD1} - V_{DD2}$ (M1 on) and V_{DD2} (M2 on).

When V_{G2} is zero, the supply V_{DD2} must be higher than the threshold of M3 or it will not turn on, and to drive a heavy load V_{DD2} must be substantially larger than the threshold. The effect of V_{DD2} being too low can be seen in the simulation in figure 6.13 for V_{DD2} = 1.0 V—the increased rise time is a consequence of the insufficient gate overdrive because the threshold $V_{T(M3)}$ is around 0.6 V.

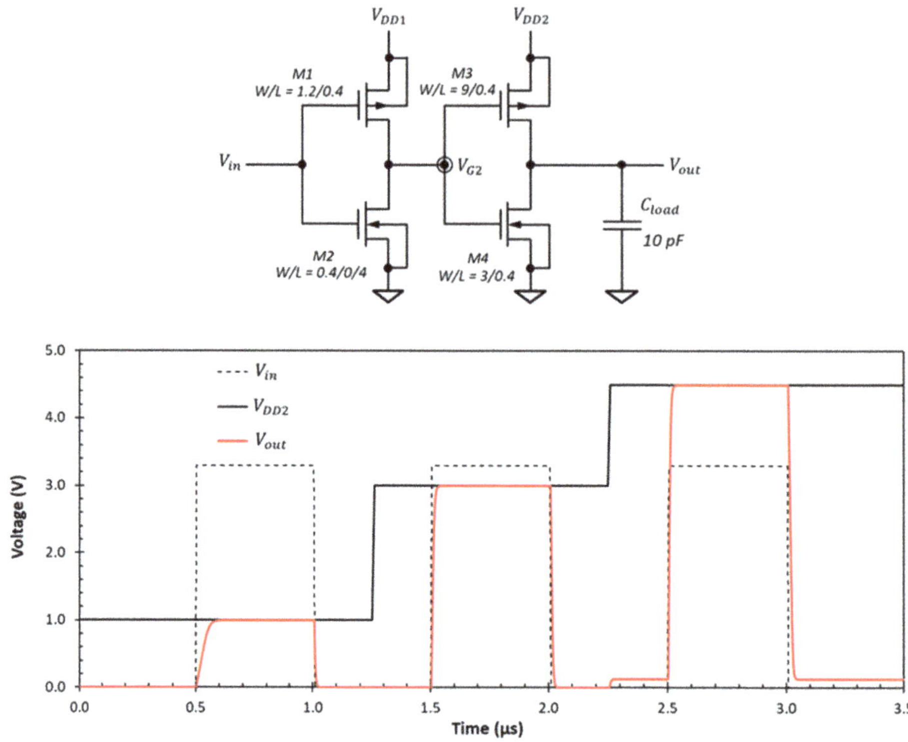

Figure 6.13. Non-inverting row driver (top) and its SPICE simulation (bottom) for V_{DD1} = 3.3 V and V_{DD2} increasing in steps from 1.0 to 3.0 V and 4.5 V.

As V_{DD2} increases to $V_{DD2} > V_{DD1} + V_{T(M3)}$ the transistor M3 cannot be turned off and continuously conducts—this is seen in the simulation in figure 6.13 for V_{DD2} = 4.5 V. This leakage is undesirable, therefore the maximum V_{DD2} should be limited to below $V_{DD1} + V_{T(M3)}$. For $V_{T(M3)} < V_{DD2} < V_{DD1} + V_{T(M3)}$ the circuit in figure 6.13 is a usable voltage shifting row driver. For higher output voltages different circuits which avoid the continuous conduction of M3, such as the classic CMOS level shifter [4], can be used instead.

In addition to the symmetrical driver in figure 6.13, which has nearly equal rise and fall times, drivers with reduced slew rate on the falling edge are also used. Lower slew rate reduces the capacitive coupling from the gate of the reset transistor to the sense node, resulting in a higher voltage at the sense node after reset. Lower slew rate at the transfer gate is used to reduce charge spill-back at large signals in 4T pixels.

A popular way to reduce the negative slew rate a CMOS driver is to use a NMOS transistor with reduced drive strength, achieved by choosing an appropriate aspect ratio W/L, and usually W < L. This is effective but becomes fixed in the layout and the slew rate cannot be changed. A better, but more complex solution is to insert a current sink in series with the output transistor. In figure 6.14 the transistor M5, part of the 10:1 current mirror M5–M6 is used to limit the slew rate by restricting the

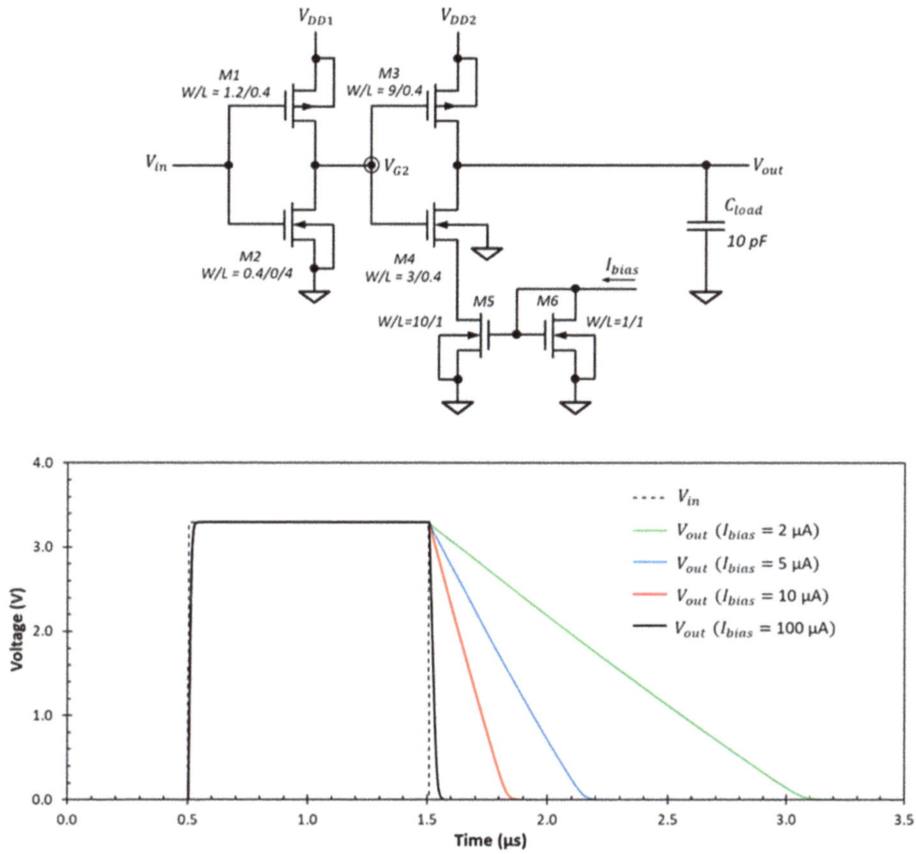

Figure 6.14. Row driver with controlled slew rate of the falling edge (top) and its SPICE simulation (bottom).

current through M4. By selecting the bias current I_{bias} the slew rate can be changed at will, as the simulation in figure 6.14 demonstrates.

The row drivers in figures 6.13 and 6.14 switch between two levels—ground and one externally supplied voltage. The same functionality can be realised with analogue switches, and a driver with more than two voltage levels can be built with the analogue multiplexer described in section 6.1.7.

6.1.6 Pixel addressing

Usually pixels are addressed on a row-by-row basis so that an entire row is read out at once, with its outputs stored in the column CDS circuitry. The stored voltages are accessed sequentially by selecting a column at a time to connect to the output or to an ADC, until the signals from the entire row are read out.

Figure 6.15 shows a typical timing diagram to read out a 4T sensor with N columns. The control signals and the addresses can be supplied to the sensor or generated internally. The column circuitry can be segmented into several blocks with

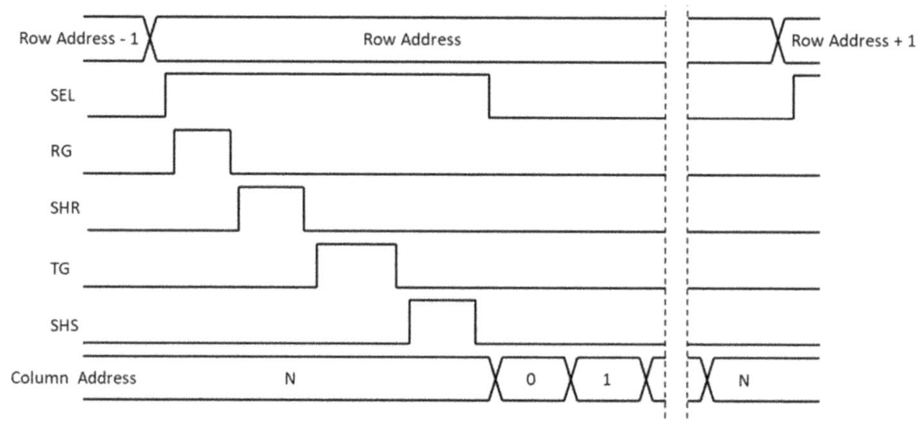

Figure 6.15. Timing diagram for whole sensor readout.

common addressing but separate outputs. In this way, all the column blocks can be read out in parallel, and the readout time of the sensor can be shortened.

Only the simplest image sensors (containing just a few tens of pixels) can have the control and the address lines available as external connections. Any sensor bigger than that needs to implement on-chip circuitry for row and column selection due to the exceedingly large number of individual connections—a CIS with the modest 1000 rows would require 1000 external connections just for SEL and would be completely impractical. Implementing on-chip row and column addressing is quite easy, and virtually all CIS include it in various forms.

The circuits used for row and column selection are called address decoders. These are logic circuits that generate enable signal only on one of their outputs ('one-hot' enable) and have much smaller number of inputs than outputs. Address decoders can be implemented as either combinational or sequential logic.

Combinational logic is the simplest and the most straightforward choice to build address decoders. To select only one row or column at a time, only one output of the address decoder can be at logic 1 (high) for any combination of inputs. As an example, the logic level diagram of a straight-binary 4-output address decoder in figure 6.16 shows how the outputs change with the input address. The SEL outputs can connect directly to the select gates of a pixel row.

The operation shown in figure 6.16 can be easily built with simple AND and NOT gates as shown in figure 6.17. In modern CMOS processes these gates are small and it is straightforward to lay them out on a pitch matching the pixel height. Decoders with more outputs can be built by using multi-input AND gates instead of 2-input ones, each preceded by the necessary combination of NOT gates and direct connections to the address lines. The circuit in figure 6.17 is particularly well suited for automated generation with CAD software which can insert NOT gates or wires in the appropriate places in the layout.

Very often the decoder works in Gray code to reduce current fluctuations and noise. In Gray code only one bit changes from one address to the next [5], while in

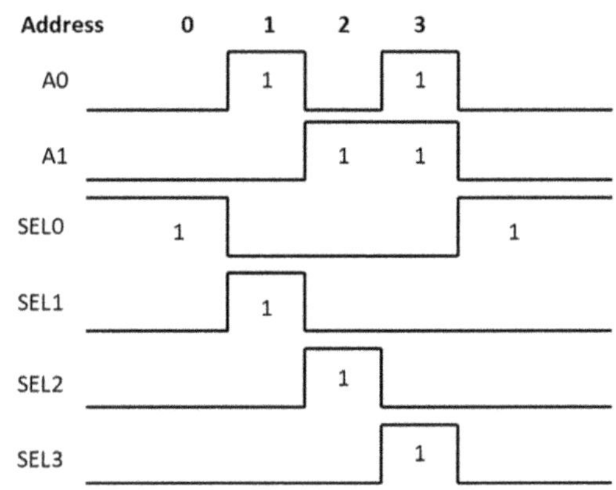

Figure 6.16. Logic diagram for a binary 4-output address decoder with two inputs A0 and A1.

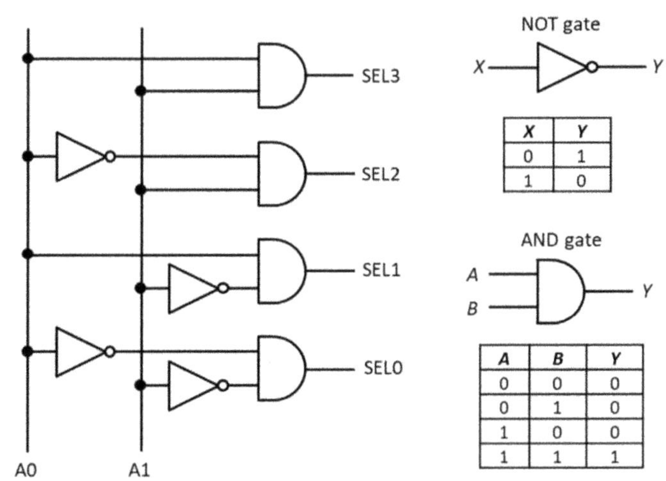

X	Y
0	1
1	0

A	B	Y
0	0	0
0	1	0
1	0	0
1	1	1

Figure 6.17. Schematic of a 4-row address decoder and the truth table of its constituent elements.

straight binary the number of changing bits varies from one to all of them. Incrementing the address in binary creates a predictable pattern in the current draw from the decoder, which could induce a similar pattern in the image due to the coupling through the power supply lines and the substrate. By working in Gray code this parasitic pattern is heavily suppressed.

Combinational logic decoders take N inputs to select 2^N addresses, which saves a huge number of inputs to the sensor but supplying the address can still take many tens of pads on-chip. By using sequential logic such as serial-in, parallel-out shift registers and latches [5] the number of input pads can be reduced to just two or three. This allows a very large number of addresses to be decoded while keeping the number of input pads the same, at the expense of higher logic complexity.

The disadvantage of using a shift register as an address decoder is the time it takes to shift in the new bit stream. For example, it would take 200 ns to shift in 10 bits (for a maximum of 1024 addresses) using a 50 MHz clock. This is rarely a limitation for the row decoding due to the relatively long time to read out a row, which can be tens or hundreds of microseconds. The shift-in time can become a bottleneck for the column addressing because of the much faster rate of the address change. In this case counters can be used because they can increment (or decrement) the address as fast as several nanoseconds. With counters the addresses are consecutive, therefore arbitrary and region-of-interest (ROI) readout are much more difficult to implement.

And finally, to completely avoid supplying any addresses or control signals to generate them, they can be created internally. Most CIS with digital outputs generate their pixel control signals and addressing as part of the readout timing sequence. The setup of the readout is done with numerous configuration registers accessible via SPI or I^2C interfaces.

6.1.7 Analogue switches and multiplexers

Analogue switches are the bread and butter of CMOS electronics and are widely used in image sensors. A simple NMOS transistor works well as a switch, provided that its gate–source voltage is always higher than the threshold and is found in every pixel. As discussed in chapter 1, a switch comprising complementary NMOS and PMOS transistors is far more versatile and can work with input voltages spanning between ground (substrate) and the supply. Due to its increased complexity, it is almost exclusively used outside the pixel, such as in the CDS circuit in figure 6.9.

A stand-alone CMOS switch, shown in figure 6.18, includes an inverter and consists of four transistors in total. The inverter is normally built with minimum size transistors and is supplied with the same voltage as the switch transistors. The control input and its inverse must swing between substrate and V_{DD} if the switch is to

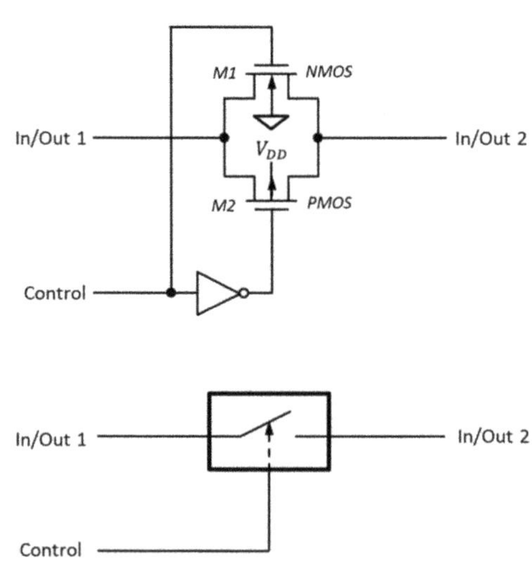

Figure 6.18. Analogue switch with complementary MOS transistors (top) and its schematic symbol (bottom).

operate for inputs between 0 V and V_{DD}. When the control input is high, both M1 and M2 are turned on; the NMOS M1 does most of the work at low input voltages and the PMOS takes over when the input goes near the supply voltage V_{DD}. When the control input is low, both transistors are off, and the switch is an open circuit.

The switch is symmetrical, and the input and the output are interchangeable. As shown in chapter 1, the series resistance of the analogue switch can be significant (thousands of Ohms) and is nonlinear, therefore care must be taken to ensure that it does not cause undesired effects. An analogue switch connecting a capacitor to a voltage is safe because as the capacitor charges up the current through it falls to zero, and so does the voltage drop across the switch. The switch resistance increases the settling time but does not affect the final voltage.

By using two basic analogue switches as in figure 6.19, a circuit that can switch between two arbitrary voltage levels can be made. At any time only one of the switches is on, and in electrical terms this is known as single pole double throw (SPDT) switch operation, or 2:1 multiplexer. During the rise and the fall time of the control signal both switches can conduct momentarily, which shorts the two inputs through the switch resistance. If this is to be avoided, the control must be split into two individual signals, so that one switch is completely off before the other turns on. The SPDT switch in figure 6.19 consists of a minimum of six transistors: two in each switch and two in the inverter, as only one inverter is needed and it can be shared between the switches.

As mentioned in section 6.1.5, a row driver can be built using analogues switches, and the SPDT switch in figure 6.19 is perfectly suitable. The two input voltages V_1 and V_2 can be provided externally and the control signal can come from an address decoder. The voltage at the output changes between V_1 and V_2, which are arbitrary

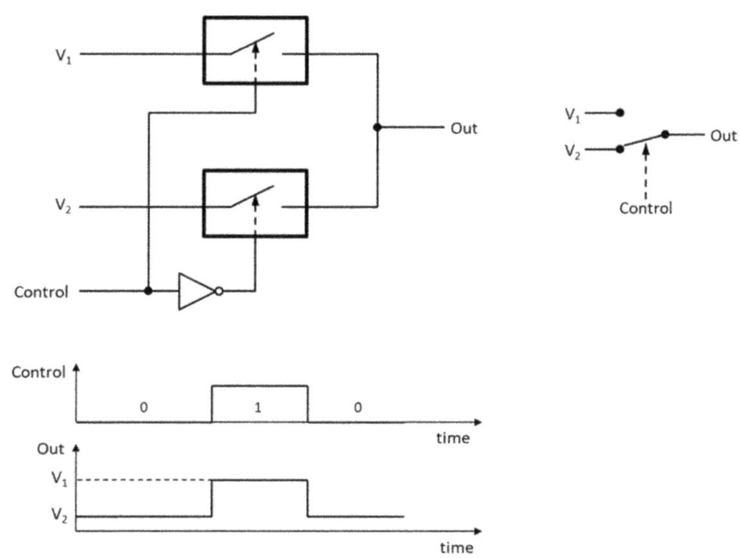

Figure 6.19. SPDT analogue switch with its mechanical equivalent (top) and its operation (bottom).

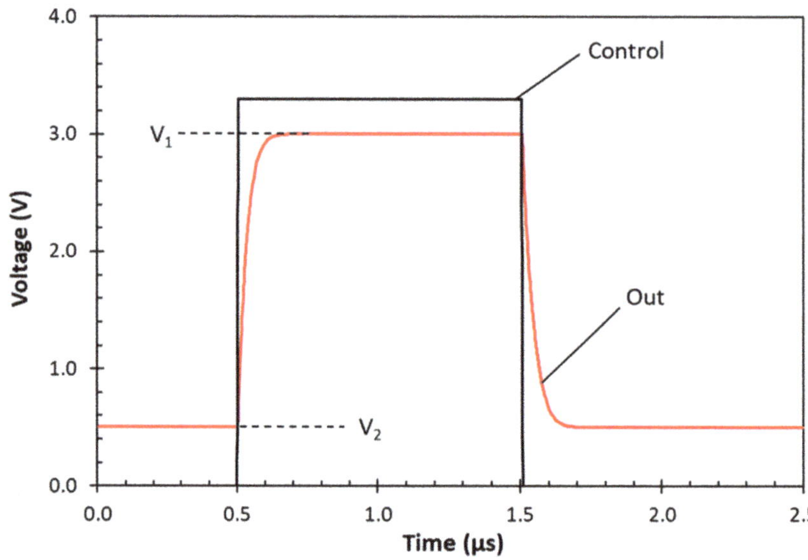

Figure 6.20. SPICE simulation of the switch in figure 6.19 for V_{DD} = 3.3 V, V_1 = 3.0 V and V_2 = 0.5 V with 10 pF load capacitor connected to the output. The aspect ratio of the switch transistors is W L^{-1} = 0.4/0.4 μm for the NMOS and W L^{-1} = 1.6/0.4 μm for the PMOS.

voltages between ground and the positive supply. Figure 6.20 shows a simulation of such a circuit used as a row driver, loaded with a 10 pF capacitor. Here the voltage levels are chosen as 0.5 and 3.0 V, but if they are ground and the supply, this operation is functionally identical to the driver in figure 6.13.

Combining an address decoder with many analogue switches results in the last block before the output amplifier in figure 6.1—the analogue multiplexer. Shown in figure 6.21, each output of the address decoder controls a single analogue switch. For every digital input to the address lines A only one output Y is high, which turns on the corresponding switch. With N digital inputs 2^N voltages can be multiplexed. This is the required functionality to route the signals from the column CDS to the output amplifier, one column at a time. In sensors with differential outputs two banks of switches must be used, controlled by the same address decoder.

6.1.8 Output amplifier

CIS with analogue output must be able to drive an off-chip load at the desired readout rate. The load predominantly consists of the parasitic capacitance of the chip pad, the package, the PCB tracks and the input capacitance of any external amplifier. Normally, an on-chip buffer with unity gain is used.

A source follower as in figure 6.3 could do the job of an output amplifier in principle. Provided that the balance of NMOS–PMOS transistors has been followed to avoid the reduction of the output swing, and some signal loss is acceptable due to the gain being less than one, a source follower could be a simple solution.

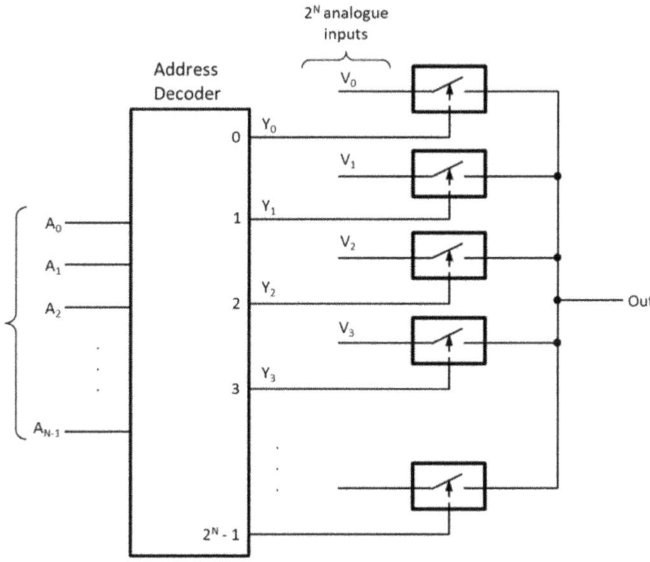

Figure 6.21. Analogue multiplexer.

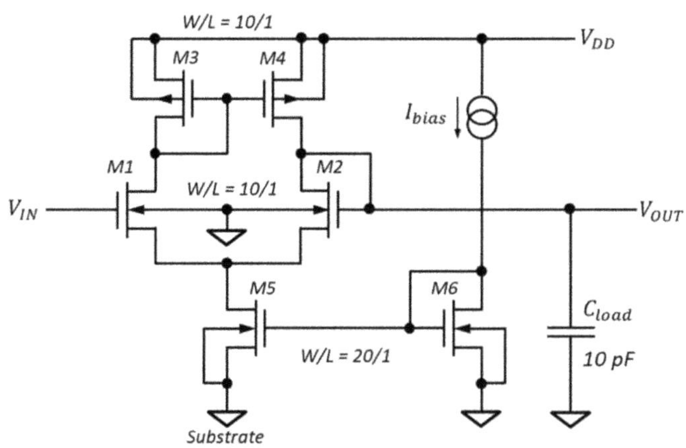

Figure 6.22. 5T buffer-amplifier.

Very often, a buffer with unity gain and low output impedance is needed. A very popular buffer architecture is the 5T differential amplifier with current mirror load [6], shown in figure 6.22. Due to the 100% negative feedback from the output (the drains of M2 and M4) to the inverting input (the gate of M2) the circuit's gain is nearly one. Unlike the source follower, with the 5T buffer the DC offset between the input and the output is only a few millivolts, due to the mismatch between M1 and M2. This is thanks to the large open-loop gain of the differential amplifier (here around 100), which forces the output to closely follow the input. The 5T buffer is a major improvement over the simple source follower, but the price to pay is the

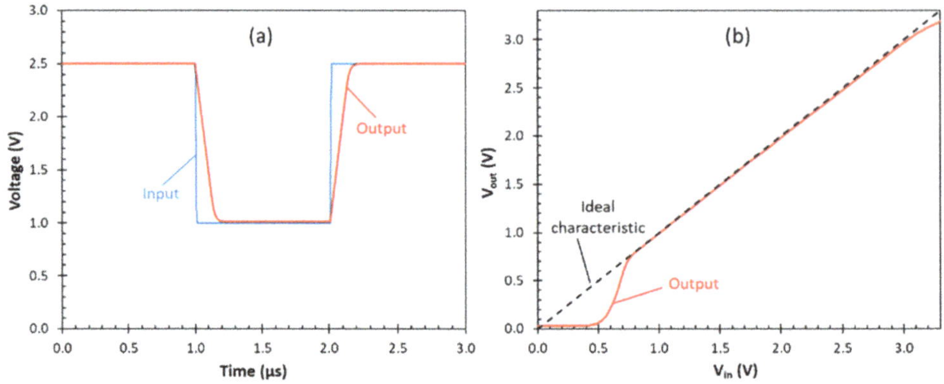

Figure 6.23. Simulated transient response of the 5T buffer in figure 6.22 for $I_{bias} = 100$ µA and $V_{DD} = 3.3$ V (a); electrical transfer function (b).

increased complexity, power dissipation and area. Since there are only a handful of output buffers per chip, the disadvantages are outweighed by the increased performance.

The simulation in figure 6.23(a) shows the transient response of the circuit in figure 6.22 implemented with transistors in a 0.18 µm CMOS process. The capacitor C_{load} represents the off-chip load. With $I_{bias} = 100$ µA the settling time is 250 ns and the bandwidth around 8 MHz. The output remains within 1% linearity for input voltages between 0.8 and 3.0 V (figure 6.23(b)), which is the common-mode input range with $V_{DD} = 3.3$ V. The output impedance is around 2 kΩ, therefore the subsequent off-chip amplifier must have much higher input impedance.

For more demanding applications, two-stage CMOS operational amplifiers [6] can be used to provide greater output drive capability and higher bandwidth.

Using some of the building blocks described in this section, and following the block diagram in figure 6.2 (but without a column buffer or an amplifier), we can draw the complete schematic of the readout chain as in figure 6.24. And yes, it does fit into one side of A4.

This schematic can be very useful when tracing the signal from the sense node to the output. In some sensors it is possible to fix the row and the column address and turn on either the signal or the reset S&H switch. In this way it becomes possible to observe the pixel signal with an oscilloscope and see its operation during reset and charge transfer.

6.2 Off-chip electronics

6.2.1 General requirements

CMOS image sensors are complex devices with numerous analogue and digital inputs and outputs, and usually require several supply voltages. Stable, low noise supply is required by the analogue circuits, while the digital supply is less critical. In addition, there could be several input bias voltages (with little or no power consumption) and bias currents that set the working conditions for the internal circuitry.

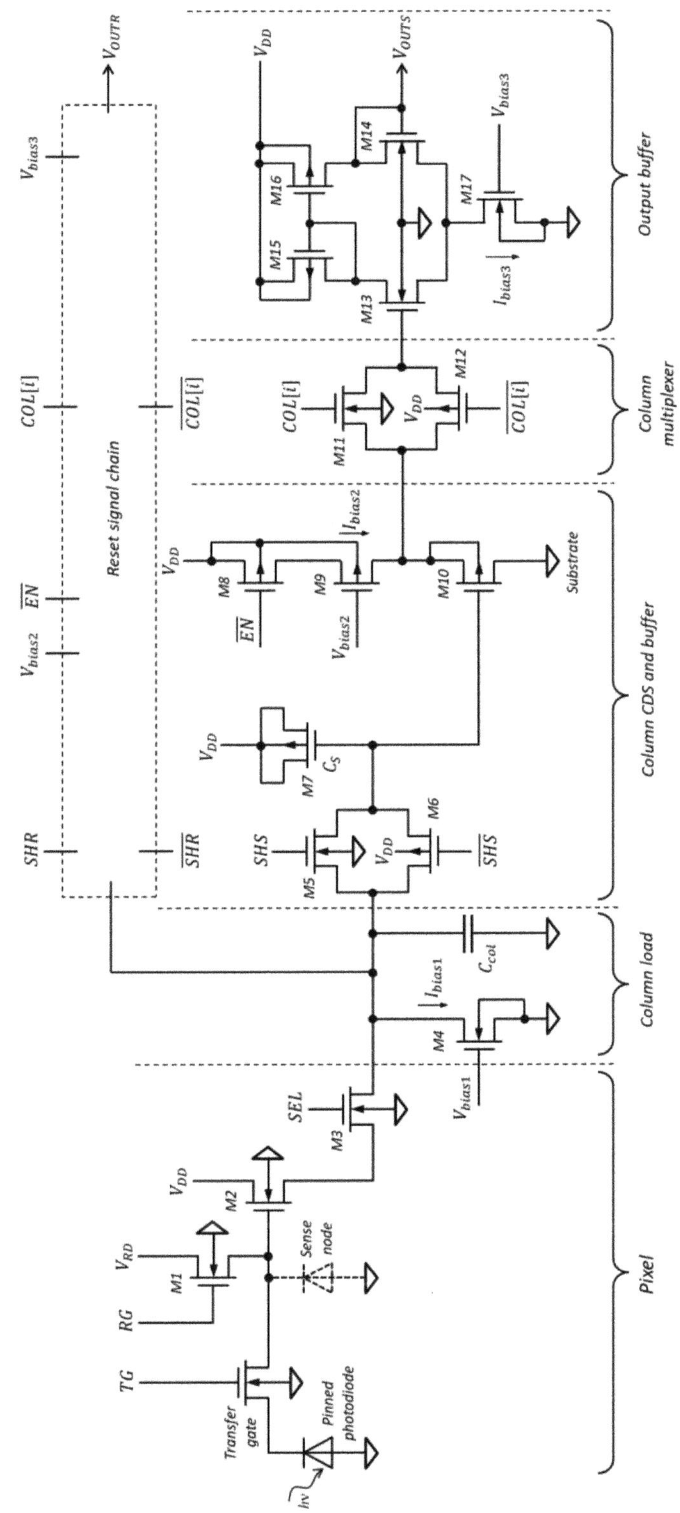

Figure 6.24. Complete readout chain schematic of a 4T image sensor without a column amplifier and with differential CDS and output.

Digital input/output (I/O) can take many forms:

- Low voltage CMOS logic with levels matching the 1.2–3.3 V supplies for image data transfer (for example using the parallel digital video port (DVP) interface), sensor configuration via SPI or I^2C interface, and also for input clocks, trigger and other control signals,
- LVDS, SLVS or CML physical layers for high speed image data transfer,
- MIPI with C-PHY (single-ended), D-PHY, or M-PHY (differential) physical layers for high speed image data transfer.

LVDS, CML, SLVS and MIPI interfaces can transmit data at many gigabits per second.

A system capable or operating an image sensor can have a block diagram like figure 6.25. Usually a field programmable gate array (FPGA) or an application specific integrated circuit (ASIC) at the heart of the system takes care of all the digital I/O, controls the power supply to the CIS (including power sequencing), provides current and voltage bias via digital-to-analogue convertors (DACs), and for sensors with analogue readout controls the analogue-to-digital converters (ADCs) and reads the digitised image signals.

Many CIS for the consumer and the industrial markets have no analogue I/O at all because the image is digitised on-chip, and all voltage and current bias is generated on-chip too. Scientific CIS tend to have simpler digital circuitry with emphasis on pixel performance and offer greater flexibility in operation. This is why many have analogue outputs and externally supplied biasing.

Sensors with internal ADCs are not considered here because they offer little chance of seeing what the output looks like. By providing digital output all this functionality is hidden from the user and the sensor becomes a 'black box' with digital-in, digital-out interface.

6.2.2 Signal amplifiers

In sensors with analogue outputs the image signals normally need to be buffered and amplified before digitisation, so that the signal is matched to the input range of the ADC. The on-chip output buffers are rarely capable of directly driving an ADC or the

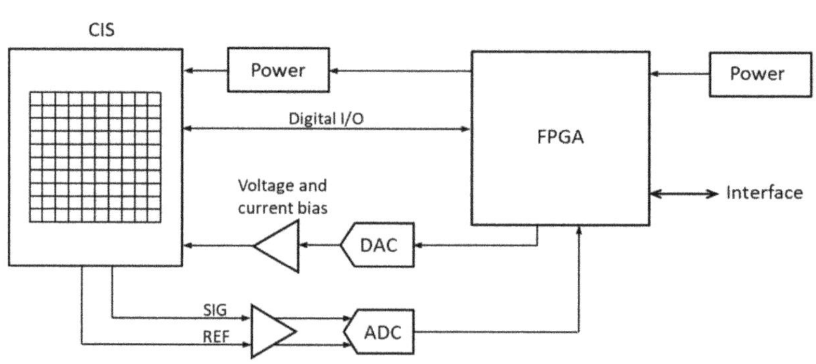

Figure 6.25. General block diagram for power, control and readout of a CIS with analogue outputs.

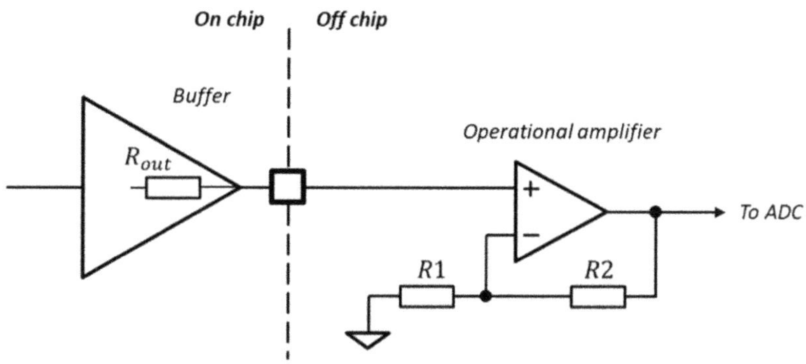

Figure 6.26. Signal amplifier for a CIS with single-ended output.

capacitance of a cable. Their output impedance R_{out} can be high and therefore the amplifier must have much higher input impedance, so that the signal is not attenuated.

Sensors with single-ended outputs can be easily served by a non-inverting, high input impedance operational amplifier (opamp) as in figure 6.26. The signal gain is given by

$$G = 1 + \frac{R_2}{R_1} \tag{6.2}$$

There is a huge choice of opamps suitable for this application. The considerations for the selection of the opamp are:

- **Bandwidth:** normally chosen at 2–3 times the pixel rate (and it depends on the settling accuracy, see example 6.2). Bear in mind that the bandwidth is given by the opamp's gain-bandwidth product (GBWP) divided by the gain in equation (6.2).
- **Slew rate:** make sure that the slew rate allows the settling time to be achieved.
- **Input and output range:** input range including ground and rail-to-rail outputs are preferred for low voltage supplies. Input range including the supply is normally not needed because the sensor's output is well below the power rail.
- **Noise:** low input-referred voltage noise (as a guide, below $5\ \text{nV}/\sqrt{\text{Hz}}$) and low input current noise. A noise calculation as in example 6.2 is a good idea.
- **Supply current:** a general requirement is to have low power consumption. Aim for less than 1 mA per opamp.

The amplifier in figure 6.26 is DC-coupled, therefore both the signal and the DC voltage at the output are amplified by the same factor. As an example, for an amplifier with a gain of two, supplied by 5 V, the output signal in figure 6.10 would bring the circuit's output close to the supply rail. An opamp with rail-to-rail output would be useful in this case, but it is clear that the gain cannot be much higher. To use this circuit as a unity gain buffer the resistor R_1 should not be soldered, and R_2 is zero.

For sensors with differential outputs we need either a difference (differential-in—single-ended out, figure 6.27(a)) or a differential (differential-in—differential-out, figure 6.27(b)) signal amplifiers. Both circuits eliminate the DC offset at the chip's

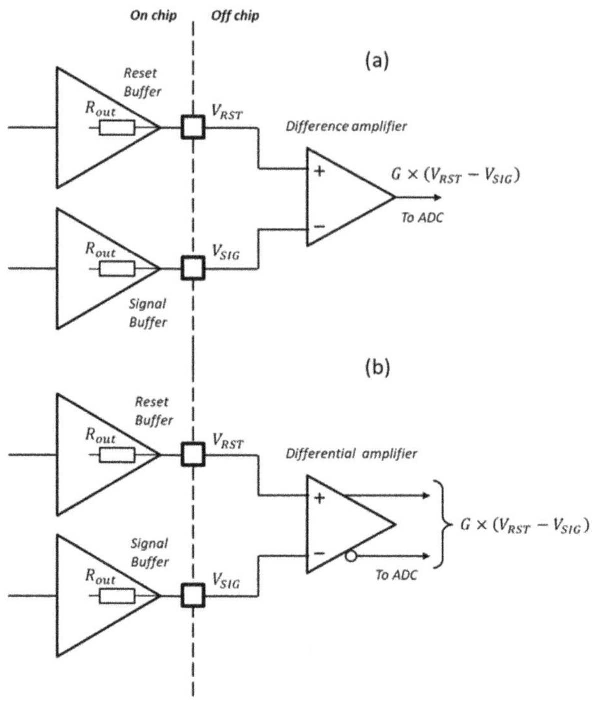

Figure 6.27. Difference (a) and differential amplifier (b) for CIS with analogue outputs.

outputs and pass to the ADC the difference between the input signals amplified by the gain G. Notice the polarity at the output: we have $V_{\mathrm{RST}} - V_{\mathrm{SIG}}$ so that the ADC signal increases with increasing optical illumination[2].

Many ADCs have differential inputs and are very well suited for the arrangement in figure 6.27(b). The differential amplifier in figure 6.28 has been designed to have high input impedance and low noise and is intended to drive differential ADCs; therefore both the input and the output are differential. The differential gain is $G = 1 + 2R_1/R_2$ (provided that $R_1 = R_3$) and the common-mode gain is unity [7]. The output signal is given by:

$$V_{\mathrm{OUT\_RST}} - V_{\mathrm{OUT\_SIG}} = \left(1 + \frac{2R_1}{R_2}\right)(V_{\mathrm{IN\_RST}} - V_{\mathrm{IN\_SIG}}) \quad (6.3)$$

At the output the low-pass filter consisting of R_4, R_5 and C_4 provides differential filtering with cut-off frequency of $1/4\pi R_4 C_4 = 8$ MHz. For differential signals C_4 can be considered as two capacitors in series, each with capacitance of $2C_4$, which halves the cut-off frequency compared to the familiar value of $1/2\pi RC$.

Figure 6.29 shows a SPICE simulation of the amplifier in figure 6.28 for a 1.5 V step change of IN_SIG corresponding to an image signal, and a sensor reset level

[2] If we did $V_{\mathrm{SIG}} - V_{\mathrm{RST}}$ the polarity should be flipped somewhere else, e.g. in software.

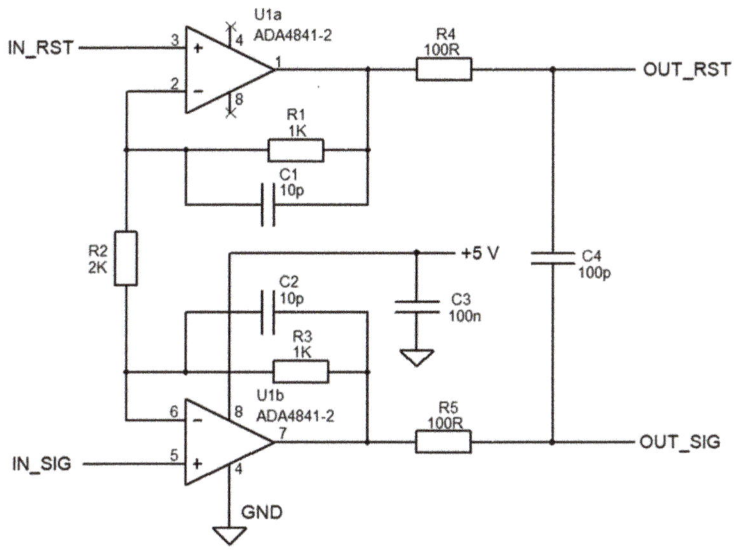

Figure 6.28. Low noise differential-in, differential-out amplifier.

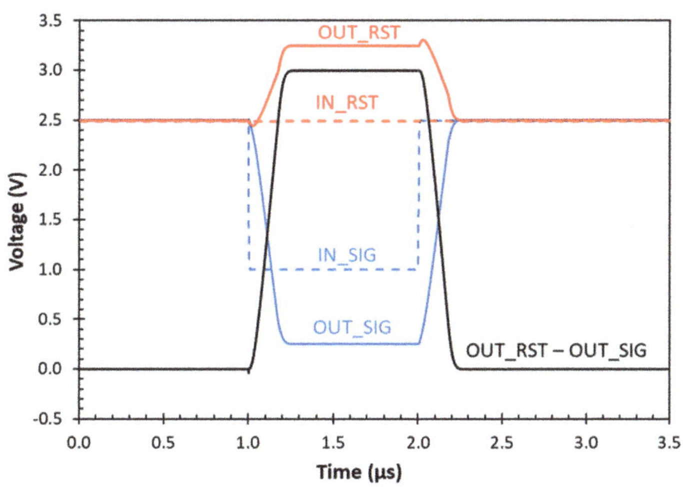

Figure 6.29. SPICE simulation of the circuit in figure 6.28, showing a differential gain of 2.

(IN_RST) of 2.5 V. The circuit uses operational amplifiers with rail-to-rail outputs which allows the single +5 V supply. Notice how OUT_RST goes up despite the input IN_SIG decreasing and IN_RST not moving. This is because the amplified output difference is centred around the mid-point of the two input voltages, equal to (2.5 + 1.0)/2 = 1.75 V. With 3.0 V differential output the voltages OUT_RST and OUT_SIG are 3.25 and 0.25 V, correspondingly. If IN_RST is 2.0 V the output OUT_SIG will want to be –0.25 V, which is not possible because of the single supply. Therefore, care should be exercised to make sure the outputs do not saturate for a different reset level, gain and input signal.

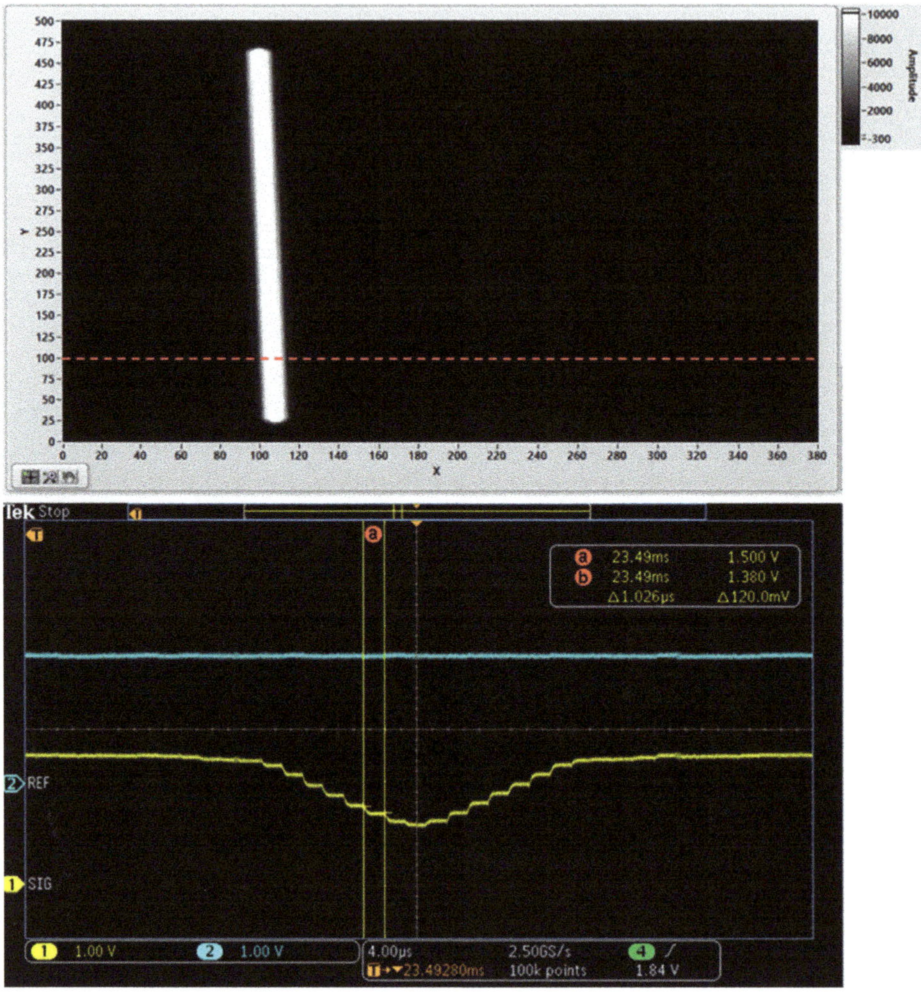

Figure 6.30. Slit illumination image (top) and the signal output from row #100 (bottom) showing consecutive pixel signals. Yellow (SIG): signal output; Cyan (REF): reset output.

Figure 6.30 shows a scope trace from an image sensor using the output amplifier in figure 6.28. An amplifier like this is widely used for sensors with differential outputs and this is why its noise performance is calculated in the following example.

Example 6.2. Calculate the noise added by the circuit in figure 6.28 for an image sensor with readout frequency of 1 Mpix s^{-1}. The input noise voltage density of the ADA4841-2 is 2.1 nV/$\sqrt{\text{Hz}}$ and the input noise current density is 1.4 pA/$\sqrt{\text{Hz}}$. Take the output impedance of the sensor's output R_{out} as 1 kΩ and the sampling accuracy $\varepsilon = 0.01\%$. Consider if this circuit will be adequate for an image sensor with quoted readout noise of 4 e^- RMS and CVF = 50 μV/e^-.

Solution: First, we need to calculate the required signal bandwidth of the circuit. The time constant for settling to 0.01% within 1 μs is given by:

$$\tau_c = \frac{1\ \mu s}{|\ln \varepsilon|} = \frac{1\ \mu s}{|\ln(0.0001)|} = \frac{1\ \mu s}{9.21} = 0.11\ \mu s$$

Therefore, the required signal bandwidth must be higher than $BW_{min} = 1/2\pi\tau_c = 1.47$ MHz. The circuit has much higher bandwidth than this, limited by the output low-pass filter R_4, R_5 and C_4 to $1/4\pi R_4 C_4 = 8$ MHz. The noise bandwidth is:

$$BW_n = \frac{1}{4 \times 2R_4C_4} = 12.5\ \text{MHz}$$

The input-referred noise voltage density is the quadrature sum of [8]:

- The opamp's input noise voltage $e_n = 2.1\ \text{nV}/\sqrt{\text{Hz}}$;
- The thermal noise of R_{out}, $\sqrt{4kTR_{out}} = 4.1\ \text{nV}/\sqrt{\text{Hz}}$;
- The thermal noise of the feedback resistors $\sqrt{4kT(R_1||R_2/2)} = 2.9\ \text{nV}/\sqrt{\text{Hz}}$;
- The noise voltage developed by the input noise current on the output resistance of the on-chip buffer $i_nR_{out} = 1.4\ \text{nV}/\sqrt{\text{Hz}}$;
- The noise voltage developed on the feedback resistors $i_n(R_1||R_2/2) = 0.7\ \text{nV}/\sqrt{\text{Hz}}$.

R_2 is divided by two because we could think of its middle being at AC ground (due to the signal being differential the middle point stays at a constant potential) and is in parallel with R_1 for AC signals. The total input-referred noise density becomes

$$e_n^{tot} = \sqrt{2.1^2 + 4.1^2 + 2.9^2 + 1.4^2 + 0.7^2} = 5.7\ \text{nV}/\sqrt{\text{Hz}}$$

The RMS noise voltage is given by e_n^{tot} multiplied by the square root of the noise bandwidth and by $\sqrt{2}$ because the circuit is differential:

$$v_n = e_n^{tot}\sqrt{2BW_n} = 5.7 \times 10^{-9} \times \sqrt{2 \times 12.5 \times 10^6} = 28.5\ \mu\text{V RMS}$$

The sensor's output noise voltage can be approximated by the product of the CVF and the noise in electrons, which gives 50 μV/e$^-$ × 4 e$^-$ = 200 μV RMS. This is much higher than the input-referred noise added by the amplifier and therefore the circuit is more than adequate. The ENC of this amplifier, for the sensor it serves, is 4 × 28.5/200 = 0.57 e$^-$ RMS, and the total system ENC is $\sqrt{4^2 + 0.57^2}$ = 4.04 e$^-$ RMS.

When the ADC has single-ended input, a conversion from differential to single-ended signal is required. The amplifier in figure 6.31 accomplishes that by adding a difference stage to the circuit in figure 6.28. It uses bipolar ±5 V supply to have greater flexibility in input signals and offsets and can drive heavier loads. The gain of the second stage, built with the opamp U2, is near unity. R9 is used to add a positive 200 mV offset to the output, needed by the ADCs to correctly capture the reference level at zero input signal. R10 is the counterpart to R9 and is used to match the non-inverting and the inverting gain.

The simulation in figure 6.32 uses the same 1.5 V input step for IN_SIG as in figure 6.29, but here IN_RST is 2.0 V instead of 2.5 V. With the same differential

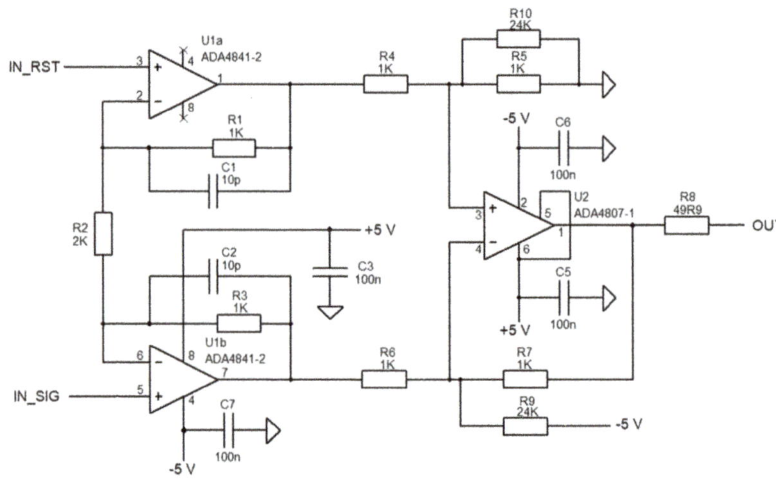

Figure 6.31. Difference amplifier for driving single-ended ADCs.

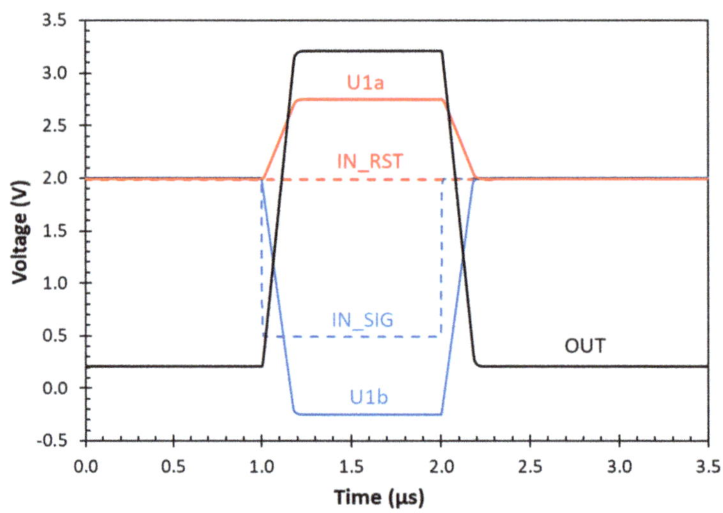

Figure 6.32. SPICE simulation of the amplifier in figure 6.31.

gain of two, the output of U1b now goes below ground, which is fine because the supply is bipolar. The output shows a 3 V signal on top of a 200 mV fixed DC offset.

6.2.3 Power supplies

Most image sensors use standard power supplies such as 3.3 and 1.8 V, and occasionally non-standard voltages. There is a great choice (see, for example the offerings of Texas Instruments and Analogue Devices) of linear low dropout (LDO) regulators capable of supplying low noise power with load currents from about 100 mA to several amps. LDOs with fixed outputs cater for the standard voltages, and those with adjustable outputs take care of the rest. Switch-mode power supplies

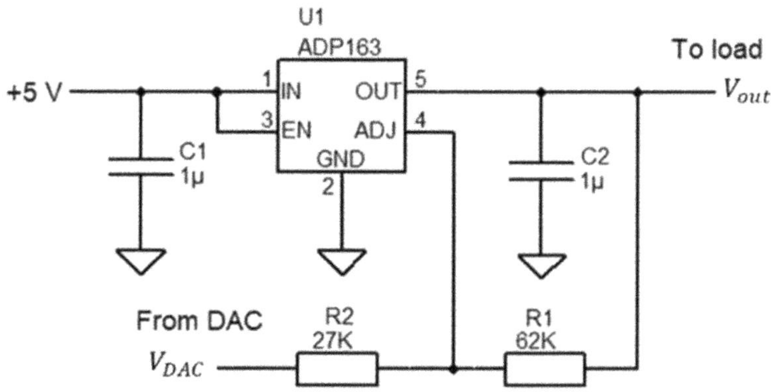

Figure 6.33. Programmable LDO power supply.

are best avoided for the sensitive analogue supplies due to their high frequency noise, but may be used for the digital circuitry.

Sometimes a programmable power supply capable of delivering hundreds of milliamps is needed for laboratory evaluation. A very convenient way to build such a supply is to use an adjustable LDO regulator under the control of a DAC.

Figure 6.33 shows a programmable power supply built around the ADP163 LDO linear regulator, capable of 150 mA output current. Currents up to 1.5 A are possible with LDOs like the LT1963, and many other parts are available (see [7], pp 614–5). The benefits of using a regulator instead of a discrete implementation are the full overload and thermal protection offered by the LDOs, and their compact size.

The feedback aims to maintain the voltage at the ADJ pin to be the same as the internal reference voltage V_{ref}. This can be written as

$$V_{\text{ADJ}} = (V_{\text{out}} - V_{\text{DAC}})\left(\frac{R_2}{R_1 + R_2}\right) + V_{\text{DAC}} = V_{\text{ref}} \tag{6.4}$$

The output voltage is therefore

$$V_{\text{out}} = V_{\text{ref}}\left(\frac{R_2 + R_1}{R_2}\right) - V_{\text{DAC}}\frac{R_1}{R_2} \tag{6.5}$$

and *decreases* as V_{DAC} is increased. ADP163 has internal reference $V_{\text{ref}} = 1.0$ V, and when $V_{\text{DAC}} = 0$ V the output voltage is 3.3 V for the values in figure 6.33. The enable (EN) can be used to turn the regulator on after the DAC voltage has been set.

The current in each supply can be conveniently monitored with a high-side current shunt monitor, connected at the input of each regulator. In the example in figure 6.34 the output voltage is 10 mV per milliamp of load current. Such circuits occupy very small space and can be extremely useful for diagnostics.

6.2.4 Bias circuits

Very often a programmable bias voltage is necessary for detailed characterisation of an image sensor. The current consumption from such supply is usually very small, but it is always decoupled with a capacitor and needs some attention.

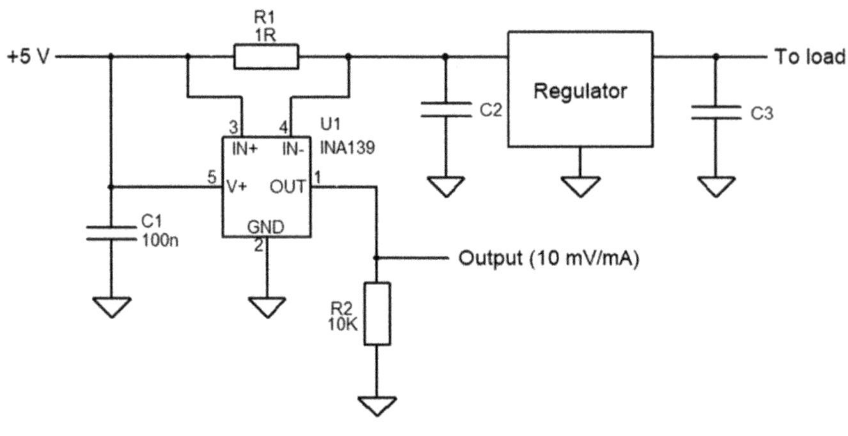

Figure 6.34. Current monitor circuit.

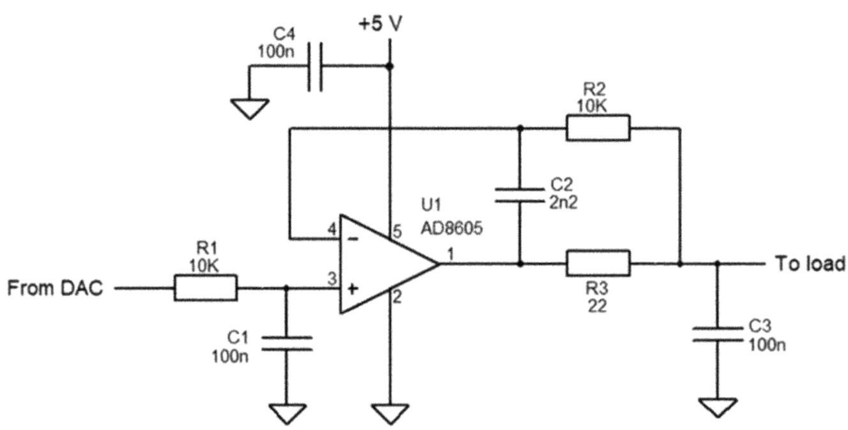

Figure 6.35. Bias circuit with capacitive decoupling.

Figure 6.35 shows a classic circuit capable of driving a bias line with its decoupling capacitor. It operates as a buffer with a gain of unity and is stable with a capacitive load, provided that $R_2C_2 \gg R_3C_3$. The chosen opamp can supply tens of milliamps at output voltages approaching the supply rails. It can also both source and sink current unlike a LDO regulator, which can only source current. Some opamps like the LM7332 can drive very large capacitances.

Many CIS require several bias currents instead of voltages, used to set the current sink of the pixel source followers, or the bias of other circuits such as amplifiers and drivers. The simplest way to provide a current input is via a resistor, as shown in figure 6.36(a) for a typical current mirror input. Knowing the voltage developed at the transistors' gates V_{bias}, the current flowing into the input is

$$I_{\text{in}} = (V_{\text{DD}} - V_{\text{bias}})/R \tag{6.6}$$

This is very simple but is subject to fluctuations in V_{DD} and V_{bias} due to its temperature dependence, and is not programmable. A more flexible way is to use a

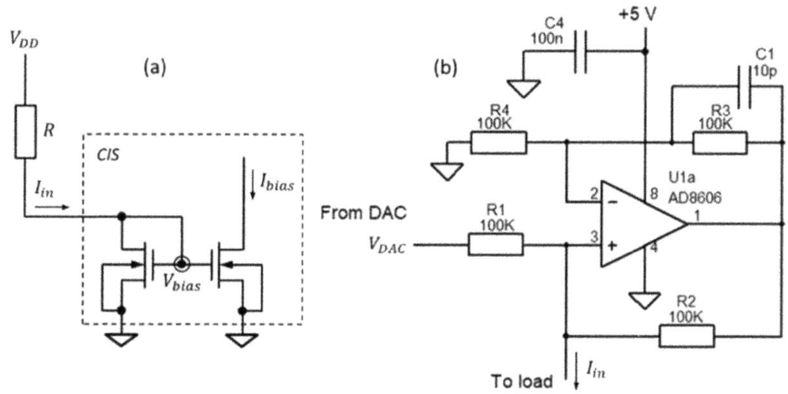

Figure 6.36. Resistor as a current source (a); Howland current source providing 10 μA per volt of DAC input (b).

variant of the Howland current source [9] in figure 6.36(b). This circuit generates current given by $I_{in} = V_{DAC}/R_1$ (provided that all resistors are the same) and is bidirectional, i.e. it can source and sink current. For the values in figure 6.36(b) the output is 10 μA per volt of the input DAC voltage, and is independent of the supply and V_{bias}, within the limits of the opamp's output.

6.2.5 Noise measurements

Often, we want to measure the noise from transistors, amplifiers or more complex circuits. Characterising MOSFETs is challenging because of their small operating currents and inability to drive any significant loads. A typical setup, shown in figure 6.37, consists of an appropriately biased test transistor, low noise amplifier (LNA) and a digitiser or a spectrum analyser. A dedicated LNA is needed because most commercially available instruments have 50 Ω inputs and too high noise.

The LNA should have low voltage noise density ($e_n \leqslant 2\text{nV}/\sqrt{\text{Hz}}$), little $1/f$ noise, bandwidth >10 MHz and gain of at least 10, so that the noise of the spectrum analyser becomes insignificant. Also, low input capacitance (C_{in}< 5 pF) and input impedance above 1 MΩ are needed so that the MOSFET is not loaded excessively and its bandwidth is not limited.

The LNA is crucial for obtaining quality measurements. It must be heavily shielded together with the device under test (DUT), and be supplied from a low noise power supply, or even a battery. Noise is an AC signal, therefore no DC precision is needed. Single-ended input can be used, which also reduces the noise by $\sqrt{2}$ compared to a differential input. Some commercial LNAs have very large input capacitance, but here we want C_{in} to be as low as possible to maximise the bandwidth of the measurement, and the product $C_{in}e_n$ must be as small as possible.

Figure 6.38 shows one LNA implementation using discrete components, featuring a gain of 20 (on a high impedance load) and bandwidth from 10 Hz to 10 MHz. Many other circuits are available, see for example [7] p 152.

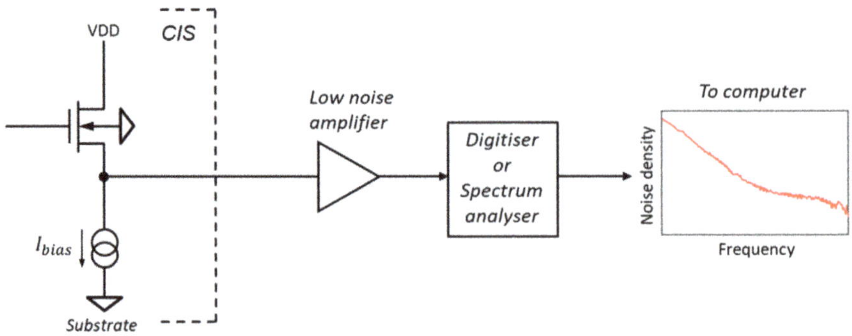

Figure 6.37. Characterisation setup for noise measurements.

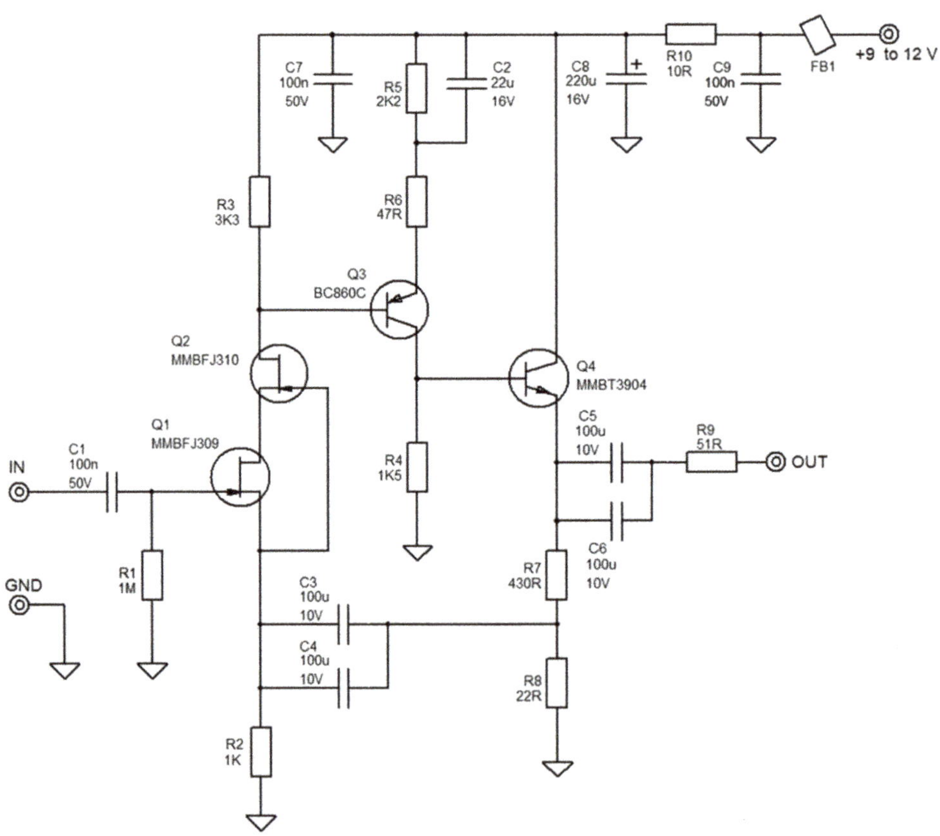

Figure 6.38. Low noise amplifier for component and circuit characterisation, based on [10].

As with every measurement, it is important that it is calibrated, here with a known noise source. Fortunately, this is already available: with the input left open, the thermal noise from the 1 MΩ input resistor R_1 should give 120.5 $nV/\sqrt{Hz}$ reading at 20 °C, which is quite close to the measurement in figure 6.39. The input-referred noise

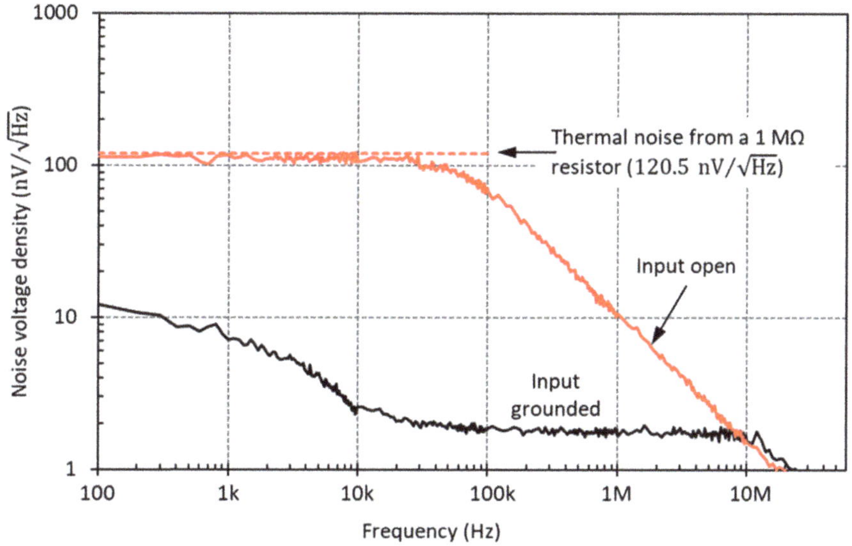

Figure 6.39. Measured input-referred noise of the circuit in figure 6.38.

of this amplifier, measured with the input grounded, is slightly below 2 $nV/\sqrt{Hz}$ away from the $1/f$ knee.

The input capacitance C_{in} can be estimated from the noise spectrum with an open input by using that C_{in} and R_l are connected in parallel, creating a low-pass filter for the thermal noise from R_l. The cut-off frequency f_c of the input noise is approximately 60 kHz, therefore $C_{in} = 1/2\pi f_c R_l = 2.6$ pF. No batteries are used, just a high-quality benchtop 12 V linear power supply from a respectable brand.

A very good LNA can also be built with the JFET-input opamp ADA4817-1 [11], which has very low input capacitance (1.3 pF) and 4 $nV/\sqrt{Hz}$ thermal noise density above 50 kHz.

Chapter summary

1. Column amplifiers using capacitive feedback can be used to provide selectable signal gain before the CDS.
2. Column CDS using two storage capacitors implements the double sampling method, is relatively simple but effective, and provides a differential sensor output.
3. Row drivers should be able to drive large capacitance and implement level shifting and controlled slew rate.
4. External column and row addressing in small sensors can be designed with combinational logic.
5. Analogue multiplexers, controlled by the column address, are used to select the stored signal from the column CDS for readout via an on-chip buffer.

6. Off-chip output signal amplifiers should be designed so that they do not degrade the noise performance of the sensor and can cope with the desired pixel rate.
7. Low noise power, bias and current inputs can be generated with a wide choice of commercial linear regulators and opamps.
8. Noise characterisation of transistors and sensors requires low noise amplifiers with high impedance and low capacitance input.

References

[1] Fowler B 2011 Single photon CMOS imaging through noise minimization *Single-Photon Imaging* ed P Seitz and A Theuwissen (Berlin: Springer) pp 159–95

[2] Boukhayma A, Peizerat A and Enz C 2016 A sub-0.5 electron read noise VGA image sensor in a standard CMOS process *IEEE J. Solid-State Circuits* **51** 2180–91

[3] Ivory J, Stefanov K D and Holland A D 2020 Mitigating charge spill-back induced image lag with a multi-level transfer gate pulse in PPD image sensors *Proc. of SPIE, 11454*

[4] Kaeslin H 2008 *Digital Integrated Circuit Design, From VLSI Architectures to CMOS Fabrication* (Cambridge: Cambridge University Press)

[5] Wakerly J F 2001 *Digital Design: Principles and Practices* 3rd edn (Englewood Cliffs, NJ: Prentice-Hall)

[6] Allen P E and Holberg D R 2012 *CMOS Analog Circuit Design* (Oxford: Oxford University Press)

[7] Horowitz P and Hill W 2015 *The Art of Electronics* 3rd edn (New York: Cambridge University Press)

[8] Jung W 2005 *Op Amp Applications Handbook* (Oxford: Newnes)

[9] AN-1515 A Comprehensive Study of the Howland Current Pump, Texas Instruments (https://www.ti.com/lit/an/snoa474a/snoa474a.pdf)

[10] Jefferts S R 1989 A very low-noise FET input amplifier *Rev. Sci. Instrum.* **60** 1194–6

[11] Shen S and Yuan J 2020 1/fγ low frequency noise model for buried channel MOSFET *IEEE J. Electron Devices Soc.* **8** 126–33

www.ingramcontent.com/pod-product-compliance
Ingram Content Group UK Ltd.
Pitfield, Milton Keynes, MK11 3LW, UK
UKHW051901100926
8556IPUK00005B/719

9 780750 332330